World Geomorphological Landscapes

Series Editor

Piotr Migoń, Institute of Geography and Regional Development, University of Wrocław, Wrocław, Poland

Geomorphology – 'the Science of Scenery' – is a part of Earth Sciences that focuses on the scientific study of landforms, their assemblages, and surface and subsurface processes that moulded them in the past and that change them today. Shapes of landforms and regularities of their spatial distribution, their origin, evolution, and ages are the subject of geomorphology. Geomorphology is also a science of considerable practical importance since many geomorphic processes occur so suddenly and unexpectedly, and with such a force, that they pose significant hazards to human populations. Landforms and landscapes vary enormously across the Earth, from high mountains to endless plains. At a smaller scale, Nature often surprises us creating shapes which look improbable. Many geomorphological landscapes are so immensely beautiful that they received the highest possible recognition – they hold the status of World Heritage properties. Apart from often being immensely scenic, landscapes tell stories which not uncommonly can be traced back in time for millions of years and include unique events. This international book series will be a scientific library of monographs that present and explain physical landscapes across the globe, focusing on both representative and uniquely spectacular examples. Each book contains details on geomorphology of a particular country (i.e. The Geomorphological Landscapes of France, The Geomorphological Landscapes of Italy, The Geomorphological Landscapes of India) or a geographically coherent region. The content is divided into two parts. Part one contains the necessary background about geology and tectonic framework, past and present climate, geographical regions, and long-term geomorphological history. The core of each book is however succinct presentation of key geomorphological localities (landscapes) and it is envisaged that the number of such studies will generally vary from 20 to 30. There is additional scope for discussing issues of geomorphological heritage and suggesting itineraries to visit the most important sites. The series provides a unique reference source not only for geomorphologists, but all Earth scientists, geographers, and conservationists. It complements the existing reference books in geomorphology which focus on specific themes rather than regions or localities and fills a growing gap between poorly accessible regional studies, often in national languages, and papers in international journals which put major emphasis on understanding processes rather than particular landscapes. The World Geomorphological Landscapes series is a peer-reviewed series which contains single and multi-authored books as well as edited volumes.

World Geomorphological Landscapes – now indexed in Scopus®!

Jasper Knight · Stefania Merlo · Andrea Zerboni
Editors

Landscapes and Landforms of the Central Sahara

Editors
Jasper Knight
School of Geography, Archaeology
and Environmental Studies University
of the Witwatersrand
Johannesburg, South Africa

Stefania Merlo
McDonald Institute for Archaeological Research
University of Cambridge
Cambridge, UK

Andrea Zerboni
Dipartimento di Scienze della Terra "A. Desio"
Università degli Studi di Milano
Milan, Italy

ISSN 2213-2090 ISSN 2213-2104 (electronic)
World Geomorphological Landscapes
ISBN 978-3-031-47159-9 ISBN 978-3-031-47160-5 (eBook)
https://doi.org/10.1007/978-3-031-47160-5

This Springer imprint is published by the registered company Springer Nature Switzerland AG
The registered company address is: Gewerbestrasse 11, 6330 Cham, Switzerland

Paper in this product is recyclable.

Series Editor Preface

Landforms and landscapes vary enormously across the Earth, from high mountains to endless plains. At a smaller scale, Nature often surprises us by creating shapes which look improbable. Many physical landscapes are so immensely beautiful that they have received the highest possible recognition—they hold the status of World Heritage properties. Apart from often being immensely scenic, landscapes tell stories which not uncommonly can be traced back in time for tens of million years and include unique events. In addition, many landscapes owe their appearance and harmony not solely to natural forces. For centuries, or even millennia, they have been shaped by humans who modified hillslopes, river courses and coastlines and erected structures which often blend with the natural landforms to form inseparable entities.

These landscapes are studied by Geomorphology—'the Science of Scenery'—a part of Earth Sciences that focuses on landforms, their assemblages, the surface and subsurface processes that moulded them in the past and that change them today. The shapes of landforms and the regularities of their spatial distribution, their origin, evolution and ages are the subject of research. Geomorphology is also a science of considerable practical importance since many geomorphic processes occur so suddenly and unexpectedly, and with such a force, that they pose significant hazards to human populations and not uncommonly result in considerable damage or even casualties.

To show the importance of geomorphology in understanding the landscape and to present the beauty and diversity of the geomorphological sceneries across the world, we have launched a new book series *World Geomorphological Landscapes*. It aims to be a scientific library of monographs that present and explain physical landscapes, focusing on both representative and uniquely spectacular examples. Each book will contain details on geomorphology of a particular country or a geographically coherent region. This volume covers not a country, but a region divided into a number of countries—the central Sahara. It epitomises global desert landscapes, being both extremely diverse, remote, difficult to reach and challenging to do research, although our understanding of even very remote and seemingly hostile lands has recently significantly increased due to access to remote sensing techniques and data sources. The central Sahara hosts fascinating bedrock landscapes, including several inscribed as World Heritage properties, dune fields, desert lakes or their vestiges and striking valley systems, largely inherited from previous geological periods. It also contains a fantastic geomorphological record of major environmental changes which have affected the contemporary deserts during the Pleistocene. Last but not least, they show a range of human adaptations to desert conditions, illustrating very close linkages between landforms, geomorphic processes, cultures and economy. Chapters contained in this book offer various glimpses into this unique geomorphic environment.

The *World Geomorphological Landscapes* series is produced under the scientific patronage of the International Association of Geomorphologists—a society that brings together geomorphologists from all around the world. The IAG was established in 1989 and is an independent scientific association affiliated with the International Geographical Union and the International Union of Geological Sciences. Among its main aims are to promote geomorphology and to foster dissemination of geomorphological knowledge. I believe that this lavishly illustrated series, which sticks to the scientific rigour, is the most appropriate means to fulfil these aims and to serve the geoscientific community. To this end, my great thanks go to the editors of the volume, Jasper Knight, Stefania Merlo and Andrea Zerboni, who attempted a difficult task to coordinate the book that involved many authors from different countries and ensured that the final product meets expectations. I am grateful to all individual contributors who agreed to participate in this project and delivered such interesting stories to read. Sadly, the current political situation in many Saharan countries makes on-site visits to see these marvellous landscapes hardly possible. The book thus not only fills the gap in knowledge and extends our understanding of desert landscapes, but also reveals a world beyond the reach of most of us.

Wrocław, Poland Piotr Migoń

Contents

Introduction to the Sahara: The Romance and Mystery of the Desert

1

Jasper Knight, Stefania Merlo and Andrea Zerboni

Abstract

The Sahara as a whole is a contested space with different meanings and identities generated both from within and imposed by external actors. This chapter discusses the nature, background and history of this contested space, focusing on its physical landscape and evolution, and the ways in which varied human activities have been shaped by, and in turn have shaped, its range of landscapes and multiple identities.

Keywords

Sahara Desert · Identity · Sense of place · Colonialism · Tuareg

1.1 Introduction

The physical landscape represented by the Sahara region of North Africa occupies a key space geographically and in the public imagination (Murphy 2013) (Fig. 1.1). It is a landscape of romance and mystery but also of misconception. In part this arises from the Sahara's large area, its general inaccessibility and hostile climatic environment, but also from its long and varied human history (Goudie 2019). General descriptions of the Sahara and its physical environments have been provided in several studies and are drawn from here. However, these descriptions refer to the whole Sahara region and do not always capture the physical variability present within the region. For many, the Sahara is a region of sand dunes whereas, in reality, dunes and sand sheets cover only 28% of the land surface compared to 43% covered by bedrock mountains and 10% by rock surfaces and hamadas (Goudie 2013, p. 36). Further, the presence of mountain massifs > 3000 m in elevation, such as the Hoggar, Tibesti, Ennedi and Aïr massifs, split the central Sahara into distinctive semi-connected basins. The names and meanings of these massifs and basins are inscribed by different tribal and language groups in those areas (Gearon 2011, p. xii), and the nature of this varied topography and the different substrates are likely to have influenced historical migration patterns and thus the ways in which different peoples have made use of and valued space (Lecocq 2015; Lydon 2015; Braun and Passon 2020). The Saharan landscape as a whole combines both monotonous uniformity extending over vast distances, and unexpected local diversity in rock outcrops, wadi channels and oases peppering this landscape, and with a distinctive human imprint. Of the Libyan sand dunes, the British engineer and pioneer of aeolian sediment transport studies Ralph Bagnold remarked:

> Here, instead of finding chaos and disorder, the observer never fails to be amazed at a simplicity of form… In places vast accumulations of sand… move inexorably, in regular formation… growing, retaining their shape… in a manner which, by its grotesque imitation of life, is vaguely disturbing to an imaginative mind. Elsewhere the dunes are cut to another pattern – lined up in parallel ranges, peak following peak in regular succession like the teeth of a monstrous saw…. (Bagnold 1941, p. xix)

1.1.1 What and Where is the Sahara?

The transliterated word *Sahara* has an Arabic root, derived from ارحص meaning desert, and has been used in a geographical sense from at least the ninth century AD (Gearon

J. Knight (✉)
School of Geography, Archaeology & Environmental Studies, University of the Witwatersrand, Johannesburg 2050, South Africa
e-mail: jasper.knight@wits.ac.za

S. Merlo
McDonald Institute for Archaeological Research, University of Cambridge, Downing Street, Cambridge CB2 3ER, UK
e-mail: sm399@cam.ac.uk

A. Zerboni
Dipartimento di Scienze della Terra "Ardito Desio", Università degli Studi di Milano, 20133 Milan, Italy
e-mail: Andrea.Zerboni@unimi.it

J. Knight et al. (eds.), *Landscapes and Landforms of the Central Sahara*, World Geomorphological Landscapes, https://doi.org/10.1007/978-3-031-47160-5_1

Fig. 1.1 Tableau of some key **a–c** physical and **d–f** human elements of the Central Sahara landscape, Fezzan region, SW Libya. *Photos* Stefania Merlo

2011). Thus the term "Sahara Desert" is a somewhat of a tautology. Likewise, the distinctive areas of different sand seas, bedrock uplands and depressions means that there is no single uniform landscape or regional identity of the Sahara that adequately captures its diversity. The spatial area covered by the Sahara is also under debate, and the Sahara can be defined from different disciplinary viewpoints. The most common approach is the region delimited by the 100 mm yr^{-1} isohyet (Harrison Church et al. 1964). However, in particular the southern boundary of the Sahara is very much a broad Sahelian zone 150–200 km wide (Mainguet 1983) which may be better defined on a geomorphic rather than climatic basis. This geomorphic approach defines the southern Sahara limit as the location where sand surfaces are stabilised (i.e. are no longer moving by aeolian sediment transport), which is generally the position of the 200 mm yr^{-1} isohyet (Mainguet 1983). Based on the area encompassed by the 200 mm yr^{-1} isohyet, desert size therefore varies interannually according to rainfall amount and distribution (Tucker et al. 1991; Tucker and Nicholson 1999). Much of the general literature, including atlases, cite an area of around 7–9 million km^2. Based on vegetation mapping from satellite imagery, where vegetation is controlled by rainfall, desert area values for the period 1980–1997 average 9,150,000 km^2 but vary by up to 16% from this figure in any one year (Tucker and Nicholson 1999) (Fig. 1.2). Increased desert area in 1984, for example, was linked to the strong Sahelian drought experienced in that year. On the longer term, the Sahara Desert has expanded overall during the twentieth century by 11–18%, but this value is dependent on the timeframe of analysis and also varies geographically across North Africa (Thomas and Nigam 2018). A different approach to identifying the extent of the desert environment was put forward by Foody (2001) who examined normalised difference vegetation index (NDVI) values for the southern Sahara Sahel region and then used

Fig. 1.2 Changes in Sahara Desert area, 1980–1997, based on NDVI mapping from satellite imagery. Graph drawn from data presented in Tucker and Nicholson (1999)

an (unstated) threshold value in order to identify a boundary between desert and Sahel. This approach has utility because a defined threshold can be automatically applied consistently across a region based on remote sensing data, and can map seasonal changes in the desert–Sahel interface. Foody (2001) found a strong relationship between interannual changes in desert area and rainfall ($r = -0.84$), and migration of this desert–Sahel interface is driven by increases (decreases) in vegetation by increased (decreased) rainfall.

Definition of the Sahara as a region can also take place according to the cultural identities of its constituent peoples (MacEachern 2007; Mattingley et al. 2007; Lecocq 2015). Identifying cultural identity/ies is, however, problematic as the region has been affected by different episodes of invasion, migration and trade, involving different ethic, language and religious groups, over the last 2000 years or so (Gearon 2011). These shifting cultural imprints in the Sahara and in North Africa more generally challenge the colonial viewpoint of what is meant by region and identity in the Sahara, prevalent throughout the 19th and early twentieth centuries (Cleaveland 1998; MacEachern 2007). Today, this is still being played out in the construction of a national or regional identity in the Western Sahara, for example (Murphy 2013; Campos-Serrano and Rodríguez-Esteban 2017). Thus, the Sahara remains a contested and multidimensional space. The central Sahara region covered in this book focuses geographically on the most arid centre of the Sahara and thus of North Africa, and includes the area around the shared borders of Mauritania, Mali, Algeria, Niger, Libya and Chad, and covering an area of ~3.5 million km^2 (shown in Fig. 1.2). The central Sahara does not include all of these countries' territories: for example, the Mediterranean coastal fringes of Algeria and Libya are not included, but this book does not adopt a strict geographical or climatic definition of the region.

1.2 Saharan Climate and Landscapes

General physical properties (climate, landforms, evolution) of the Sahara as a whole have been described in summary form in several texts (e.g. Cloudsley-Thompson 1984; Lancaster 1996, pp. 212–213; Goudie 2013, pp. 303–307). The most detailed book available on Saharan landscapes, specific landforms and their evolution is by Busche (1998), written in German (no English translation is available). A commonly used climatic descriptor of the Sahara is that of annual rainfall. A potential problem however is that weather stations are sparsely distributed (Nicholson 1979) and the high interannual variability of rainfall means that long records are required in order to derive any meaningful viewpoint of average rainfall. Satellite data are today used routinely to evaluate climate patterns across the region, in particular dust and other aerosols in response to synoptic circulation patterns (e.g. Ashpole and Washington 2013; Hobby et al. 2013). There is less emphasis, however, on building long-term climate datasets from either in situ observational or reanalysis data.

Climatic properties of the Sahara can also be based on calculation of climatic aridity. This refers to the ratio P/PET, in which P is mean annual precipitation and PET is mean annual potential evapotranspiration (UNESCO 1979). The latter is strongly related to mean annual air temperature and relative humidity. Climate records from the region support a more humid Mediterranean coastal fringe (Merabti et al. 2018), but the central Sahara itself is classified as hyperarid (precipitation < 100 mm yr^{-1}; P/PET < 0.03). Based on these problems with weather station records, satellite data are now used for climatological purposes. Kelley (2014) used Tropical Rainfall Measuring Mission (TRMM) satellite data for the period 1998–2012 to identify spatial and temporal patterns of rainfall across North Africa, in particular in western Egypt/eastern Libya. Seasonality of rainfall strongly relates to precipitation source, especially in more humid fringes of the region, but seasonality is more subdued in drier areas.

The precipitation regime of the Sahara is generally low (hyperarid) but is highly variable with episodic and localised torrential rain events. Patterns of present annual rainfall show very low values in the central Sahara (<20 mm yr^{-1}) with steep precipitation gradients in particular through the Atlas mountain range from Morocco to Tunisia. Precipitation and relative humidity also increase from east to west through the Sahara (UNESCO 1979). Snow falls occasionally over the northern Sahara mountains. Recent weather records also show that the Sahara is one of the world's hottest regions, with a maximum recorded temperature of 51.3 °C recorded in September 1979 in Algeria (RADP 2010). A minimum temperature in Algeria of −13.8 °C was recorded in January 2005 (ibid). The temperature regime is highly variable, and diurnal temperature range may be a control on thermal weathering (cataclastism).

Today's landscapes of the central Sahara reflect both contemporary and past climate, and different physiographic regions have been identified in the central Sahara (Mainguet 1983). The eastern part of the region comprises bedrock massifs, plateaus and intervening basins. Sand seas are present within many of these basins and are a well-known landscape element of the Sahara. The western part of the region is broadly flat with a slight westward dip. Volcanic intrusions provide the basis for isolated uplands of the Hoggar and Aïr highlands. Geomorphological features typically found in the Sahara are described in Table 1.1, based on transliterated local Arabic words and their geomorphic meanings.

Table 1.1 Geomorphic meaning of some place-name terms used in the Sahara, transliterated from the Arabic (after Salem and Busrewil 1980, p. xi)

Feature name	Geomorphic/topographic meaning
Awaynat	Spring
Bahr	Sea, lake
Baltat	Depression, mudflat
Birkah	Intermittent lake
Dur	Hills or hilly area surrounding a plain
Emi	Mountain massif
Erg	Sand sea
Gilf	Escarpment
Graras	Endorheic basin into which wadis flow
Hamada	Bare bedrock plateau
Hatiyat	Depression, oasis, valley
Hawa	Cave, sinkhole
Jabal	Mountain
Kaf	Cliff, mountain, peak, ridge, spur
Khurmah	Pass
Naqqazah	Terrace
Qarah	Low, flat-topped hill
Qararah	Depression, valley
Reg (serir, ténéré)	Gravel plain
Sabkha	Saline salt marshes or salt flats
Tadrart	Mountain, massif
Tassili	Barren plateau
Wadi (oued)	Intermittent river channel/valley
Zahr	Escarpment, plateau

Not all place-name terms are included, and the transliterated spelling may vary between sources or in different publications

1.3 Peoples and Cultures of the Sahara

The literature dealing with the peoples of the Sahara, their histories, interactions, traditions and values spans the disciplines of archaeology, anthropology, cultural studies and historical geography and is thus somewhat dispersed. Some general texts also describe the longer term and regional-scale development of human activity in the Sahara (e.g. Austen 2010; Braun and Passon 2020; Scheele 2020). One useful proxy for past human occupation of the Sahara and use of its resources is from the archaeological record, where material archaeological evidence (stone-wall settlements, rock art, palaeoecological records) attests to different phases of human activity (e.g. Mattingley et al. 2007; Merlo et al. 2013). Such studies, constrained by radiocarbon dating, have been able to establish the relationships between phases of human activity and climate, such as the timing of occupation/abandonment of settlements and how environmental resources are used by populations at different times (e.g. Cremaschi and Zerboni 2009; Ruan et al. 2016; Scheele 2020). This includes the "Green Sahara" within the African Humid Period extending from the late glacial to early Holocene, in which increased rainfall resulted from changes in monsoon rainfall penetration into the region (de Menocal 2015). An important developing area of Saharan research refers to cultural landscapes and the meanings and identities ascribed to landscape features (rock outcrops, water sources), in particular through rock art and stone-built structures (di Lernia 2006; di Lernia and Tafuri 2013; Barnett and Guagnin 2014; Honoré 2019). This evidence can link landscape physical properties to their social, political, economic and cultural contexts (e.g. Brass 2019) (Fig. 1.1).

Culturally, different groups are found in different areas, with for example Berbers in southern Algeria, Tuareg in southern Sahara, Tibu in the Tibesti region and Moors in the Western Sahara (Gearon 2011). This pattern is the outcome of historical events of migration, occupation and interactions between different groups both within the Sahara, and by Islamic, Ottoman and Arabic influences from outside the region (Cleaveland 1998; Lecocq 2015; Braun and Passon 2020). Thus, the peoples and cultures of the Sahara are not amorphous, and their trading networks and sociocultural interactions have shaped the development of Saharan cities, farming patterns and styles and their associated non-material cultures (Mattingley and Sterry 2013; Knight 2022). A summary of human occupation in the Sahara to more recent times is given by McDougall (2019), describing European imperial expansion in the region during the nineteenth century, mainly by France, which was seen as a *mission civilatrice* (civilising mission). Explorers' accounts, such as by Mungo Park and Wilfred Thesiger, focused on the customs and practices of Saharan peoples, viewed through a western colonial lens (Lupton 1979; Fookes et al. 2019), and 19th-century fiction and newspaper accounts are replete with heroic imperialist stories (Livingstone 2018). The Sahara has also been a place of colonial speculation: in the 1870s, a plan by French engineers was proposed to create an inland sea of 13,280 km^2 (Anon 1879) by flooding low-lying areas of north-east Algeria. Mackenzie (1877), arguing in favour of the proposal, outlined "the desirability and feasibility of opening it up to Commerce and Civilization" (p. iii). Further, "it would improve rather than hurt the climate of Europe, while it would fertilize Northern Africa" (p. xiii). Plympton (1886) examined the physical and climatic impacts of such a proposal.

1.3.1 The Colonial Gaze

The Sahara, its peoples, landscapes and their meanings, has most commonly been viewed in the western, 20th-century literature through a colonial lens (Lecocq 2015; Domínguez-Mujica et al. 2018). Further, the present governance and physical infrastructure of the region strongly reflects its colonial inheritance (Davis 2017). Exploration and colonial expansion by different European powers has meant that primary sources from government and military records are archived in different places in Europe and Africa, and not widely available. Much of the extant literature today is based on only a partial analysis of these records, or based on secondary sources, and primary evidence from indigenous knowledge systems and the views and values of local people, today or in the past, has been less commonly studied (Love 2017; Warscheid 2018; McDougall 2019). However, these sources are vital because they can inform on pre- and post-colonial worlds, values and cultures. A well-known example is Timbuktu, Mali, which has been a cosmopolitan centre of migration, trading and scholarship since at least the twelfth century, and influenced by repeated episodes of political occupation and cultural imprints including from Islamic and Arabic peoples (De Villiers and Hirtle 2007; Krätli and Lydon 2010; Braun and Passon 2020). This cultural palimpsest is shown through the World Heritage-inscribed status of the urban centre of Timbuktu, the urban street pattern, Islamic design motifs, and historical and religious manuscript libraries (Park 2010; Warscheid 2018; Baumanova et al. 2019). Today, the mythologising of Timbuktu in the geographic and popular imagination (Heffernan 2001; Andersson 2019) reflects the combination of physical and human environmental properties and processes, set in a historical context. A post-colonial interpretation of space in the Sahara emphasises the continuity of culture and history across the region (Cleaveland 1998; Lecocq 2015; Lydon 2015; Domínguez-Mujica et al. 2018; Barker 2019), and the dissonance between this and present international boundaries and national identities (Murphy 2013; Campos-Serrano and Rodríguez-Esteban 2017).

1.4 Contemporary Issues and Future Research Directions in the Sahara

The contemporary Sahara is less physically accessible than it was in the past (up to as late as the 1970s), largely because of safety concerns due to political instability, insurgence and religious extremism in the region. As a consequence, access for tourists and researchers in many countries is restricted and dangerous. Although this is problematic for field-based research, remote sensing methods can be successfully deployed to give new insight into regional-scale landscape and geology patterns, including mineral resources and groundwater (Peña and Abdelsalam 2006; Lamri et al. 2016; Lezzaik and Milewski 2018), and palaeodrainage patterns (Abdelkareem and El-Baz 2015). Exploitation of oil and gas resources, and groundwater pumping for agriculture and developing cities, means that extensive subsurface geological data exist for some areas such as southern Libya, but this is largely not available to researchers for commercial reasons. Associated with this resource exploitation, however, is pollution and soil/water contamination, and this is a significant environmental concern (Fayiga et al. 2018). In addition, threats to archaeology and built heritage are posed by political unrest, terrorism, urbanisation and climate change (di Lernia 2015). In summary, these physical and human elements of Saharan landscapes are vulnerable to a range of contemporary processes associated with climate change, development and political context. Future pure and applied research to safeguard the unique landscapes of the Sahara is needed. This book

Fig. 1.3 Landscapes and landforms of the central Sahara region, as discussed in this book, mainly come from locations within the red circle, including the regions shared by Algeria, Libya, Chad, Niger and Mali. The locations of specific sites discussed in some chapters are numbered according to chapter number: 5—Ennedi, 10—Ubārī Sand Sea, 12—Lake Chad, 13—Bodélé Depression. Other chapters discuss sites from across the region (base map: Bing Maps, reproduced under Creative Commons license)

explores elements of this multifaceted landscape, including its physical properties, Quaternary evolution, meanings and landscape–human interactions.

1.4.1 Structure of This Book

This book presents both overlies and detailed case studies of central Sahara landscapes and landforms and their wider connections to past and present climate, environments, human occupation and heritage. The broad region of interest is enclosed within the red circle in Fig. 1.3. Most chapters in this book examine specific environments and themes with locations across the central Sahara; other chapters describe specific localities that are also marked in Fig. 1.3. In detail, Chapters 2, 3, 4, 5, 6, 7, 8, 9, 10, 11, 12, 13, 14 and 15 deal mainly with landscape geology, geomorphology and landscape evolutionary history. Chapters 16, 17, 18, 19, 20, 21, 22 and 23 deal mainly with the intersectionality of the Saharan landscape with the human world, both past and present. Each chapter of this book is written by international experts, and all chapters were externally peer-reviewed as well as by the book editors and the book series editor Professor Piotr Migoń. We thank all authors and reviewers for their contributions to this book.

Acknowledgements We thank Jennifer Fitchett for comments on a previous version of this chapter.

References

Abdelkareem M, El-Baz F (2015) Evidence of drainage reversal in the NE Sahara revealed by space-borne remote sensing data. J Afr Earth Sci 110:245–257

Andersson R (2019) The Timbuktu syndrome. Soc Anthropol/anthropol Sociale 27:304–319

Anon (1879) Flooding the Sahara. Nature 19:509

Ashpole I, Washington R (2013) A new high-resolution central and western Saharan summertime dust source map from automated satellite dust plume tracking. J Geophys Res Atmos 118:6981–6995

Austen RA (2010) Trans-Saharan Africa in world history. Oxford University Press, New York, p 157

Bagnold RA (1941) The physics of blown sand and desert dunes. Methuen, London, p 264

Barker G (2019) Libyan landscapes in history and prehistory. Libyan Stud 50:9–20

Barnett T, Guagnin M (2014) Changing places: rock art and Holocene landscapes in the Wadi Al-Ajal, south-west Libya. J Afr Archaeol 12:165–182

Baumanova M, Smejda L, Rüther H (2019) Pre-colonial origins of urban spaces in the West African Sahel: street networks, trade, and spatial plurality. J Urban Hist 45:500–516

Brass M (2019) The emergence of mobile pastoral elites during the middle to Late Holocene in the Sahara. J Afr Archaeol 17:53–75

Braun K, Passon J (eds) (2020) Across the Sahara: tracks, trade and cross-cultural exchange in Libya. Springer, Cham

Busche D (1998) Die Zentrale Sahara. Justus Perthes Verlag, Gotha, p 284

Campos-Serrano A, Rodríguez-Esteban JA (2017) Imagined territories and histories in conflict during the struggles for Western Sahara, 1956–1979. J Hist Geogr 55:44–59

Cleaveland T (1998) Islam and the construction of social identity in the nineteenth-century Sahara. J Afr Hist 39:365–388

Cloudsley-Thompson JL (ed) (1984) Key Environments: Sahara Desert. Pergamon Press, Oxford, p 348

Cremaschi M, Zerboni A (2009) Early to Middle Holocene landscape exploitation in a drying environment: Two case studies compared from the central Sahara (SW Fezzan, Libya). C R Geosci 341:689–702

Davis MH (2017) "The Transformation of Man" in French Algeria: economic planning and the postwar social sciences, 1958–62. J Contemp Hist 52:73–94

de Menocal PB (2015) End of the African humid period. Nature Geosci 8:86–87

De Villiers M, Hirtle S (2007) Timbuktu. In: The Sahara's Fabled City of Gold. Walker & Company, New York, p 302

di Lernia S (2006) Cultural landscape and local knowledge: a new vision of Saharan archaeology. Libyan Stud 57:5–20

di Lernia S (2015) Save Libyan archaeology. Nature 517:547–549

di Lernia S, Tafuri MA (2013) Persistent deathplaces and mobile landmarks: the Holocene mortuary and isotopic record from Wadi Takarkori (SW Libya). J Anthropol Archaeol 32:1–15

Domínguez-Mujica J, Andreu-Mediero B, Kroudo N (2018) On the trail of social relations in the colonial Sahara: a postcolonial reading. Soc Cult Geogr 19:741–763

Fayiga AO, Ipinmoroti MO, Chirenje T (2018) Environmental pollution in Africa. Environ Develop Sustain 20:41–73

Foody QM (2001) Monitoring the magnitude of landcover change around the southern limits of the Sahara. Photogram Eng Remote Sens 67:841–847

Fookes P, Waring D, Lee M (2019) Wilfred Thesiger and the Umm as Samim, Oman. Geol Today 35:221–227

Gearon E (2011) The Sahara. A cultural history. Oxford University Press, Oxford, p 264

Goudie AS (2013) Arid and semi-arid geomorphology. Cambridge University Press, Cambridge, p 454

Goudie A (2019) Desert exploration in North Africa: some generalisations. Libyan Stud 50:59–62

Harrison Church RJ, Clarke JI, Clarke PJH, Henderson HJR (1964) Africa and the Islands. Longman, London, p 494

Heffernan M (2001) "A dream as frail as those of ancient Time": the in-credible geographies of Timbuctoo. Environ Planning D Soc Space 19:203–225

Hobby M, Gascoyne M, Marsham JH, Bart M, Allen C, Engelstaedter S, Fadel DM, Gandega A, Lane R, McQuaid JB, Ouchene B, Ouladichir A, Parker DJ, Rosenberg P, Ferroudj MS, Saci A, Seddik F, Todd M, Walker D, Washington R (2013) The fennec automatic weather station (AWS) network: monitoring the saharan climate system. J Atmos Ocean Technol 30:709–724

Honoré E (2019) Prehistoric landmarks in contrasted territories: rock art of the Libyan Desert massifs. Egypt. Quat Int 503:264–272

Kelley OA (2014) Where the least rainfall occurs in the Sahara Desert, the TRMM radar reveals a different pattern of rainfall each season. J Clim 27:6919–6939

Knight J (2022) Presence, absence, transience: the spatiotemporalities of sand. Geographies 2:657–668

Krätli G, Lydon G (eds) (2010) The trans-Saharan book trade. Manuscript culture, Arabic literacy and intellectual history in Muslim Africa. Brill Books, Leiden, p 424

Lamri T, Djemaï S, Hamoudi M, Zoheir B, Bendaoud A, Ouzegane K, Amara M (2016) Satellite imagery and airborne geophysics for geological mapping of the Edembo area, Eastern Hoggar (Algerian Sahara). J Afr Earth Sci 115:143–158

Lancaster N (1996) Desert environments. In: Adams WM, Goudie AS, Orme AR (eds) The physical geography of Africa. Oxford University Press, Oxford, pp 211–237

Lecocq B (2015) Distant shores: a historiographic view on trans-Saharan space. J Afr Hist 56:23–36

Lezzaik K, Milewski A (2018) A quantitative assessment of groundwater resources in the Middle East and North Africa region. Hydrogeol J 26:251–266

Livingstone JD (2018) Travels in fiction: baker, Stanley, Cameron and the adventure of African exploration. J Victorian Cult 23:64–85

Love PM (2017) The colonial pasts of medieval texts in northern Africa: useful knowledge, publication history, and political violence in colonial and post-independence Algeria. J Afr Hist 58:445–463

Lupton K (1979) Mungo park, the African traveler. Oxford University Press, Oxford, p 272

Lydon G (2015) Saharan oceans and bridges, barriers and divides in Africa's historiographical landscape. J Afr Hist 56:3–22

MacEachern S (2007) Where in Africa does Africa start? Identity, genetics and African studies from the Sahara to Darfur. J Social Anthropol 7:393–412

Mackenzie D (1877) The Flooding the Sahara: an account for the proposed plan for opening central Africa to commerce and civilization from the north-west coast, with a description of Soudan and Western Sahara, and notes on ancient manuscripts, &c. Sampson Low, Marston, Searle, & Rivington, London, p 287

Mainguet M (1983) Tentative mega-geomorphological study of the Sahara. In: Gardner R, Scoging H (eds) Mega-Geomorphology. Clarendon Press, Oxford, pp 113–133

Mattingley DJ, Sterry M (2013) The first towns in the central Sahara. Antiquity 87:503–518

Mattingley D, Lahr M, Armitage S, Barton H, Dore J, Drake N, Foley R, Merlo S, Salem M, Stock J, White K (2007) Desert migrations: people, environment and culture in the Libyan Sahara. Libyan Stud 38:115–156

McDougall EA (2019) Saharan peoples and societies. Oxford Research Encyclopedia, African History. https://doi.org/10.1093/acrefore/9780190277734.013.285

Merabti A, Meddi M, Martins DS, Pereira LS (2018) Comparing SPI and RDI applied at local scale as influenced by climate. Water Res Manag 32:1071–1085

Merlo S, Hakenbeck S, Balbo AL (2013) Desert migrations project XVIII: the archaeology of the northern Fazzan. A preliminary report. Libyan Stud 44:141–161

Murphy AB (2013) Territory's Continuing Allure. Ann Assoc Am Geogr 103:1212–1226

Nicholson SE (1979) Statistical typing of rainfall anomalies in Subsaharan Africa. Erdkunde 33:95–103

Park DP (2010) Prehistoric Timbuktu and its hinterland. Antiquity 84:1076–1088

Peña SA, Abdelsalam MG (2006) Orbital remote sensing for geological mapping in southern Tunisia: implication for oil and gas exploration. J Afr Earth Sci 44:203–219

Plympton GW (1886) Flooding the Sahara. Science 7:542–544

RADP (République algérienne démocratique et populaire) (2010) Seconde communication nationale de l'Algerie sur les changements climatiques a la CCNUCC. Projet GEF/PNUD 00039149. Ministère de l'Aménagement du Territoire et de l'Environnement, Algier, p 202

Ruan J, Kherbouche F, Genty D, Blamart D, Cheng H, Dewilde F, Hachi S, Edwards RL, Régnier E, Michelot J-L (2016) Evidence of a prolonged drought ca. 4200 yr BP correlated with prehistoric settlement abandonment from the Gueldaman GLD1 cave, Northern Algeria. Clim past 12:1–14

Salem MJ, Busrewil MT (eds) (1980) The geology of Libya, vol III. Academic Press, London

Scheele J (2020) Connectivity and its discontents: the Sahara—Second face of the Mediterranean? Zeit Ethnol 145:219–236

Thomas N, Nigam S (2018) Twentieth-century climate change over Africa: seasonal hydroclimate trends and Sahara desert expansion. J Clim 31:3349–3370

Tucker CJ, Nicholson SE (1999) Variations in the size of the Sahara Desert from 1980 to 1997. Ambio 28:587–591

Tucker CJ, Dregne HE, Newcomb WW (1991) Expansion and contraction of the Sahara Desert from 1980 to 1990. Science 253:299–300

UNESCO (1979) Map of the world distribution of arid regions. UNESCO, Paris, p 54

Warscheid I (2018) The Islamic literature of the precolonial Sahara: sources and approaches. Hist Compass 16:e12449. https://doi.org/10.1111/hic3.12449

Geology and Long-Term Landscape Evolution of the Central Sahara

2

Jasper Knight

Abstract

The geological history of the central Sahara region extends over ~3.5 billion years and encompasses tectonics, climate and sediment system responses to their varied forcings. The main elements of the geological setting of the central Sahara are summarised in this chapter, including volcanism, sedimentary basin development and meteorite impacts. Long-term geological history sets the scene for geomorphic development during the Cenozoic (last 66 million years) and the Quaternary (last 2.58 million years) and provides the physical framework for today's Saharan landscapes.

Keywords

Cenozoic · Climate change · Mantle swells · Tectonic history · Volcanism · Weathering

2.1 Introduction

The central Sahara occupies a vast region (~4.5 million km^2 examined here) of North Africa and encompasses a wide range of geologies, extending from Precambrian to Quaternary in age, and including intrusive and extrusive igneous, metasedimentary and sedimentary rocks (Fig. 2.1). The varied rock types and long geological history of the central Sahara region have implications for how climatic forcing over 10^9–10^{11}-year timescales has impacted on geomorphological processes of weathering and erosion, and therefore on landscape development. This means that today's Saharan physical landscapes have a strong geologic as well as climatic imprint. Recent overviews of the geology and geologic evolution of Africa as a whole are given by Summerfield (1996), Schlüter (2008) and van Hinsbergen et al. (2011). The geology of the Sahara has been described by Williams (1984) and Fabre (2005), and with respect to the Cenozoic by Busche (1998, pp. 17–68), Guiraud et al. (2005) and Swezey (2009). There is also specific detail on the geology of Libya, a key area of oil and gas exploration, by Conant and Goudarzi (1967), Salem and Busrewil (1980) and Tawadros (2001). Equivalent summaries for many other countries or North African regions are not readily available. A history of geological mapping in Africa is outlined briefly by Schlüter (2008, pp. 9–11), and a summary of the tectonic context and history of the African continent is provided by Burke and Gunnell (2008) and van Hinsbergen et al. (2011).

Although there are several regional-scale studies on Saharan geology (Klitzsch 1994; Fabre 2005; Schlüter 2008), there is a lack of detailed field data both spatially and by time period across the Sahara. This may be due in part to the size and relative inaccessibility of the region, for transport and security reasons. In addition, although there have been industry investigations of subsurface geology (using gravity and seismic methods) and borehole logging for oil and gas field development, these data are often commercially sensitive and may not be readily available to researchers. Field data on Saharan geology may also lack up to date analytical information such as on geochemistry and geochronology. As a consequence, remote sensing technologies have been most commonly used to examine macroscale stratigraphy and tectonic evolution at a regional scale (Ballantine et al. 2005), but to do this effectively ground-truthing and field geochemical and stratigraphic data are also needed. Thus, Saharan geology as a topic of investigation presents for researchers a mixture of both great opportunities and significant constraints.

In detail, this chapter describes (1) the large-scale distribution of rock types and their ages across the central Sahara region; and (2) the tectonic setting of the African continent,

J. Knight (✉)
School of Geography, Archaeology & Environmental Studies, University of the Witwatersrand, Johannesburg 2050, South Africa
e-mail: jasper.knight@wits.ac.za

J. Knight et al. (eds.), *Landscapes and Landforms of the Central Sahara*,
World Geomorphological Landscapes, https://doi.org/10.1007/978-3-031-47160-5_2

with respect to its plate tectonic and orogenic history, and with specific reference to the movement of mantle swells thought to be contributors to long-term epeirogenic uplift and subsequent landscape denudation. (3) Specific elements of Saharan geology are then examined, which are volcanism, sedimentary basins and meteorite impacts. Finally, (4) geologic and environmental changes during the Cenozoic and Quaternary are then described, which is important because these contribute to today's topographic and sediment patterns and macroscale landscapes and landforms.

2.2 Macroscale Geologic Patterns and Geological Development of the Central Sahara

A generalised regional-scale geologic map of the central Sahara is shown in Fig. 2.1. Cenozoic-age sediments dominate the land surface throughout the region, especially within topographic basins where surface sediments accumulate, mainly within sand seas (see Knight and Merlo, this volume). These surface sediments (which may be tens

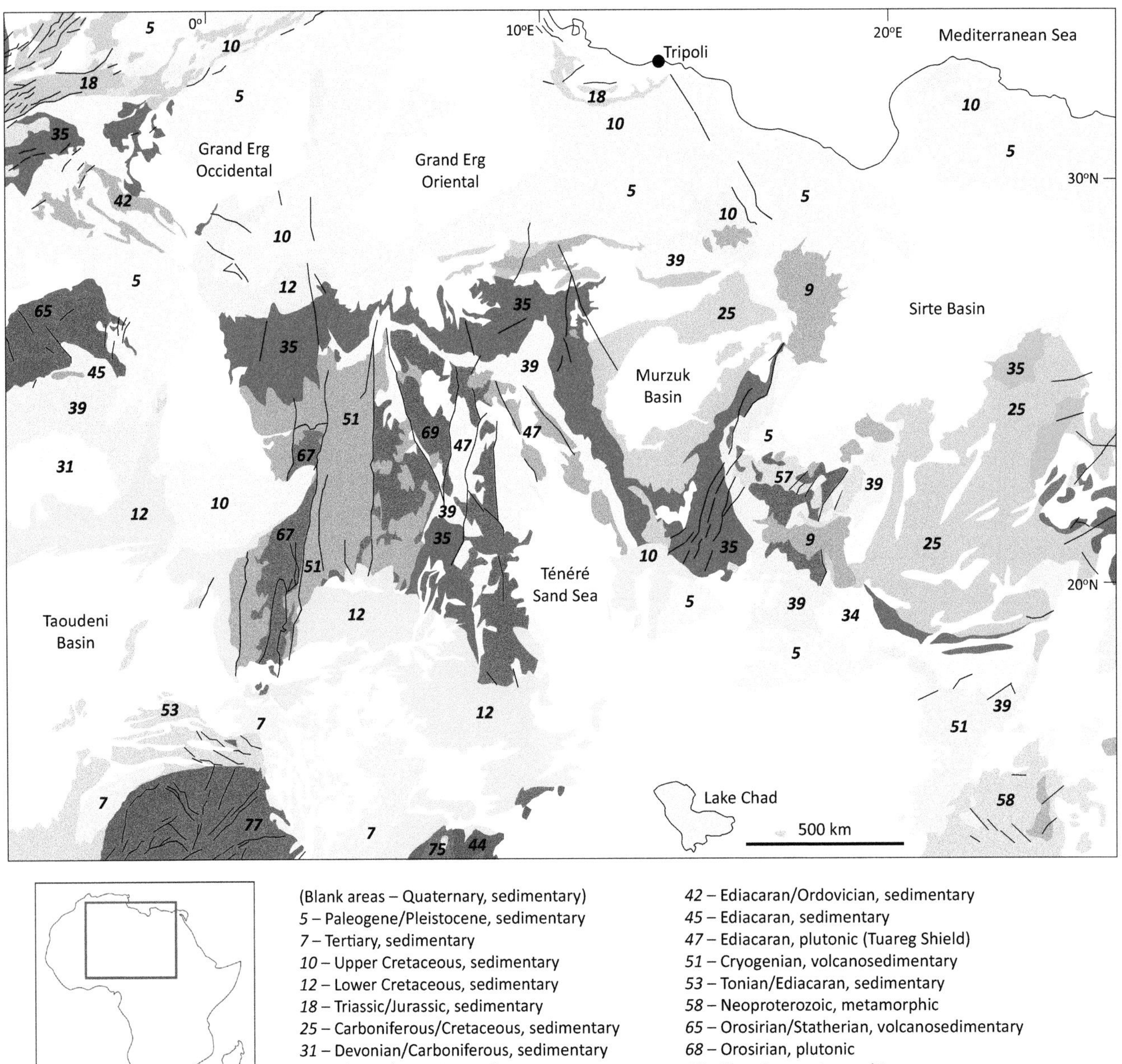

Fig. 2.1 Regional geological map of the central Sahara region (redrawn and simplified after BRGM 2016). Continuous black lines mark the position of faults

of metres thick) therefore conceal the nature of underlying bedrock which remains largely unknown (Senosy et al. 2013). Areas of igneous and metamorphic blocks are commonly demarcated by parallel faults that reflect terrane assembly and associated metamorphism, and often form topographic highs (Guiraud et al. 2005). This is consistent with the emergence of these areas over time as a consequence of low surface denudation rates. Basement rocks underlying the central Sahara region are Archaean in age, and represent accreted elements of three cratons that are present across North Africa, which are the West African, East Sahara and Uweinat-Chad cratons (Schlüter 2008). These cratons include a range of igneous and metamorphic/metasedimentary rocks that span some 3 billion years' (Ga) duration (Key 1992) and are separated by sheared or sutured margins where mountain building and volcanism has taken place (Black et al. 1994; Liégeois et al. 2003). The major fault zones and rifts across North Africa were mapped by Guiraud et al. (2005, their Fig. 3). Fault zones commonly frame the upland blocks of Tibesti, Aïr and Hoggar. The East Sahara and Uweinat-Chad cratons are separated by the Tibesti orogenic belt (Key 1992). There is a north–south sense of shear between the West African and the East Saharan cratons (Nkono et al. 2018); the Uweinat-Chad craton has south-eastward lateral movement. Thus, many areas of the central Sahara are underlain by this old eroded basement. Successive phases of crustal development are recorded in proxy by isotopic studies of detrital zircons within river systems that inform on changes in river catchment source areas over time. Results show isotopic peaks at 2.7, 2.1–1.8, 1.2–1.0, 0.7–0.5 and 0.3 Ga (Iizuka et al. 2013), and these periods correspond to key phases in the tectonic development and breakup of Gondwana (e.g. Torsvik and Cocks 2011).

Today, no Precambrian rocks are exposed in the central Sahara region although elements of intracratonic shield rocks likely underlie North Africa (Torsvik and Cocks 2011). These correspond to two distinctive tectonosedimentary cycles (Fabre 2005). Evidence for tectonic assembly during the Pan-African Orogeny at ~600–530 Ma is shown by the presence of slab-like accreted terranes of the Tuareg shield that are stacked against each other such as in the Hoggar massif (Black et al. 1994; Amara et al. 2017). During the Cambrian, lowland terrestrial to shallow marine conditions existed in the Sahara region, recorded by the deposition of fluvial to marine sandstones and, in places, coarser arkoses and greywackes. Early Cambrian glacial sediments exist over the West African craton, with marine carbonates and ash beds (Guiraud et al. 2005). Crustal deformation, rifting and volcanism took place around the West African craton at this time. During the Ordovician, fluvial to marine sandstones were deposited across much of the region, reflecting marine transgression. Gondwana was then affected by extensive glaciation during the late Ordovician while the palaeo-South Pole was located over the Sahara region, and at least three episodes are known to have affected the region. Evidence for this is the presence of striations, glacial valleys, erratics and glacigenic sediments preserved as diamictites (see Ghienne and Le Heron, this volume). Both before and after glacial episodes, shelf sandstones and carbonates were deposited in marginal areas in the Rheic Ocean. Climate warming in the Silurian, in association with Gondwana continental drift, led to sea-level transgression and deposition of graptolitic shales including the hydrocarbon source rocks of the Tanezzuft Formation (discussed below), and then by fluvial marine sandstones in subsiding basins. Tethys rifting took place from the end of the Silurian onwards and was associated with coeval sea-level fall. In the Devonian, transgression across north-west Africa and under a subtropical climate led to deposition of marginal evaporites and marine shales, and fluvial sandstones in more continental areas, such as the Tadrart Formation (discussed below). Tectonic reactivation of lineaments took place throughout the Devonian. Warm conditions and deposition of limestones continued into the Carboniferous and especially in North Africa through to the end of the Carboniferous. Continental collision of Gondwana with Laurussia (forming the Pangaea megacontinent) took place in the early Carboniferous with the Hercynian orogeny in the late Carboniferous. This latter event was associated with uplift in the central Sahara and a dominance of folding, transpression and shortening in north-west Africa. This also affected the north–south fault zones through the Hoggar, for example. Glaciation took place during the late Carboniferous/early Permian but few rocks of this age are known in the region. The opening of Neotethys at the end of the Permian is associated with volcanic activity in East Africa and some locations in North Africa. Rifting and some transgression in north-west Africa continued through the Triassic. Rifting in East Africa was associated with magmatic activity extending into southern Egypt. In the Jurassic, rifting conditions continued, mainly in north-west Africa in the Atlas Mountains, and marine deposition took place in marginal areas. In the Cretaceous, more extensive rifting took place throughout the region, and as the African Plate split from the Arabian and South American plates. Continental deltaic deposition took place in the central Sahara in the early Cretaceous, followed by transgression and carbonate deposition from the north of the region and into the Chad and Niger basins (Azil and Ait Ouali 2021). Events in the Cenozoic and Quaternary are reviewed below.

2.3 Tectonics and Topography

A tectonic overview of Africa as a whole is given by Burke and Gunnell (2008) and key stages in the assembly of Africa were summarised by Schlüter (2008). The Pan-African orogeny united different cratons into Gondwana and Pangaea, followed by breakup during the Jurassic and Cretaceous. A significant driver of continental-scale dynamics and land surface geomorphology is the role of mantle swells, which are areas of lithospheric uplift that result from shallow convection within the upper mantle. Burke and Gunnell (2008, their Fig. 2.2) identified eight swell locations within the central Sahara region (Tripoli, Sawda, Haruj, Tibesti, Uweinat, Ahaggar, Adrar, Aïr), and these are all associated with locations where middle to late Cenozoic volcanism is recorded, in between the more stable cratonic blocks (Ashwal and Burke 1989). Land surface uplift from mantle swells or by volcanism is commonly influenced by structural lineaments, which act as locations of differential shear between the structural blocks (Nkono et al. 2018).

The topography and relief of the Sahara region is variable: upland areas (summits and plateaus) reach 3400 m asl at Tibesti, 3000 m at Jebel Marra, 2900 m at Hoggar and 2000 m at Aïr, and these highlands are separated by basins. The macroscale geomorphology of the region is discussed by Mainguet (1983). This includes African Surfaces

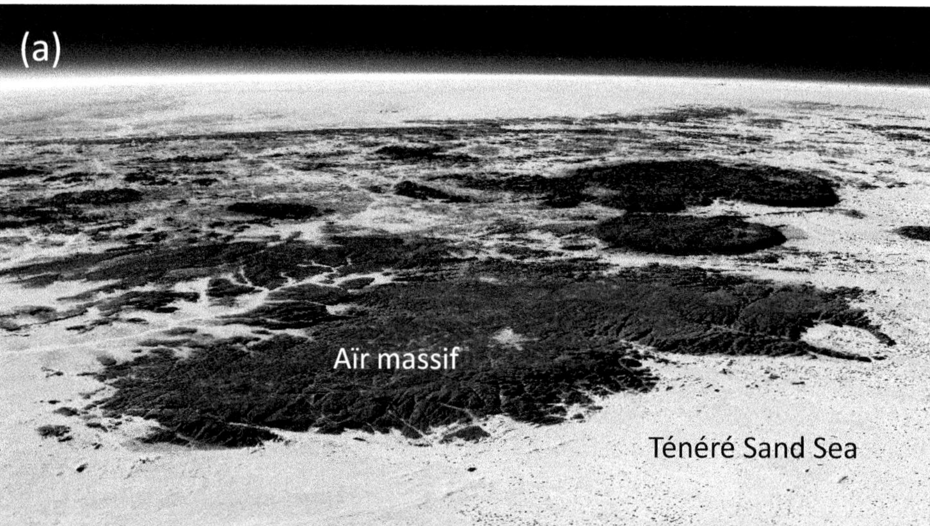

Fig. 2.2 Examples of volcanic provinces of the central Sahara. **a** Oblique view to the north-west of volcanics of the Aïr massif (source: http://earth.imagico.de), **b** Tarso Toussidé volcano, Tibesti massif (source: NASA Earth Observatory). Reproduced under Creative Commons License

that correspond to peneplanated erosion surfaces that are of regional extent (Guillocheau et al. 2018). Several composite surfaces are identified across Africa, the most prolonged phase of development of which took place around 180–130 Ma during the late Jurassic and early Cretaceous. The relationship of these surfaces to periods of land surface uplift is not certain: this may vary from a deterministic forcing–response relationship, to a lagged and transient relationship caused by a slower land surface response and redistribution of erosional products. Existing dating and elevational evidence suggests that Africa surfaces are composite features (Partridge and Maud 1987). In northern Africa, the African surfaces are commonly marked by laterites or occasionally bauxites that develop on stable, long-lived surfaces subject to long-term chemical weathering (Burke and Gunnell 2008). Evidence for planated surfaces tends to come from areas outside of the central Sahara itself, in Ethiopia and Egypt, where phases of peneplanation are linked to activity of the Afar plume. Regionally, a significant unconformity at 30 Ma marks a period of global sea-level fall, and this also provided accommodation space for sedimentary basin infilling such as in the Chad and Congo basins.

2.4 Volcanism

Emplacement of igneous intrusions took place episodically throughout the geological history of the central Sahara (described above), and this was most commonly related to rifting (e.g. Paquette et al. 1998; Nkono et al. 2018) and evidence for volcanism found on the surface dates mainly from the Paleozoic and Cenozoic (Williams 1984; Elshaafi and Gudmundsson 2017). There are also more recent volcanoes associated with intraplate mantle plumes, and the locations and eruptive history of individual volcanic provinces reflect the position and strength of these plumes. Key centres of volcanism include the Hoggar, Tibesti, Jebel Marra and Aïr massifs. Palaeozoic volcanism is located outside of the stable West African and Congo cratons. By contrast, volcanism during the Cenozoic was associated with north–south fracture zones present through Algeria and Libya (Keppie et al. 2011). This was also accompanied by significant uplift of about 500 m in the Jebel Marra and 1000 m in the Tibesti and Hoggar massifs (Williams 1984). The calculated volumes of igneous products associated with these massifs are 3000 km^3 for Tibesti and 8000 km^3 for Jebel Marra (Williams 1984).

Volcanic development of the Hoggar massif is particularly complex because it comprises accreted units from the margin of the West African craton during the Pan-African Orogeny (Paquette et al. 1998; Liégeois et al. 2003; Amara et al. 2017). Radiometric dates on granodiorite and granite from the Hoggar massif correspond to different accretion phases within this orogenic period (Liégeois et al. 2003), and emplacement of these plutons then led to mineralisation from circulating fluids (Aissa and Marignac 2017). Late Miocene to recent alkali basalts and trachytes are also present, and geochemical analysis suggests a mantle plume origin (Dupuy et al. 1993). Volcanism in Tibesti dates from the Oligocene onwards. Active craters and rhyolitic lava slopes in this area were described by Grove (1960) and mapped by Permenter and Oppenheimer (2007). Here, the volcanic province covers an area of 59,400 km^2, with an average caldera diameter of 12 km, and flank slopes of 10–30° (Permenter and Oppenheimer 2007). The volcanic complex comprises wide and relatively flat shield volcanoes with basaltic lava fields and surficial tephra deposits, with ignimbrite sheets (Deniel et al. 2015) (Fig. 2.2). Different volcanic phases are also identified based on superposition relationships and degree of weathering, showing that the location of volcanic activity has migrated across the area over time, and this may match the direction and pace of migration of the African Plate over an underlying mantle plume. The chronology of volcanism here is (after Deniel et al. 2015): (1) plateau flood basalt volcanism at 12–8 Ma during the Miocene with the migration of feeder dikes from north-east to south-west over time; (2) development of composite calderas in the late Miocene and along a central NNE–SSW fault zone; (3) development of three major ignimbrite volcanic complexes from the late Miocene to middle Pleistocene as a result of doming and then collapse of caldera basins; (4) deposition of basaltic surface lava flows and tephras, mainly in the period 7–5 Ma; and (5) more subdued volcanic activity in pre-existing caldera basins, extending to the present day. This broad chronology is similar to lava shields elsewhere in the central Sahara region (Keppie et al. 2011; Elshaafi and Gudmundsson 2017).

2.5 Sedimentary Basins

Different sedimentary basins formed at different times in the central Sahara as a result of different forcing factors (Guiraud et al. 2005). Epeirogenic sag basins such as in north-east Libya developed as a result of lithospheric downwarping. Basins also developed as a result of rifting, such as the petroleum-rich Sirte Basin of northern Libya (Bellini and Massa 1980; van Houten 1983; van der Meer and Cloetingh 1993). Interior (endorheic) basins developed inland located between raised bedrock massifs. Palaeozoic basins were affected by the Caledonian (Middle Silurian to Lower Devonian) and Hercynian orogenies (Middle to Upper Carboniferous), resulting in faulting and extension, limited volcanism, and basin downwarping. Global tectonic and climatic changes were accompanied by changes in sea level that affected coasts, variously inundating or uplifting

Saharan margins and resulting in coastal and marine sedimentation. This took place from the Triassic onwards in embayments or gulfs demarcated by an archipelago of existing highlands, which resulted in large-scale sedimentary systems becoming developed across these eroded land surfaces. The subdued topography of the Sahara has meant that sedimentation systems and the directions of sediment fluxes that follow drainage and slope patterns have been responsive to uplift and erosion cycles (Williams 1984) and therefore tectonic processes. The sedimentary history of the Sahara is therefore linked to many different aspects of tectonism, including plate movement, plumes and mantle swells, and emplacement of igneous bodies. The dominance of sandstone, mudstones, siltstones and limestones as the major sedimentary rocks recorded within the Sahara is testimony to the reworking of weathered products by water, and thus is climatically controlled. As a consequence, arid phases tend to be relatively quiescent with respect to regional sediment deposition, with limited reworking by water or wind. Thus, the extensive surficial sediments forming the Saharan sand seas can also be considered as relict products of long-term weathering (see Knight, this volume).

Sedimentary basins of the Sahara have been mainly examined from the context of development of hydrocarbon source rocks (Conant and Goudarzi 1967; Turner 1980). The nature, history and sediments of two significant Saharan basins (Murzuq and Sirte) are now described, as illustrative of sedimentary processes and environments during the geological development of the Sahara, and as an expression of sedimentary response to basin subsidence and the creation of relief.

The Murzuq Basin in Libya is a triangular-shaped subsiding basin with an area of 320,000 km^2 (Bellini and Massa 1980). It developed in the East Saharan Craton during the Paleozoic and was affected by multiple extensional tectonic phases related to the Caledonian and Hercynian orogenies. Shallow marine and fluvial sediments were deposited on basement Precambrian rocks during the Cambrian, followed by uplift in the Lower Ordovician (Ramos et al. 2006) (Fig. 2.3). This area was strongly affected by, and retains a significant erosional and depositional record of, late Ordovician (Hirnantian) glaciation (Moreau 2011). These effects of glaciation were influenced by bedrock structures (faults, tectonic blocks). The development of tectonic highs (e.g. Tihemboka Arch, Brak-Bin Ghanimah High, Gargaf Arch) separated the Murzuq Basin from adjacent basins. The relative interplay of epeirogenic uplift and basinal subsidence from the Silurian to Jurassic led to different phases of terrestrial to shallow marine sedimentation across the Murzuq Basin (Le Heron et al. 2013), with sediments sourced from different areas (Morton et al. 2011).

In detail, a major unconformity corresponding to land surface incision was developed in response to the Pan-African Orogeny, and this marks the bottom of the basin infill (Guiraud et al. 2005) (Fig. 2.3). The base of this infill (Cambrian) comprises cross-bedded sandstones and breccias of the Mourizidie Formation and similar sandy cross-beds of the overlying Hassouna Formation. Fossiliferous Ordovician deposits attest to a marine depositional setting, and shallowing upwards into the Silurian. The upper Silurian/lower Devonian has a rich flora biodiversity from the Acacus to Tanezzuft formations, and these represent distinctive sedimentary cycles of sandstones to siltstones and shales. A period of emergence, folding, erosion and development of palaeosols followed the Silurian. Devonian strata (Tadrart Formation, Ouan Kasa Formation) are marine-influenced sandstones to mudstones with abundant plant debris. During the Carboniferous, transgression across much of the region resulted in deposition of fossiliferous carbonates, shales and calcareous sandstones, and variations

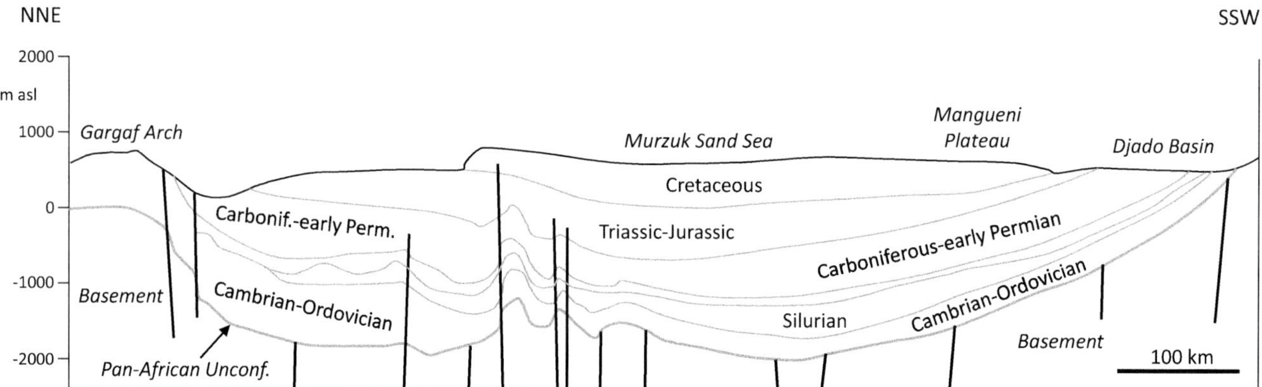

Fig. 2.3 Schematic cross section through the Murzaq Basin (after Guiraud et al. 2005). Brown lines represent unconformities. Thick vertical lines represent faults

in water depth from lagoonal to shelf settings. Gypsum and stromatolites are also present in some levels. The Jurassic to Cretaceous Messak Sandstone (formerly "Nubian Sandstone" in this area) represents deposition in braided fluvial channels in a slowly subsiding basin (Lorenz 1980). These channels were separated by ponded areas in which clays were deposited. These sediments were unconformably overlain by late Cretaceous to Paleogene carbonates. This generalised stratigraphy varies somewhat spatially, likely as a result of differential subsidence across the basin. The burial depths of Silurian source rocks such as the Tanezzuft Formation have led to maturation of hydrocarbons (Galushin and Eloghbi 2014) which are today being exploited by oil and gas wells.

The Sirte Basin (north-east Libya) contains a range of tectonically controlled platforms and basins with 1500–7000 m thickness of sediment with ages from Cambrian to Cenozoic (Tawadros 2001; Gumati 2021). The Sirte Basin is best known as one of the world's most significant hydrocarbon reservoirs, with source rocks deposited mainly in the Silurian, late Cretaceous and Eocene (Macgregor 1996). The distribution of faults and horst/graben blocks has split the basin into a number of sub-basins and troughs that developed, and thus became separated from each other, at different times. The chronology of geological events within the Sirte Basin is given by Guiraud et al. (2005, their Fig. 48). Overall, pre-graben, syn-graben and sag basin stages have been recognised, but these are poorly constrained in time and space (Gumati and Nairn 1991). The later pre-graben history of the Sirte Basin comprises mantle upwelling and consequent doming and thinning of the overlying crust in the period 140–100 Ma (early Cretaceous) above a mantle hotspot (van Houten 1983). Thinning and doming likely took place in a number of episodes, but these may have varied in timing and magnitude across the basin. The pre-rift stage most likely terminated in the Albian (mid-Cretaceous, ~100 Ma). The syn-rift stage through structural collapse of the Sirte Arch took place likely through the Upper Cretaceous and Eocene (~100 to 34 Ma) (van der Meer and Cloetingh 1993). This rifting was related to slab subduction down the Hellenic Trench, culminating during the early Eocene (Capitanio et al. 2009). The rifting phase was characterised by crustal extension with syn-rift infilling by a range of sediments including sandstones (quartzites), shales, evaporites and carbonates. Stratigraphic analysis of these syn-rift sediments reveals the phasing relationships between sediment infill and tilting, and faulting of basement rocks (Gras 1998; Gumati 2021). Post-rift downwarping by thermal subsidence took place mainly during the Eocene (~56 to 34 Ma). Well log data show that subsidence in the Sirte Trough reached 2085 m, whereas adjacent horsts show <1000 m of subsidence (Gumati and Nairn 1991).

2.6 Meteorite Impact Structures

Several dozen meteorite impact structures are known across Africa (Koeberl 1994; Reimold and Koeberl 2014), and because of their spatial scale (from around 350 m to several hundred km in diameter) remote sensing methods have been commonly used to identify and map them (e.g. Koeberl et al. 2005; Schmieder and Buchner 2007; Lobpries and Lapen 2019). Geomorphically, impact structures are broadly circular with a bowl-shaped interior and raised outer rim which reflect the lithospheric response to impact followed by mantle compression, uplift and collapse. Examples from North Africa are shown in Fig. 2.4. Aorounga (northern Chad) is a circular impact structure 12.6 km in diameter (Koeberl 1994) that has a surrounding ring ridge that is 100 m higher than the interior depression (Fig. 2.4a). Based on different radar datasets, the Aorounga structure appears to comprise several sets of ridges and depressions, arranged in a concentric pattern, with a maximum diameter of 16 km (Reimold and Koeberl 2014). Today, the structure is dissected by longitudinal desert dunes (Koeberl et al. 2005). The Gweni-Fada structure (Fig. 2.4b) appears to be slightly asymmetric in topography with a local relief of 280 m (Koeberl et al. 2005) and with a total impact crater width of ~21 to 23 km. Based on structural and geophysical data, both of these structures are highly eroded (Reimold and Koeberl 2014). The age control on individual impact events is provided by isotopic radionuclide dating. The ability to date these structures using this method arises because the very high-impact velocities, and thus high pressure and temperature conditions that result, allow for rock recrystallisation and the formation of unusual rock types such as tektites (Koeberl 1994, 2002). Impact structures also have a clear lithostratigraphic and thus age context with respect to overlying impact breccias or other sedimentary rocks. For the examples shown in Fig. 2.4, Gweni-Fada has an age of <345 Ma, Tenoumer of 2.5 ± 0.5 Ma, and Aorounga of 10 ka (but this age is disputed; Reimold and Koeberl 2014, p. 92) (Reimold and Gibson 2009). Even older impacts are recorded in Africa, such as Vredefort (South Africa) at >2 Ga. The preservation in the landscape of evidence for such ancient events attests to both the role of these high magnitude events in shaping surface topography and subsurface lithospheric structure, and the high preservation potential of the variety of geological evidence for these events (Koeberl 1994). The presence of bare bedrock surfaces also more readily enables remote sensing tools to be used, such as different spectral band combinations (van Gasselt et al. 2017; Lobpries and Lapen 2019), or Shuttle Radar Topography Mission (SRTM) data to produce a Digital Elevation Model (DEM) of the land surface (Koeberl et al. 2005; Schmieder and Buchner 2007)

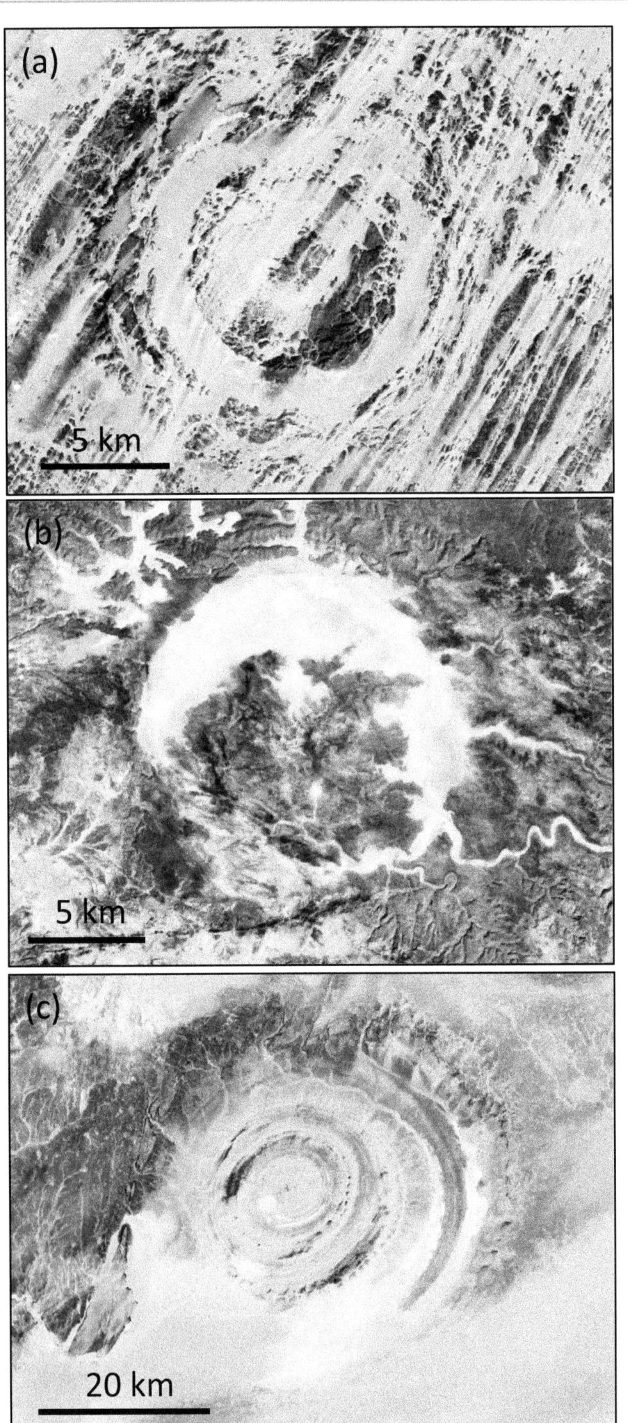

Fig. 2.4 Examples of meteorite impacts in North Africa. **a** Aorounga, Chad (19° 06′N, 19° 15′E), **b** Gweni-Fada, Chad (17° 25′N, 21° 45′E), **c** Tonoumer, Mauritania (22° 55′N, 10°24′E). *Source* Google Earth

which can be used to infer impact size, shape and age. In the central Sahara region, post-impact weathering, over long time scales, has in some cases amplified local relief with depressions in the landscape becoming filled with sand (e.g. Fig. 2.4b) and has also influenced the position of watersheds and drainage patterns.

2.7 Cenozoic Climates and Environments

Saharan landscape development in response to climate change during the Palaeogene and Neogene within the Cenozoic Era (66–2.58 Ma) sets the scene for the impacts of higher-frequency climate changes associated with global glacials and interglacials during the Quaternary (2.58 Ma to present). A detailed summary of the Cenozoic evolution of the Sahara is given by Fabre (2005, his Chap. 9) and Swezey (2009), and this evolution reflects the interplay between climate, tectonics and geomorphological processes (Mainguet 1983; Sepulchre et al. 2009). A reduction in uplift, volcanism and rifting, in addition to regional aridification towards the end of the Neogene, contributed to land surface stability, meaning that most geomorphic elements of the central Saharan landscape were in place before the onset of the Quaternary. As such, the macroscale landscapes of today's central Sahara can be considered as relict features.

Different phases of Cenozoic sedimentation in different Saharan basins are shown in Fig. 2.5. Magmatic activity and extensional rifting took place in some basins throughout the Cretaceous and into the Paleocene. Marine-influenced environments dominated during the Paleocene and Eocene, associated with high sea levels and hot climatic conditions when the African Plate was located south of its present position (Swezey 2009). This period corresponds to the Late Paleocene Thermal Maximum (Fig. 2.5). In the Eocene, marine depositional conditions continued with the development of a carbonate platform, supported by the development of phosphates and lignite at this time. Fluvial deposits and evaporites were present in areas marginal to the marine embayment. A fall in Eocene sea levels and coeval uplift of the Atlas Mountains and regional folding and faulting led to development of a prominent unconformity that marks an absence of depositional signatures or with quiescent sedimentary conditions through the Oligocene, except for localised fluvial deposition in marginal areas (Guiraud et al. 2005). In the Miocene, fluvial deposition dominated throughout the central Sahara region in response to sea-level rise and increased humidity. Evidence for Saharan palaeohydrology at this time is discussed by Knight (this volume). Higher fluvial sediment availability may have been driven by mountain uplift. Changes in the direction of fluvial drainage during the Miocene are related to tectonism as well as changes in basinal accommodation space. In some places, lacustrine, overbank deposits and ferruginous duricrusts also developed during the Miocene, especially outside of the main river channel systems (Sghari 2014). Upper Miocene sediments are more variable as the climate became more arid, culminating in peak aridity during the Messinian Salinity Crisis (~5.9 to 5.3 Ma) associated with decreased sea level in the Mediterranean basin and leading to increased incision of inflowing rivers. Volcanic activity

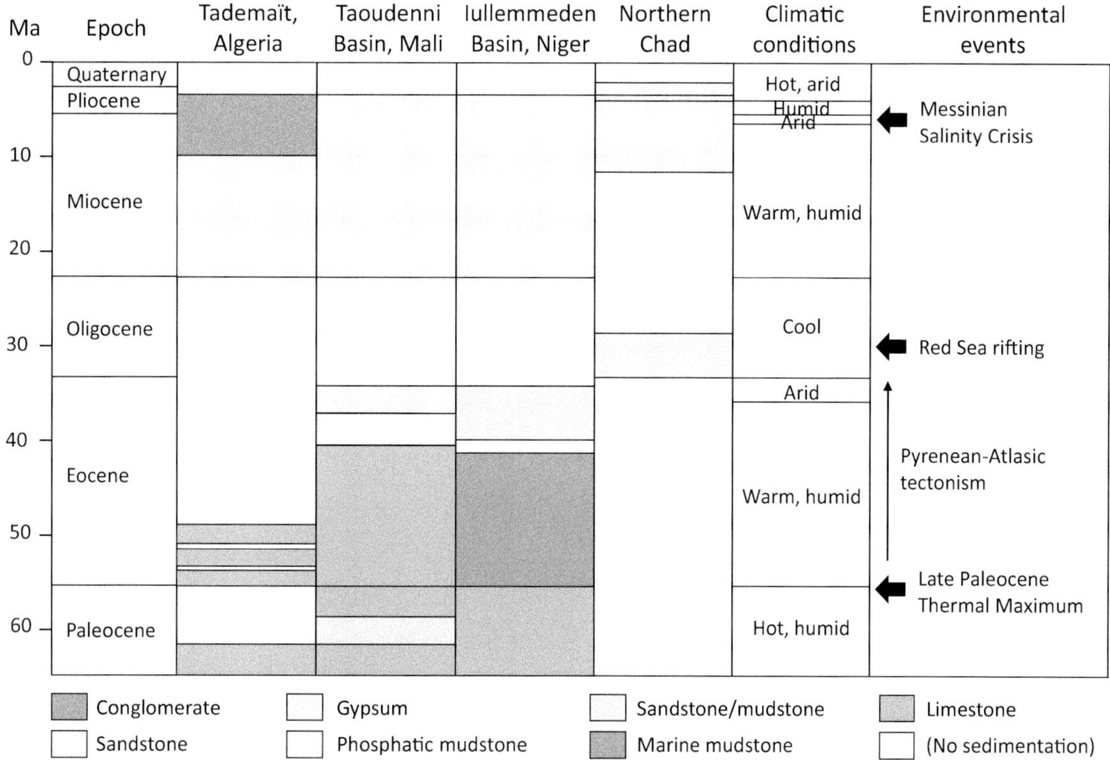

Fig. 2.5 Examples of the different depositional environments and lithologies found in different central Sahara sedimentary basins during the Cenozoic (redrawn from Swezey 2009, with addition information from Guiraud et al. 2005)

also took place extensively throughout the region during the Miocene, including in the Hoggar, Aïr, Jebel el Haruj, Tibesti and Darfur areas. During the Pliocene, sandstone and mudstone deposition continued and longer-term aridity became established across North Africa. This aridification and establishment of the sand seas present within the central Sahara today took place as a result of the coincidence of the northward movement of the African Plate to dryer latitudes, and strengthening of the easterly jet stream in response to Tibetan Plateau uplift (see Knight and Merlo, this volume). Such tectonic activity resulted in dryer subsiding air in east Africa; and development of continental glaciers in Antarctica and later in the northern hemisphere which strengthened trade wind circulation and decreased tropical sea surface temperatures, resulting in lowered precipitation values (Williams 1984). Increased aridity in the central Sahara from the late Cenozoic and into the Quaternary dried up lake basins and rivers and facilitated blown sand and dust generation. Therefore, the development of today's Saharan sand seas is the culmination of prolonged aridity, mainly from the Miocene onwards. This aridification also resulted in the development of chemical weathering products, both within surface sediments (forming duricrusts, see below) and on bare rock surfaces, forming hamadas and desert pavements (see Knight, this volume).

2.7.1 Development of Duricrusts

Cenozoic weathering and erosion and long-term chemical weathering led to the development of cap rock duricrusts at the land surface. Processes contributing to development of duricrusts in Africa is described by Faniran and Jeje (1983, their Chap. 8) and Nash and McLaren (2007). These duricrusts are genetically linked to deep subsurface weathering within the rock profile. The nature of duricrusts in the central Sahara region has not been fully described in terms of duricrust type, chemistry and evolutionary history. However, weathering of sandstones leads to mobilisation of iron and silicate minerals and thus development of different duricrusts, which can enhance differential weathering rates over space.

The properties and distribution of duricrusts of different types and origins found in the northern Sahara region are discussed by Goudie (1973). Calcrete is the most common type found in the Sahara, and silcrete has also been identified in some areas such as Fezzan (Libya) although its distribution is less well known. Calcrete thickness reaches a maximum of 10–20 m with individual induration hardpans in the range 10–35 cm. Calcrete reaches 5–10 m in thickness in Algeria (Goudie 1973). The processes contributing to calcrete formation in the Sahara have not been studied in detail. Their distribution on plateau surfaces (Fig. 2.6)

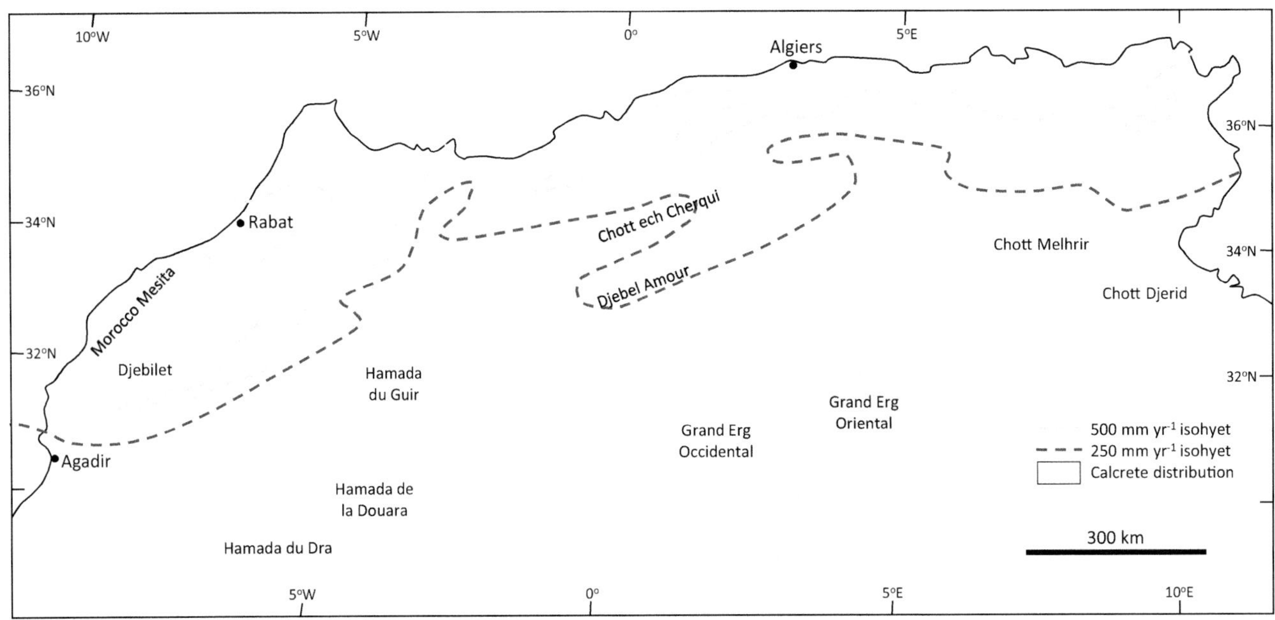

Fig. 2.6 Distribution of calcrete in north-west Africa (redrawn from Goudie 1973, p. 77)

suggests a combination of in situ chemical precipitation and secondary development by the accumulation of blown dust (e.g. Horta 1980). Different calcrete members are identified regionally in northern Libya, associated with *terra rossa* soils and formed during cooler glacials (Tawadros 2001, p. 302). For example, the mean composition of calcrete from North Africa was given by Goudie (1973, his Table 5b) as 75% $CaCO_3$ and 11% SiO_2, but with very wide ranges of individual values. Abdullah et al. (2017) used U/Th dating of calcretised lake sediments in the Murzuq Basin, Libya, to suggest formation during a pre-Pleistocene to middle Pleistocene dry period. In the Maghreb, thick calcrete beds formed during the Eocene are found in association with regionally extensive malacological faunas that have affinities with southern Europe and the Mediterranean (Bensalah et al. 1991). The thickness and spatial extent of duricrusts exert a strong control on landscape denudation because these chemically hardened surfaces tend to reduce erosion rates and thus preserve palaeosurfaces in the landscape. For example, K–Ar dating of chemical weathering products from an African Surface at ~2290 m asl in Eritrea yields ages of 58.6 ± 2.3 and 59.4 ± 2.3 Ma which dates the surface to the late Paleocene (early Cenozoic) (Perelló et al. 2020). This type of evidence can therefore help constrain the age of any stable land surface, and thus the nature of landscape development during the Cenozoic. Secondary karstic dissolution of calcrete can also occur, and this has been noted in the Sahara region (Horta 1980). Surface weathering and lateritisation during the late Cretaceous, forming hematite, goethite and gibbsite, have also been noted in the south-east Egyptian desert (El Desoky et al. 2020).

2.8 Quaternary Climate Changes and Their Impacts in the Central Sahara

Palaeoenvironmental records for the Quaternary period (2.58 Ma to present) are relatively uncommon from the central Sahara region. In terms of number of records, these tend to be either from the Neogene and dealing specifically with palaeohydrology and palaeoriver systems (see Knight, this volume), or from the late Holocene. There are few records that span the last few glacial–interglacial cycles and, as such, the Quaternary of the Sahara is not well resolved. Broadly speaking, glacials were dry and interglacials were wet in the Sahara, as a result of changes in atmospheric circulation patterns (determining the position of the intertropical convergence zone and monsoon rain belts), and that these were orbitally controlled (Kutzbach et al. 2020). However, shorter pluvial phases also occurred within glacials and dryer phases within interglacials, especially during climate transitions. As today, locations closer to ocean moisture sources were also wetter than more continental areas. These general climatic characteristics fit with much longer and continuous Quaternary records from speleothems in southern Arabia (Nicholson et al. 2020). This also highlights that palaeoclimatic and palaeoenvironmental Quaternary records of the Sahara, as elsewhere, have to be considered in a spatial and temporal context, and in responding to topographic controls of the local area (e.g. Knight and Fitchett 2021).

A Middle Pleistocene sedimentary record (last 400 kyr) from the Murzuq Basin is one of the longest available in the region and shows wackestone and packstone units interbedded with conglomerates and lacustrine limestones (Geyh and Thiedig 2008). These latter units, which are of regional extent and contain coquina beds, attest to lake formation during pluvial Quaternary phases at 420, 320, 250 and 130 ka, constrained by U/Th dating. ^{13}C and ^{18}O isotopic records suggest that the intensity of these wetter periods decreased over time. Cave sand with concretions, found on the floor of caves in the Fezzan, southern Libya, are dated by the luminescence method to 90–60 kyr, corresponding to Marine Isotope Stages (MIS) 4 and 5a–c (Cremaschi 1998). It is notable that no sediment accumulation then occurred until the Holocene, presumably due to regional aridity. Luminescence dating of dune and palaeolake shoreline deposits in this region shows wetter periods during MIS 5c, d and in the early/middle Holocene (Drake et al. 2018) with a possible earlier lacustrine period at 420 ka (Armitage et al. 2007). Wetter conditions and higher groundwater table position in southern Egypt at 115 ka (MIS 5e), 102 ka (MIS 5c,d), 80 ka (MIS 5a) and 41 ka (MIS 3) and also supported lake development at these times (Nicoll 2018). Speleothem growth also took place intermittently during these wetter periods (El-Shenawy et al. 2018).

Dryer periods during the Quaternary are marked by sand dune reactivation in the southern Saharan Sahel. Potential earlier phases of dune activity are not able to be readily identified by luminescence dating where sand grains have been later reworked and their exposure signals zeroed, but ground-penetrating radar techniques can be used to show the internal structures of subsurface sediments (e.g. Drake et al. 2018). Available luminescence ages on sand dunes from the Sahel zone show latest phases of dune activity during MIS 2 (~20 to 15 ka) especially in Nigeria and Mauritania, the lateglacial (~12 to 9 ka) in Niger and Mauritania, the early Holocene (~10 to 6 ka) in Sudan, the mid-Holocene (~7 to 5 ka) in Nigeria, and the last 1000 years in Sudan and Mali (Bristow and Armitage 2016). Swezey (2001) showed that different dated dune phases are found in different regions of the Sahara. Stratigraphic, dating and geochemical evidence shows that phases of dust export from the central Sahara took place during short dry episodes coincident with North Atlantic Heinrich events (Ehrmann et al. 2017; Torfstein et al. 2018) (see Knight, this volume). This may also reflect increased seasonality in rainfall through the lateglacial period, as revealed by $\delta^{18}O$ analysis of bovine tooth enamel (Reade et al. 2018). Dryer land surface conditions across the region during Heinrich events also led to changes in the position of the Sahara–Sahel boundary and the spread of sand dunes into former Sahelian areas (Collins et al. 2013). It is also notable in the Saharan palaeorecord that there is very little biological (pollen) evidence preserved, if it was indeed even present except in much wetter areas such as the Atlas Mountains (Tabel et al. 2016). High chemical weathering during wetter phases may have destroyed such evidence, although phytoliths are preserved (Parker et al. 2008).

Holocene records are far more numerous and somewhat more diverse as they include lake and sabkha environments (Pachur and Hoelzmann 2000). These records show wetter conditions during the early Holocene that correspond to the later part of the African Humid Period (see Knight, this volume) with more arid conditions prevailing after around 7000 BP. A sabkha record from north-east Niger shows lake conditions existed from 10.6 to 7.3 ka, forming laminated diatomite, with high organic content and low minerogenic content (Brauneck et al. 2013). This was then followed by increased clastic and then sand input after 7.3 ka, indicative of drying out of the sabkha environment. Bubenzer et al. (2007) mapped ^{14}C ages for the onset of dryer conditions in Egypt and showed that arid conditions were initiated earlier in the north (nearest to the Mediterranean) at 7 ka and then proceeded southwards towards the interior by 5 ka. Pachur and Hoelzmann (2000) likewise identified different phasing relationships of climate transitions across the Sahara from ~35 ka onwards. Figure 2.7 illustrates the palaeoclimatic context of these records in the central Sahara region. Note that climatic records during the Africa Humid Period (spanning the lateglacial and Holocene) are examined in more detail elsewhere (Knight, this volume).

Quaternary climatic and environmental changes have implications for macroscale landscape development, through uplift and denudation cycles that give rise to preserved palaeosurfaces, through climate-driven periods of variable weathering and erosion rates, and through phases of remobilisation of loose surface sediments by gravity, wind and water. Regional aridity through the late Holocene, and during earlier Cenozoic phases, has given rise to relatively high preservation potential, evidenced through relicts of old land surfaces and features being retained in the landscape (see discussion in Knight and Zerboni 2018). In the central Sahara, this includes duricrusts, hamadas and desert pavements (Knight, this volume), landslides and alluvial fans (Knight, this volume), and development of sand seas (Knight and Merlo, this volume).

2.9 Discussion and Future Outlook

The long geological history of the central Sahara region encompasses tectonic, volcanic and sedimentary events, and the expression of these events is influenced by climate and sea-level changes. The presence of outcrops of rock types of different ages attests to the longevity and relict nature of the Saharan landscape, but these outcrops are strongly modified by geomorphological processes that give rise to the distinctive landforms found in the region today.

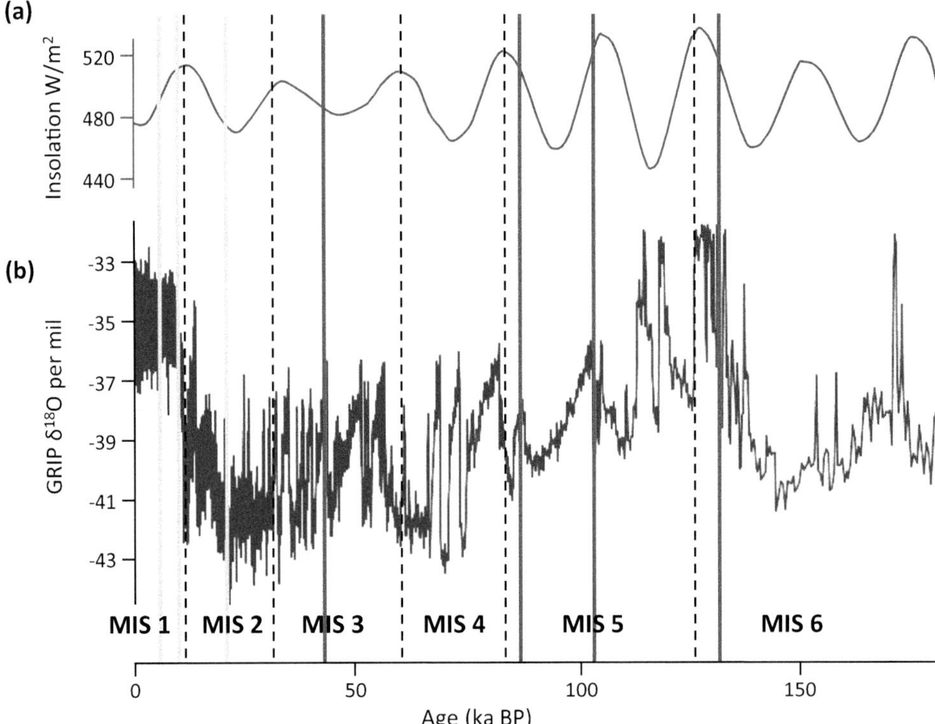

Fig. 2.7 Summary of the climatic context in the central Sahara region for the Quaternary period 180 ka to present (adapted from Ehrmann et al. 2017). **a** June insolation at 30° N, **b** Greenland GRIP $\delta^{18}O$ record. MIS stage boundaries are shown by the dotted vertical black lines. Blue vertical lines represent wet phases, and gold vertical lines represent dry phases (lines not to scale)

Although many different landscape features throughout the central Sahara have their origin in late Cretaceous uplift and peneplanation (Mainguet 1983), climate and environmental changes during the Cenozoic and in particular during the Quaternary have had most impact on landscape development. This includes Miocene aridity that led to the development of sand seas and thus the Sahara's distinctive deserts. Research on regional geology and geomorphology of the Sahara today commonly uses remote sensing methods, because of problems in safely accessing such a vast area. Remote sensing has both advantages and limitations but does not negate the need for field observations and sampling. However, new hyperspectral capabilities and increased spatial resolution of new remote platforms suggest many new opportunities for geological research in the region. This includes identifying sites of mineral and ore resources, tracking palaeochannels through heavy mineral anomalies, reconstructing sediment transport pathways and provenances, and mapping spatial variations in grain size and mineralogy over sand seas. These represent exciting areas of future research. Many locations in the central Sahara also represent sites of significant geological and geomorphic interest (see Anhaeusser et al. 2016), such as the massifs of Hoggar, Tassilis, Tadrart Acacus and Tibesti, and sandstone arches of the Ennedi. This also includes the preserved archaeological imprint of past human activity such as cemeteries, fortifications and rock art. This demonstrates the links between geology, geomorphology and human occupation in the Sahara.

References

Abdullah M, Betzler C, Frechen M, Sierralta M, Thiedig F, El Chair M (2017) Significance of calcretes for reconstruction of the Fezzan Megalake (Murzuq Basin, SW Libya). Z Deutsch Gesell Geowissen 168:199–209

Aissa D-E, Marignac C (2017) Controls on gold deposits in Hoggar, Tuareg Shield (Southern Algeria). J Afr Earth Sci 127:136–145

Amara M, Hamoudi M, Djemaï S, Bendaoud A, Dufréchou G, Jessell WM, Boubekri H, Ouzegane K, Guemmama M, Machane D (2017) New insight of the geological structures and tectonic framework of Ahnet and northwestern part of Tin Zaouatine terranes (western Hoggar, Algeria) constraints from aeromagnetic, gamma ray, and remote sensing data. Arab J Geosci 10:396. https://doi.org/10.1007/s12517-017-3060-7

Anhaeusser CR, Viljoen MJ, Viljoen RP (eds) (2016) Africa's top geological sites. Struik Nature, Cape Town, p 312

Armitage SJ, Drake NA, Stokes S, El-Hawat A, Salem MJ, White K, Turner P, McLaren SJ (2007) Multiple phases of North African humidity recorded in lacustrine sediments from the Fazzan Basin, Libyan Sahara. Quat Geochronol 2:181–186

Ashwal LD, Burke K (1989) African lithospheric structure, volcanism, and topography. Earth Planet Sci Lett 96:8–14

Azil A, Ait Ouali R (2021) Lithostratigraphy and evolution of the Lower Cretaceous Basins, in Western Saharan Atlas, Algeria. J Afr Earth Sci 175:104112. https://doi.org/10.1016/j.jafrearsci.2021.104112

Ballantine J-AC, Okin GS, Prentiss DE, Roberts DA (2005) Mapping North African landforms using continental scale unmixing of MODIS imagery. Remote Sens Environ 97:470–483

Bellini E, Massa D (1980) A stratigraphic contribution to the Palaeozoic of the southern basins of Libya. In: Salem MJ, Busrewil MT (eds) The geology of Libya, vol I. Academic Press, London, pp 3–56

Bensalah M, Benest M, Truc G (1991) Continental detrital deposits and calcretes of Eocene age in Algeria (South of Oran and Constantine). J Afr Earth Sci 12:247–252

Black R, Latouche L, Liégeois JP, Caby R, Bertrand JM (1994) Pan-African displaced terranes in the Tuareg shield (central Sahara). Geology 22:641–644

Brauneck J, Mees F, Baumhauer R (2013) A record of early to middle Holocene environmental change inferred from lake deposits beneath a sabkha sequence in the Central Sahara (Seggedim, NE Niger). J Paleolimnol 49:605–618

BRGM (2016) Geological map of Africa, 1:10 000 000 scale. CGMW-BRGM, Paris

Bristow CS, Armitage SJ (2016) Dune ages in the sand deserts of the southern Sahara and Sahel. Quat Int 410:46–57

Bubenzer O, Besler H, Hilgers A (2007) Filling the gap: OSL data expanding ^{14}C chronologies of Late Quaternary environmental change in the Libyan Desert. Quat Int 175:41–52

Burke K, Gunnell Y (2008) The African erosion surface: a continental-scale synthesis of geomorphology, tectonics, and environmental change over the past 180 million years. GSA Memoir 201, Geological Society of America, Boulder CO, p 66

Busche D (1998) Die Zentrale Sahara. Justus Perthes Verlag, Gotha, p 284

Capitanio FA, Faccenn C, Funiciello R (2009) The opening of Sirte basin: result of slab avalanching? Earth Planet Sci Lett 285:210–216

Collins JA, Govin A, Mulitza S, Heslop D, Zabel M, Hartmann J, Röhl U, Wefer G (2013) Abrupt shifts of the Sahara-Sahel boundary during Heinrich stadials. Clim past 9:1181–1191

Conant LC, Goudarzi GH (1967) Stratigraphic and tectonic framework of Libya. Bull Am Assoc Petrol Geol 51:719–730

Cremaschi M (1998) Late Quaternary geological evidence for environmental changes in south-western Fezzan (Libyan Sahara). In: Cremaschi M, Di Lernia S (eds) Wadi Teshuinat—Palaeoenvironment and Prehistory in south-western Fezzan (Libyan Sahara). CNR, Roma-Milano, pp 13–48

Deniel C, Vincent PM, Beauvilain A, Gourgaud A (2015) The Cenozoic volcanic province of Tibesti (Sahara of Chad): major units, chronology, and structural features. Bull Volcanol 77:74. https://doi.org/10.1007/s00445-015-0955-6

Drake NA, Lem RE, Armitage SJ, Breeze P, Francke J, El-Hawat AS, Salem MJ, Hounslow MW, White K (2018) Reconstructing palaeoclimate and hydrological fluctuations in the Fezzan Basin (southern Libya) since 130 ka: a catchment-based approach. Quat Sci Rev 200:376–394

Dupuy C, Dostal J, Chikhaoui M (1993) Trace element and isotopic geochemistry of Cenozoic alkali basaltic lavas from Atakor (Central Sahara). Geochem J 27:131–145

Ehrmann W, Schmiedl G, Beuscher S, Krüger S (2017) Intensity of African humid periods estimated from Saharan dust fluxes. PLoS ONE 12:e0170989. https://doi.org/10.1371/journal.pone.0170989

El Desoky HM, Shahin TM, El-Leil IA, Shafea EA (2020) Geology and mapping of laterites, South Eastern Desert, Egypt: based on field and ASTER data approach. Geol J 55:4252–4264

Elshaafi A, Gudmundsson A (2017) Distribution and size of lava shields on the Al Haruj al Aswad and the Al Haruj al Abyad volcanic systems, Central Libya. J Volcanol Geotherm Res 338:46–62

El-Shenawy MI, Kim S-T, Schwarcz HP, Asmerom Y, Polyak VJ (2018) Speleothem evidence for the greening of the Sahara and its implications for the early human dispersal out of sub-Saharan Africa. Quat Sci Rev 188:67–76

Fabre J (ed) (2005) Géologie du Sahara occidental et central. Musée royal de l'Afrique centrale, Tervuren, Belgium, p 572

Faniran A, Jeje LK (1983) Humid tropical geomorphology. Longman, London, p 414

Galushin YI, Eloghbi S (2014) Thermal history of the Murzuq Basin, Libya, and generation of hydrocarbons in its source rocks. Geochem Int 52:486–499

Geyh MA, Thiedig F (2008) The Middle Pleistocene Al Mahrúqah Formation in the Murzuq Basin, northern Sahara, Libya evidence for orbitally-forced humid episodes during the last 500,000 years. Palaeogeogr Palaeoclimatol Palaeoecol 257:1–21

Goudie A (1973) Duricrusts in tropical and subtropical landscapes. Clarendon Press, Oxford, p 174

Gras R (1998) Statistical analysis of syn-rift sediments: an example from the Sarir Sandstone, Messlah Field, Sirte Basin, Libya. J Petrol Geol 21:329–342

Grove AT (1960) Geomorphology of the Tibesti region with special reference to western Tibesti. Geogr J 126:18–27

Guillocheau F, Simon B, Baby G, Bessin P, Robin C, Dauteuil O (2018) Planation surfaces as a record of mantle dynamics: the case example of Africa. Gondwana Res 53:82–98

Guiraud R, Bosworth W, Thierry J, Delplanque A (2005) Phanerozoic geological evolution of Northern and Central Africa: an overview. J Afr Earth Sci 43:83–143

Gumati MS (2021) Basin architecture and tectonic controls on the Early Cretaceous Sarir Sandstone reservoir, eastern Sirt Basin, Libya. J Afr Earth Sci 176:104089. https://doi.org/10.1016/j.jafrearsci.2020.104089

Gumati YD, Nairn AEM (1991) Tectonic subsidence of the Sirte Basin, Libya. J Petrol Geol 14:93–102

Horta JCDeOS (1980) Calcrete, gypcrete and soil classification in Algeria. Eng Geol 15:15–52

Iizuka T, Campbell IH, Allen CM, Gill JB, Mauyama S, Makoka F (2013) Evolution of the African continental crust as recorded by U-Pb, Lu–Hf and O isotopes in detrital zircons from modern rivers. Geochim Cosmochim Acta 107:96–120

Keppie JD, Dostal J, Murphy JB (2011) Complex geometry of the Cenozoic magma plumbing system in the central Sahara, NW Africa. Int Geol Rev 53:1576–1592

Key RM (1992) An introduction to the crystalline basement of Africa. In: Wright EP, Burgess WG (eds) Hydrogeology of crystalline basement aquifers in Africa. Geological Society of London, Special Publications 66, pp 29–57

Klitzsch E (1994) Geological exploration history of the Eastern Sahara. Geol Rundsch 83:475–483

Knight J, Fitchett JM (2021) Place, space and time: resolving Quaternary records. S Afr J Geol 124:1107–1114

Knight J, Zerboni A (2018) Formation of desert pavements and the interpretation of lithic-strewn landscapes of the central Sahara. J Arid Env 153:39–51

Koberl C (1994) African meteorite impact craters: characteristics and geological importance. J Afr Earth Sci 18:263–295

Koeberl C (2002) Mineralogical and geochemical aspects of impact craters. Min Mag 66:745–768

Koeberl C, Reimold WU, Cooper G, Cowan D, Vincent PM (2005) Aorounga and Gweni Fada impact structures, Chad: remote sensing, petrography, and geochemistry of target rocks. Meteorit Planet Sci 40:1455–1471

Kutzbach JE, Guan J, He F, Cohen AS, Orland IJ, Chen G (2020) African climate response to orbital and glacial forcing in 140,000-y simulation with implications for early modern human environments. PNAS 117:2255–2264

Le Heron DP, Meinhold G, Bergig KA (2013) Neoproterozoic-Devonian stratigraphic evolution of the eastern Murzuq Basin, Libya: a tale of tilting in the central Sahara. Basin Res 25:52–73

Liégeois JP, Latouche L, Boughrara M, Navez J, Guiraud M (2003) The LATEA metacraton (Central Hoggar, Tuareg shield, Algeria): behaviour of an old passive margin during the Pan-African orogeny. J Afr Earth Sci 37:161–190

Lobpries TA, Lapen TJ (2019) Remote sensing evidence for a possible 10 kilometer in diameter impact structure in north-central Niger. J Afr Earth Sci 150:673–684

Lorenz J (1980) Late Jurassic-Early Cretaceous sedimentation and tectonics of the Murzuq Basin, southwestern Libya. In: Salem MJ, Busrewil MT (eds) The geology of Libya, vol II. Academic Press, London, pp 383–392

Macgregor DS (1996) The hydrocarbon systems of North Africa. Marine Petrol Geol 13:329–340

Mainguet M (1983) Tentative mega-geomorphological study of the Sahara. In: Gardner R, Scoging H (eds) Mega-geomorphology. Clarendon Press, Oxford, pp 113–133

Moreau J (2011) The late Ordovician deglaciation sequence of the SW Murzuq Basin (Libya). Basin Res 23:449–477

Morton AC, Meinhold G, Howard JP, Phillips RJ, Strogen D, Abutarruma Y, Elgadry M, Thusu B, Whitham AG (2011) A heavy mineral study of sandstones from the eastern Murzuq Basin, Libya: constraints on provenance and stratigraphic correlation. J Afr Earth Sci 61:308–330

Nash DJ, McLaren S (2007) Geochemical sediments and landscapes. Blackwell, Oxford, p 465

Nicholson SL, Pike AWG, Hosfield R, Roberts N, Sahy D, Woodhead J, Cheng H, Edwards RL, Affolter S, Leuenberger M, Burns SJ, Matter A, Fleitmann D (2020) Pluvial periods in Southern Arabia over the last 1.1 million-years. Quat Sci Rev 229:106112. https://doi.org/10.1016/j.quascirev.2019.106112

Nicoll K (2018) A revised chronology for Pleistocene paleolakes and Middle Stone Age—Middle Paleolithic cultural activity at Bîr Tirfawi—Bîr Sahara in the Egyptian Sahara. Quat Int 463:18–28

Nkono C, Liégeois J-P, Demaiffe D (2018) Relationships between structural lineaments and Cenozoic volcanism, Tibesti swell, Saharan metacraton. J Afr Earth Sci 145:274–283

Pachur H-J, Hoelzmann P (2000) Late Quaternary palaeoecology and palaeoclimates of the eastern Sahara. J Afr Earth Sci 30:929–939

Paquette JL, Caby R, Djouadi MT, Bouchez JL (1998) U-Pb dating of the end of the Pan-African orogeny in the Tuareg shield: the post-collisional syn-shear Tioueine pluton (Western Hoggar, Algeria). Lithos 45:245–253

Parker A, Harris B, White K, Drake N (2008) Phytoliths as indicators of grassland dynamics during the Holocene from lake sediments in the Ubari sand sea, Fazzan Basin, Libya. Libyan Stud 39:29–40

Partridge TC, Maud RR (1987) Geomorphic evolution of southern Africa since then Mesozoic. S Afr J Geol 90:179–208

Perelló J, Brockway H, García A (2020) A minimum Thanetian (Paleocene) age for the African Surface in the Eritrean highlands, Northeast Africa. J Afr Earth Sci 164:103782. https://doi.org/10.1016/j.jafrearsci.2020.103782

Permenter JL, Oppenheimer C (2007) Volcanoes of the Tibesti massif (Chad, northern Africa). Bull Volcanol 69:609–626

Ramos E, Marzo M, de Gibert JM, Tawengi KS, Khoja AA, Bolatti ND (2006) Stratigraphy and sedimentology of the Middle Ordovician Hawaz Formation (Murzuq Basin, Libya). AAPG Bull 90:1309–1336

Reade H, O'Connell TC, Barker G, Stevens RE (2018) Increased climate seasonality during the late glacial in the Gebel Akhdar, Libya. Quat Sci Rev 192:225–235

Reimold WU, Gibson RL (2009) Meteorite impact! Springer, Berlin, p 337

Reimold WU, Koeberl C (2014) Impact structures in Africa: a review. J Afr Earth Sci 93:57–175

Salem MJ, Busrewil MT (eds) (1980) The geology of Libya, vol 3. Academic Press, London

Schlüter T (2008) Geological atlas of Africa. Springer, Berlin, p 307

Schmieder M, Buchner E (2007) Short note: the Faya basin (N Chad, Africa)—A possible impact structure? J Afr Earth Sci 47:62–68

Senosy MM, Youssef MM, Zaher MA (2013) Sedimentary cover in the South Western Desert of Egypt as deduced from Bouguer gravity and drill-hole data. J Afr Earth Sci 82:1–14

Sepulchre P, Ramstein G, Schuster M (2009) Modelling the impact of tectonics, surface conditions and sea surface temperatures on Saharan and sub-Saharan climate evolution. C R Geosci 341:612–620

Sghari A (2014) Les croûtes calcaires en Tunisie: consequence des transformations géomorphologiques majeures au Messinien-Pliocène. Z Geomorphol 58:485–508

Summerfield MA (1996) Tectonics, geology, and long-term landscape development. In: Adams WM, Goudie AS, Orme AR (eds) The physical geography of Africa. Oxford University Press, Oxford, pp 1–17

Swezey C (2001) Eolian sediment responses to late Quaternary climate changes: temporal and spatial patterns in the Sahara. Palaeogeogr Palaeoclimatol Palaeoecol 167:119–155

Swezey CS (2009) Cenozoic stratigraphy of the Sahara, Northern Africa. J Afr Earth Sci 53:89–121

Tabel J, Khater C, Rhoujjati A, Dezileau L, Bouimetarhan I, Carre M, Vidal L, Benkaddour A, Nourelbait M, Cheddadi R (2016) Environmental changes over the past 25 000 years in the southern Middle Atlas, Morocco. J Quat Sci 31:93–102

Tawadros E (2001) Geology of Egypt and Libya. Balkema, Rotterdam, p 468

Torfstein A, Goldstein SL, Stein M (2018) Enhanced Saharan dust input to the Levant during Heinrich stadials. Quat Sci Rev 186:142–155

Torsvik TH, Cocks LRM (2011) The Palaeozoic palaeogeography of central Gondwana. In: van Hinsbergen DJJ, Buiter SJH, Torsvik TH, Gaina C, Webb SJ (eds) The formation and evolution of Africa: a synopsis of 3.8 Ga of earth history. Geological Society of London, Special Publications 357, pp 137–166

Turner BR (1980) Palaeozoic sedimentology of the southeastern part of Al Kufrah Basin, Libya: a model for oil exploration. In: Salem MJ, Busrewil MT (eds) The geology of Libya, vol II. Academic Press, London, pp 351–374

van der Meer F, Cloetingh S (1993) Intraplate stresses and the subsidence history of the Sirte Basin (Libya). Tectonophysics 226:37–58

van Houten FB (1983) Sirte Basin, north-central Libya: cretaceous rifting above a fixed mantle hotspot? Geology 11:115–118

van Gasselt S, Kim JR, Choi Y-S, Kim J (2017) The Oasis impact structure, Libya: geological characteristics from ALOS PALSAR-2 data interpretation. Earth Planets Space 69:35. https://doi.org/10.1186/s40623-017-0620-8

van Hinsbergen DJJ, Buiter SJH, Torsvik TH, Gaina C, Webb SJ (2011) The formation and evolution of Africa from the Archaean to present: introduction. In: van Hinsbergen DJJ, Buiter SJH, Torsvik TH, Gaina C, Webb SJ (eds) The formation and evolution of Africa: a synopsis of 3.8 Ga of earth history. Geological Society of London, Special Publications 357, pp 1–8

Williams M (1984) Geology. In: Cloudsley-Thompson JL (ed) Key environments: Sahara Desert. Pergamon Press, Oxford, pp 31–39

Evidence for Past Glaciations

3

Jean-François Ghienne and Daniel P. Le Heron

Abstract

On three occasions, during the Cryogenian, Ordovician and early Carboniferous, the present-day Sahara was glaciated. The first of these glaciations left glacial landforms and deposits in the western part of the desert. Much more widely, expansion of large ice masses towards the present north took place multiple times during the late Ordovician and led to the development of spectacular sediment–landform assemblages (found in Mali, Mauritania, Morocco, Algeria, Libya, Niger, Benin and Chad) that allow for ice sheet dynamics to be reconstructed. Striated surfaces are widespread and are only one example of linear features developed beneath the ice: highly attenuated and elongate mega-scale glacial lineations, produced by fast-flowing ice streams, occur in multiple localities. Irregular, mounded piles of sediments associated with this glaciation may represent ice stagnation landforms, and circular-outlined phenomena such as pingos occurred beyond the ice front. Meltwater-related features include ice-marginal tunnel valleys and proglacial channel networks. The early Carboniferous record is recognised in Niger, Chad and south-western Egypt.

Keywords

Glacial surface · Glacial lineations · Palaeovalley · Channel · Sand and sandstones · Ordovician

J.-F. Ghienne (✉)
Institut Terre et Environnement de Strasbourg, CNRS - Université de Strasbourg, 67000 Strasbourg, France
e-mail: ghienne@unistra.fr

D. P. Le Heron
Department of Geology, Universität Wein, 1090 Vienna, Austria
e-mail: daniel.le-heron@univie.ac.at

3.1 Introduction

In the 1950s, a widespread erosional unconformity was recognised in Mauritania (Monod 1952; Sougy 1956), which was soon identified across the Sahara. Serge Beuf and co-workers from the Institut Français du Pétrole first published photographs of striated surfaces with the legend: "surfaces (…), which may be attributed to a glacial process". These surfaces and associated diamictites bearing exotic striated clasts were simultaneously discovered in Libya, Mauritania and Algeria, and unambiguously pointed to a major glacial event in the Lower Palaeozoic (Beuf et al. 1971, and reference therein). When it was realised that the pre-Carboniferous glacial record of South Africa was also tied to a coeval glacial record, and considering the wide distribution of glaciomarine facies across South America, Arabia and Europe, a continental-scale, end-Ordovician to early Silurian glaciation was acknowledged (Vaslet 1990). The related ice sheets are now known to have impacted large parts of western Gondwana from high to intermediate southern palaeolatitudes (Fig. 3.1). After more than 50 years of research, detailed glacial reconstructions have emerged at the scale of the Gondwana platform, in particular in those areas that correspond today to the Central Sahara. In parallel, less extensive but definitive records of the Cryogenian and Lower Carboniferous glaciations were also explored.

The geological history of Africa during the Neoproterozoic is tied to the assembly of Gondwana, which was associated with the closure of oceans resulting in sutures at the margins of stable continental blocks including the West African Craton and, to the east, the Saharan metacraton. The stable cratonic blocks were comparatively unaffected by collision, but metasediments, accretionary wedge materials and subduction-related volcanic rocks located in between were intensely deformed (Liégeois 2019). Afterwards, planation of the so-called Pan-African mountain chain occurred widely. Upon the progressively planated surface (e.g. the Infratassilian surface of southern

J. Knight et al. (eds.), *Landscapes and Landforms of the Central Sahara*, World Geomorphological Landscapes, https://doi.org/10.1007/978-3-031-47160-5_3

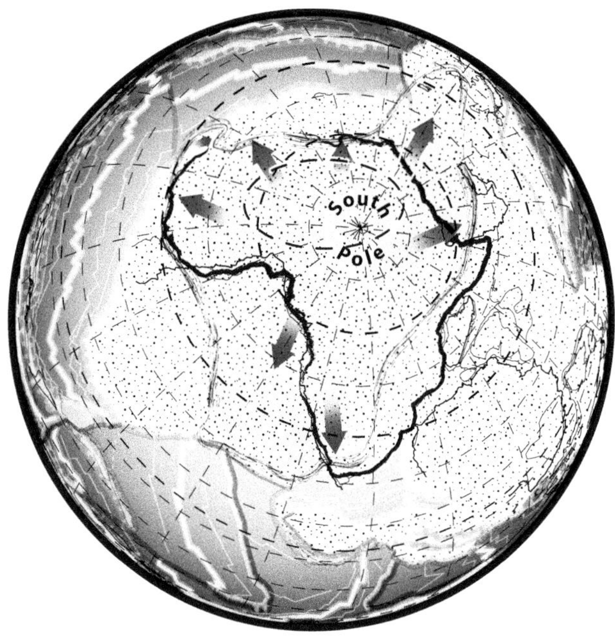

Fig. 3.1 A likely reconstruction of the end-Ordovician ice sheet (in blue, arrows for main ice flows) at its maximum extent over the Gondwana palaeocontient (dotted pattern). Central Sahara (stippled) was at that time fully glaciated. Ice reached south America, Southern Africa and parts of southern Europe (modified from Nutz et al. 2013)

Algeria; Fabre 2005), fluvial to shallow-marine sandstones and shales were laid down from the Cambrian to the Carboniferous throughout the so-called North Gondwana platform (Boote et al. 1998; Eschard et al. 2005; Ghienne et al. 2023a). In general terms, periods of sea-level rise resulted in widespread deposition of fine-grained sediments, including the postglacial shales of the early Silurian (Lüning et al. 2000). By contrast, sandstones associated with lower sea levels, including those of the late Ordovician glaciation, are typically resistant, cliff-forming strata. Thus, ancient glacial and interglacial cycles have played a major role in fashioning both cuestas and cliffs alike.

3.2 Ice Sheets in the Sahara: Cryogenian, Ordovician, and Early Carboniferous

From the late Neoproterozoic to the Carboniferous, the central Sahara was constantly lying at intermediate to high palaeolatitudes (McElhinny et al. 2003). Cryogenian glacial deposits are well exposed on the West African Craton and in Pan-African basins of the Hoggar (Deynoux et al. 2006). Remarkably, based on recent reconstructions, West Africa preserves one of the only mid- to high-latitude records of Cryogenian glaciation, with palaeogeographic reconstructions placing this region at about 60°S (Shields-Zhou et al. 2011). Although these Cryogenian deposits are unequivocally glacial (e.g. Deynoux 1985) and widespread, it is mostly the late Ordovician glaciation (Fig. 3.1) that has had a direct impact on modern day Saharan landscapes, both in exhumed forms (e.g. exquisite glacial pavements, Figs. 3.2 and 3.3) and derived or inverted landforms (Figs. 3.3 and 3.4) that result from differential weathering profiles of ancient glacial strata.

The literature on Carboniferous glaciation in North Africa is sparse, but there is excellent evidence for Visean glaciation reported to the west of the Aïr massif in northern Niger (Lang et al. 1991). This evidence includes tillites, dropstone-bearing shales, and in terms of landforms, striated surfaces and pristine glacial flutes (Fig. 3.2g). The geographic extent of the early Carboniferous glaciation, and thus the distribution of associated landforms, has not been systematically investigated. Diamictites and possible cryogenic structures have been described from the Gilf El Kebir plateau in Egypt (Klitzsch 1983) and in Libya (Pachur and Altmann 2006) but as noted previously (Le Heron et al. 2009) a glacial origin for these awaits verification. However, in Chad, multiple palaeo-ice-stream pathways and related grounded zone wedges await detailed field study and offer a promising area of investigation for future palaeo-ice sheet studies (Le Heron 2018; Kettler et al. 2023; Wohlschlägl et al. 2023) (Fig. 3.5).

3.3 The Late Ordovician Glaciation

The late Ordovician glaciation culminated in the development of a large and continuous ice sheet over North Africa and Arabia, flowing in a general NW direction (in Mauritania), a NNW to N direction (in Algeria and Libya) and a NE direction (in Saudi Arabia). From ~ 446 to 444 Ma, multiple glacial cycles witnessed the repeated recession and regrowth of the ice sheet over the palaeocontinent of Gondwana (Ghienne et al. 2007, 2023a), when North Africa occupied southern polar latitudes (Fig. 3.1). Ice sheet behaviour at this time can be interpreted with respect to its Cenozoic counterparts (Ghienne et al. 2014). If the bulk of the evidence for glaciation during the late Ordovician is confined to the Sahara, European countries contain distal glacial or glacially-related deposits as they represented peri-Gondwana terranes loosely connected to the African continent (Guttiérrez-Marco et al. 2010). Based primarily on geomorphological interpretation of satellite images, we can recognise features of fast-flowing ice streams, which are typical of ice flux behaviour in large polar ice sheets.

There is a rich heritage of glacial landform description and appreciation in the Sahara. In many respects, the golden year for landform description was 1971. In that year, a beautifully illustrated book was produced (Beuf et al. 1971). In it, a suite of different features was described all around the Tuareg Shield. This work of the Institut Français du Pétrole (IFP) underpins almost all of what is known in the Sahara on the late Ordovician glaciation.

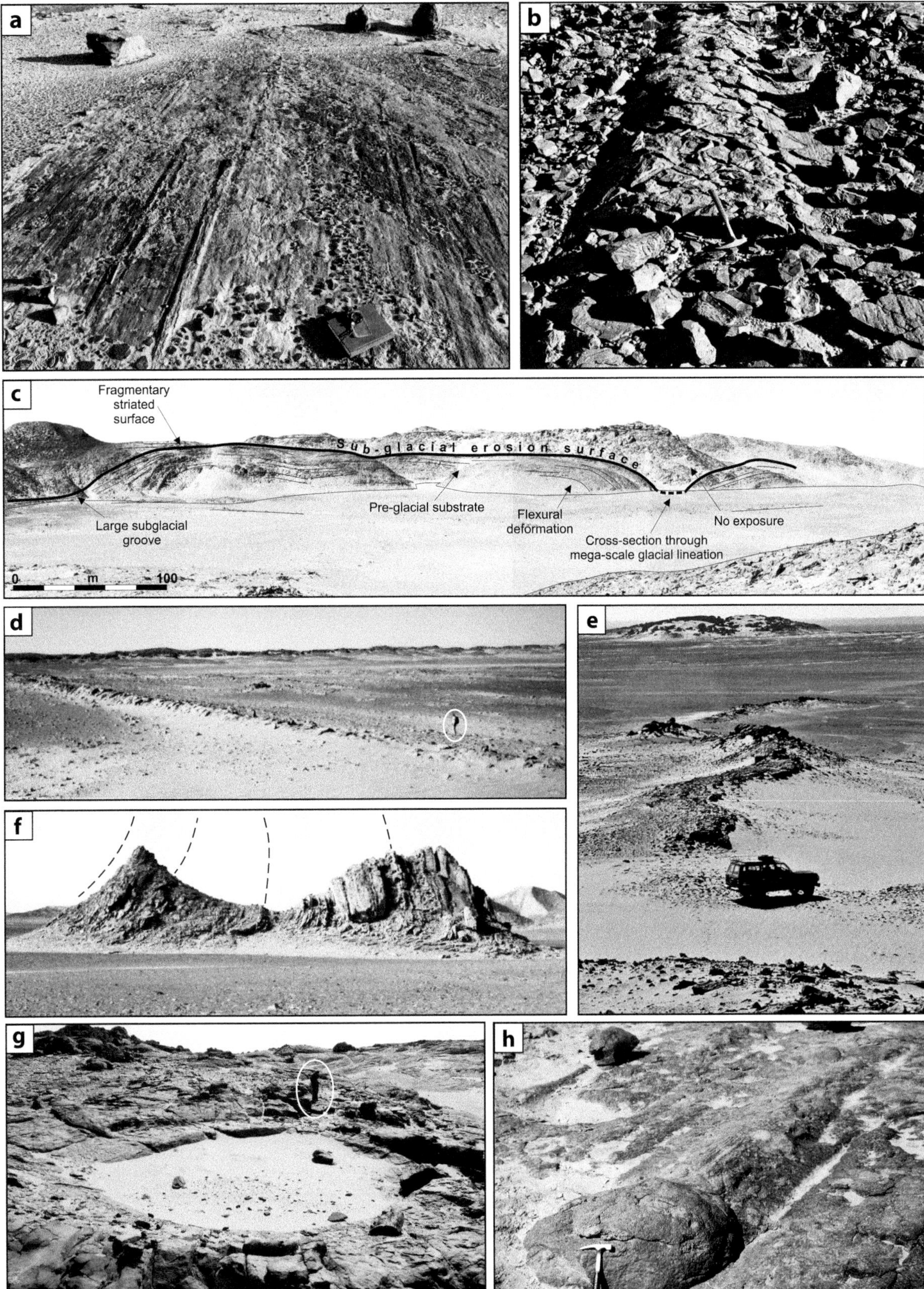

Fig. 3.2 Examples of exhumed glacial landforms. **a**, **b** Examples of exhumed subglacial pavements in SW Libya (Tihemboka area) and SE Algeria (Dider area), respectively; **c** natural section across a mega-scale glacial lineation (MSGL) showing deformation of the underlying, unconsolidated preglacial strata in the Jabal az-Zalmah (Libyan desert; Le Heron and Howard 2010); **d** MSGL formed beneath a late-glacial ice stream flowing in the Djado area (NE Niger; Denis et al. 2010) (encircled geologist for scale); **e** esker structure as a sinuous, exhumed relief form (Ghat area); **f** kettlehole structure in the Ghat-Tihemboka area (SW Libya; encircled geologist; Girard et al. 2015); **g** glaciotectonic deformation including the overturned limb of a breached and recumbent anticline (Libyan desert); **h** glacial fluting behind a granite boulder exhumed along a Lower Carboniferous glacial surface in the Aïr Massif, northern Niger (Lang et al. 1991)

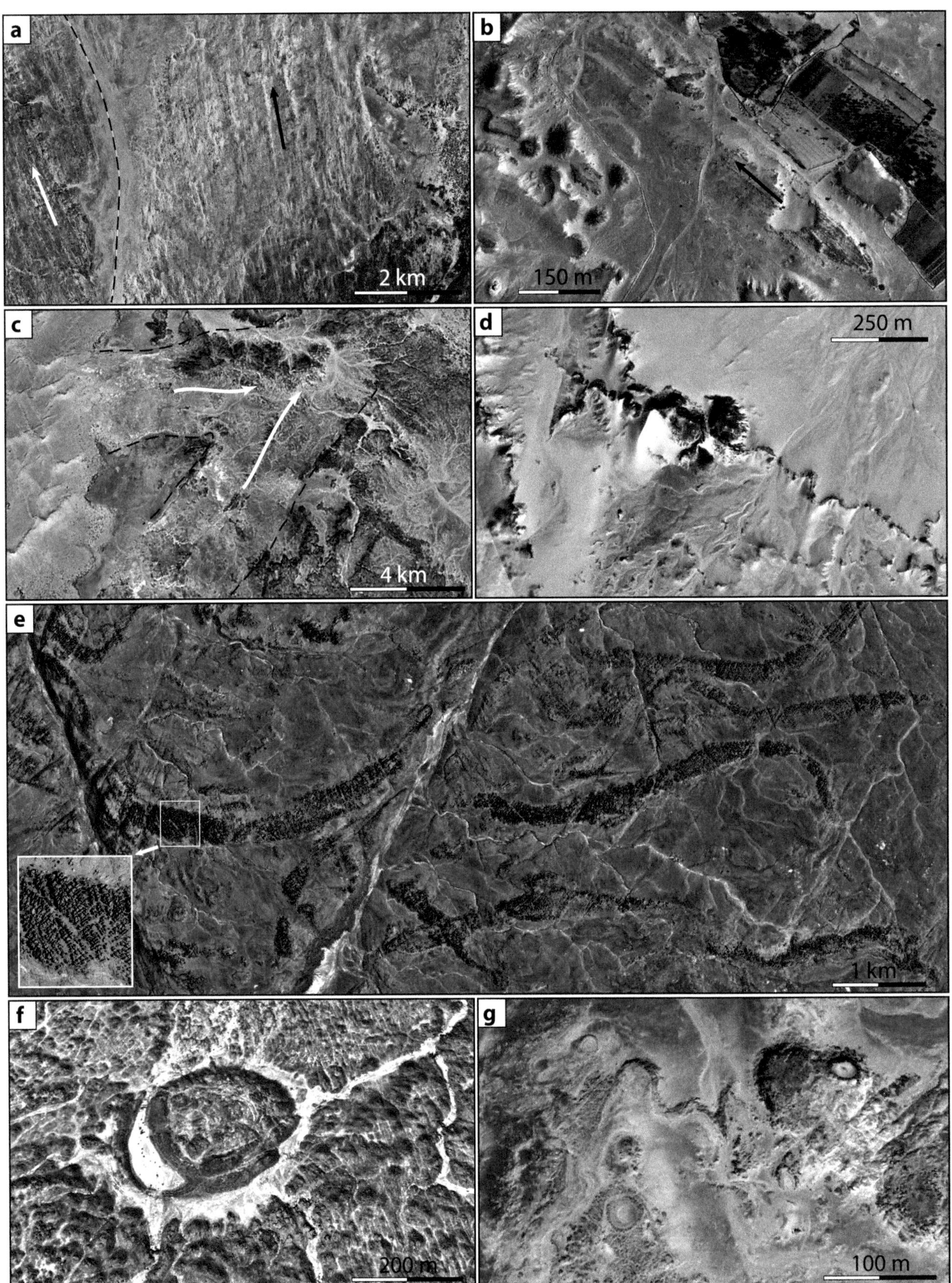

Fig. 3.3 A selection of satellite images (from Google Earth) showing exhumed and derived end-Ordovician landforms. **a** Exhumed glacial surfaces showing two groups of mega-scale glacial lineation with two distinct ice-flow orientations (arrows) close to the Algeria–Libya border (dotted line), near Al Barkat (Moreau et al. 2005); **b** a large fluted glacial lineation, with superimposed smaller-scale MSGL (west of Ghat, SW Libya); **c** confluence point between two tunnel valleys west of the Gargaf Arch (Le Heron et al. 2004); **d** inverted dendritic network of subglacial conduits in the Djado (NE Niger; Denis et al. 2007); **e** inverted network of proglacial meltwater channels in the Tassili N'Ajjer (SE Algeria, see Deschamps et al. 2013). The inset shows the orthogonal fractures and resulting corridors, which typify outcrop of channel-fill sandstones from Mauritania to Libya (see also Fig. 3.4e); **f** an exhumed pingo structure north in the Mouydir area (southern Algeria); **g** circular mini-basins in SW Libya, interpreted either as kettlehole structures or lithalsa-like features

Fig. 3.4 Landforms in late Ordovician sandstones of SW Libya deriving from intraformational features (i.e. which do not correspond originally to landforms). **a** Intraformational striae tied to an end-Ordovician subglacial, soft-sediment shear zone; **b** 'v-shaped' striae and small grooves typifying the intraformational origin contrasting with 'true' striated glacial pavements (see Fig. 3.2a); **c** intraformational striated and grooved surface; **d** sandstone column (pipe) evidencing sediment intrusion and the development of meltwater overpressures beneath the end-Ordovician ice sheet; **e** sandstone pinnacles (encircled geologist for scale) typifying exposures of massive and thick proglacial channel fills that derived from the amplification of an array of orthogonal fractures (see inset in Fig. 3.3e; Girard et al. 2015)

3.4 Late Ordovician Glacial Landforms

Landforms associated with the late Ordovician glaciation can be divided into five categories, which we describe in more detail below. These are (i) subglacial landforms, (ii) ice-marginal landforms, (iii) proglacial landforms, (iv) periglacial landforms, and (v) postglacial landforms.

3.4.1 Subglacial Landforms

In the Sahara, five main types of subglacial landform groupings occur in the end-Ordovician glacial record, often spread over wide geographical areas on barren rock surfaces. These are:

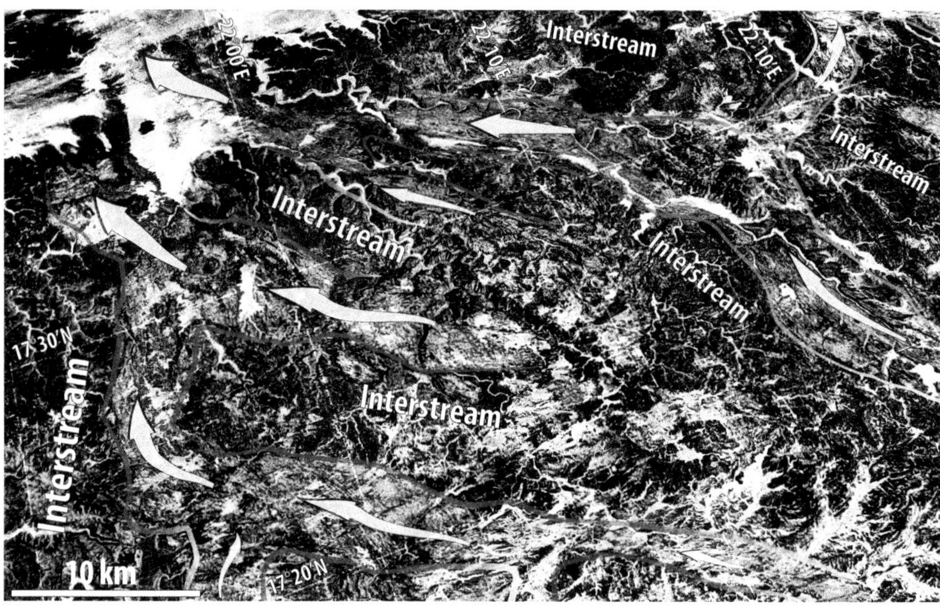

Fig. 3.5 Oblique satellite view (from Google Earth) of the exhumed network of ice-stream pathways in northern Chad, most likely of (Lower?) Carboniferous age (Le Heron 2016). Occurrences of MSGLs can be observed, which distinguish pathways from interstream areas

- Striated and grooved surfaces (true pavement vs. intraformational),
- Mega-scale glacial lineations (MSGL), giga-scale lineations and whalebacks,
- Eskers,
- Tunnel valleys, and
- Conduits and injection networks.

Much debate once centred on the precise origin of striated and grooved surfaces (Fig. 3.4a–c) because some of them typically occur in sandstone, i.e. consolidated sand, and are repeated with multiple stratigraphic occurrences over a few metres. While detailed interpretations vary, the repeated stratigraphic intervals over which they occur, 'v-shaped' cross-sections (Fig. 3.4b) and their typical association with dewatering structures which may crosscut these phenomena led to the viewpoint that most of these surfaces formed through intraformational shearing of unconsolidated sand (e.g. Deynoux and Ghienne 2004; Le Heron et al. 2005; Denis et al. 2010), rather than at the contact between an ice sheet and its bed. Nevertheless, 'stratigraphically isolated' striated surfaces, buried by diamictites or fine-grained marine sediments, are also identified, which are reminiscent of polished and grooved pavements exposed by modern retreating glaciers (Fig. 3.2a, b).

At larger scales, metric to tens-of-m scale ridges, grooves and MSGLs are common in areas such as the Tihemboka (Libya-Algeria border, Fig. 3.3a, b), at scales such that they can be resolved on satellite imagery and mapped accordingly (Moreau et al. 2005). Good examples are also recognised in the Jabal az-Zalmah (Libyan desert, Fig. 3.2c), the Djado (NE Niger, Fig. 3.2d) and north and northeastern Chad (e.g., the Ennedi Plateau, Ghienne et al. 2023b). In turn, families of MSGLs may be arranged into even larger-scale linear features (glacial ridges or giga-scale glacial lineations; Ghienne et al. 2007). Collectively, with their highly attenuated geometries and extension of km to tens of km long, these features are interpreted to have been produced by ancient ice streams that drained the ice sheet (Moreau et al. 2005; Le Heron and Craig 2008). With only a few exceptions (e.g. southern Libya; Ghienne et al. 2013), whalebacks and roches moutonnées are absent from the palaeolandform record. Most of these structures shaped into basement rocks have most likely been obliterated by more recent weathering.

The features and landforms noted above are interpreted to have formed through processes of sculpting beneath flowing ice. In addition to these, other prominent families of features are injection structures, eskers and tunnel valleys. Cut by meltwater processes beneath melting ice, they typify the late Ordovician glaciated platform as a meltwater-dominated system (Le Heron 2016). Eskers form sinuous ribbon features (Fig. 3.2e) with excellent examples exposed in the Tassili N'Ajjer plateau. Their epirelief may be primary, because meltwater channels preferentially cut up into the ice, rather than down into the underlying substrate during esker formation. Tunnel valleys, by contrast, are cut into the subglacial sediment pile. They are much larger and can measure up to a few km in width and up to tens of km long. They form either isolated structures or interwoven networks of incisions that are typically filled by up to 400 m thickness of sandstone and shale successions. They represent one of the most iconic features of the late Ordovician glaciation, with 'swarms' of tunnel valleys recognised in the Gargaf Arch in Libya (Le Heron et al. 2004) (Fig. 3.3c), and in the Tassili N'Ajjer plateau (Deschamps et al. 2013). However, the best examples in terms of landscape generation probably occur in the Western Sahara as inverted bodies of sandstone, snaking for kilometres across the Hodh and Adrar regions of Mauritania (Ghienne and Deynoux 1998). Remarkably, tunnel valleys have also been imaged at several km depth beneath the desert in the course of hydrocarbon exploration on seismic datasets (Le Heron et al. 2009).

Smaller-scale networks of dendritic to polygonal sand-filled conduits, originally cut in shale lithologies, are specifically observed in the Djado area and northern Chad, where they develop at the km scale (Denis et al. 2007) (Fig. 3.3d). In SW Libya, sandstone columns, up to 2 m in diameter (Fig. 3.4d), or walls hundreds of m long, formerly sand pipes and dykes, respectively, are commonly observed beneath glacial erosion surfaces and represent intraformational sand intrusions that were caused by subglacial overpressures (Girard et al. 2015). All of these subglacial structures have subsequently been lithified by a quartz cement and now stand proud of the surrounding desert plain.

3.4.2 Ice-Marginal Landforms

A variety of ice-marginal landforms can be recognised across the Sahara, which allows for the position of former ice fronts to be identified. The bulldozing of piles of sediment in front of an advancing ice sheet produces 'miniature mountain belts', with folds, thrust fault tectonics and shear zones. Some of the finest examples occur in Libya (northern part of the Gargaf Arch, Tihemboka area, Jabal az-Zalmah in the Libyan Desert; Fig. 3.2f) where irregular sandstone ridges 50–60 m high rise from the desert plain (Le Heron et al. 2005; Girard et al. 2015); similar structures occur in the Anti-Atlas of Morocco. Oscillation of the ice front (i.e. minor advance and retreat patterns) has in some cases produced some particularly complex relationships of intense folding and refolding. The asymmetry of the folds within such stagnation landforms provides good evidence of ancient ice-flow orientation.

3.4.3 Proglacial Landforms

Beyond the ice front, there are a few examples of exhumed kettleholes (Fig. 3.2g). These developed over melting blocks of ice initially incorporated in underlying sediments. The landforms however typifying a proglacial setting are a variety of channel systems linked to the release of meltwater. Some of these have been interpreted as representing submarine channels, whereas others have been interpreted as river channel systems. A well-known network of sinuous ribbon channels (or cordons) was mapped on the Tassili N'Ajjer plateau (Girard et al. 2012; Deschamps et al. 2013) (Fig. 3.3e). The majority are oriented SE–NW, consistent with meltwater released into channels flowing towards the NW during glacial outburst events. Most of them show lateral facies transitions, with limited basal incision. Today, channel fills appear as inverted structures in Saharan landscapes and as other end-Ordovician massive sandstone bodies. They are commonly affected by arrays of orthogonal fractures which, enhanced by weathering agents, result in impressive geometrical landscapes including checkerboard surfaces and networks of corridors (insert in Fig. 3.3e) or sets of pinnacles (Fig. 3.4e).

3.4.4 Periglacial Landscapes

One type of landscape that is rare in the Saharan ancient glacial record is that which formed under ancient periglacial conditions, i.e. those conditions representing cold climates beyond the ice front. A classic 'pingo' structure in association with some possible thermokarstic lakes is exhumed north of the Hoggar in Algeria (Beuf et al. 1971) (Fig. 3.3f), and a similar example is also known in Saudi Arabia. In southern Morocco, circular structures 10–30 m in diameter have been interpreted as small-scale depositional depressions over decaying ice bodies tied to a discontinuous permafrost, and compared to lithalsa systems (Nutz et al. 2013). Very similar structures are known in SW Libya, yet they could also be interpreted as kettleholes which acted as circular mini basins (Girard et al. 2015) (Fig. 3.3g). Le Heron et al. (2005) stressed that not all circular structures were tied to the growth and expansion of ice lenses in periglacial soils: it is suggested that the upward diapiric movement of mud beneath an overriding ice sheet could have had a similar effect.

While good examples of patterned ground are known from the Cryogenian record in Mauritania (Deynoux 1982), such structures have not been observed in the end-Ordovician record. This is intriguing when considering the large present-day high-latitude continental surfaces characterised by frozen ground. Although cannibalisation of earlier-formed periglacial features during glacial readvances may have occurred, the poor periglacial record is also possibly due to the overall depositional context: continuously buried stratal surfaces in sedimentary basins intrinsically differ from present-day non-depositional and long-exposed continental surfaces.

3.4.5 Postglacial Landforms

Deglaciation of the Gondwana continental shelves resulted in major marine flooding events. In places, subsequent fine-grained sedimentation has covered and preserved these glacial landforms that are now lying in submerged depressions. Elsewhere, and especially at the top of positive palaeorelief forms, such as giga-scale glacial lineations or thick outwash depositional systems, shallow-marine reworking occurred which modified the ancient glacial features (Moreau 2011). Resulting morphologies, such as channel fills, sand dunes or fields of wave megaripples, evolved during the early Silurian into palimpsest seascapes.

References

Beuf S, Biju-Duval B, de Charpal O, Rognon P, Gariel O, Bennacef A (1971) Les Grès du Palaéozoïque inférieur au Sahara. Editions Technip, Paris, p 464

Boote DRD, Clark-Lowes DD, Traut MW (1998) Paleozoic petroleum systems of North Africa. In: MacGregor DS, Moody RTJ, Clark-Lowes DD (eds) Petroleum geology of North Africa, vol 132. Geol Soc London Spec Publ, pp 7–68

Denis M, Buoncristiani J-F, Konaté M, Guiraud M (2007) The origin and glaciodynamic significance of sandstone ridge networks from the Hirnantian glaciation of the Djado Basin (Niger). Sedimentology 54:1225–1243

Denis M, Guiraud M, Konaté M, Buoncristiani J-F (2010) Subglacial deformation and water-pressure cycles as a key for understanding ice stream dynamics: evidence from the Late Ordovician succession of the Djado Basin (Niger). Int J Earth Sci 99:1399–1425

Deschamps R, Eschard R, Roussé S (2013) Architecture of Late Ordovician glacial valleys in the Tassili N'Ajjer area (Algeria). Sediment Geol 289:124–147

Deynoux M (1982) Periglacial polygonal structures and sand wedges in the Late Precambrian glacial formations of the Taoudeni Basin in Adrar of Mauritania (West Africa). Palaeogeogr Palaeoclimatol Palaeoecol 39:55–70

Deynoux M (1985) Terrestrial or waterlain glacial diamictites? Three case studies from the Late Precambrian and Late Ordovician glacial drifts in West Africa. Palaeogeogr Palaeoclimatol Palaeoecol 51:97–141

Deynoux M, Ghienne J-F (2004) Late Ordovician glacial pavements revisited: a reappraisal of the origin of striated surfaces. Terra Nova 17:488–491

Deynoux M, Affaton P, Trompette R, Villeneuve M (2006) Pan-African tectonic evolution and glacial events registered in Neoproterozoic to Cambrian cratonic and foreland basins of West Africa. J Afr Earth Sci 46:397–426

Eschard R, Abdallah H, Braik F, Desaubliaux G (2005) The Lower Paleozoic succession in the Tasilli outcrops, Algeria: sedimentology and sequence stratigraphy. First Break 23:27–36

Fabre J (2005) Géologie du Sahara occidental et central. Tervuren African Geoscience Collection 18, Musée Royal de l'Afrique Centrale, Tervuren, Belgique, p 572

Ghienne J-F, Deynoux M (1998) Large-scale channel fill structures in Late Ordovician glacial deposits in Mauritania, western Sahara. Sediment Geol 119:141–159

Ghienne J-F, Moreau J, Degermann L, Rubino J-L (2013) Lower Paleozoic unconformities in an intracratonic platform setting: glacial erosion versus tectonics in the eastern Murzuq Basin (southern Libya). Int J Earth Sci 102:455–482

Ghienne J-F, Desrochers A, Vandenbroucke TRA, Achab A, Asselin E, Dabard M-P, Farley C, Loi A, Paris F, Wickson S, Veizer J (2014) A Cenozoic-style scenario for the end-Ordovician glaciation. Nature Comm 5:4485. https://doi.org/10.1038/ncomms5485

Ghienne J-F, Le Heron DP, Moreau J, Deynoux M (2007) The Late Ordovician glacial sedimentary system of the West Gondwana platform. In: Hambrey MJ, Christoffersen P, Glasser NF, Hubbard B (eds) Glacial sedimentary processes and products. IAS Spec Publ 39, pp 295–319

Girard F, Ghienne J-F, Du-Bernard X, Rubino J-L (2015) Sedimentary imprints of former ice-sheet margins: insights from an end-Ordovician archive (SW Libya). Earth Sci Rev 148:259–289

Girard F, Ghienne J-F, Rubino J-L (2012) Channelized sandstone bodies ('cordons') in the Tassili N'Ajjer (Algeria & Libya): snapshots of a Late Ordovician proglacial outwash plain. In: Huuse M, Fedfern J, Le Heron DP, Dixon RJ, Moscariello A, Craig J (eds) Glaciogenic reservoirs and hydrocarbon systems. Geol Soc Lond Spec Publ 368, pp 355–380

Gutiérrez-Marco J-C, Ghienne J-F, Bernárdez E, Hacar MP (2010) Did the Late Ordovician African ice sheet reach Europe? Geology 38:279–282

Ghienne J-F, Abdallah H, Deschamps R, Guiraud M, Gutiérrez-Marco J-C, Konaté M, Meinhold G, Moussa A, Rubino J-L (2023a) The Ordovician record of North and West Africa: unravelling sea-level variations, Gondwana tectonics, and the glacial impact. In: Servais T, Harper D, Lefebvre B, Percival I (eds) A Global Synthesis of the Ordovician System: Part 2. Geol Soc Lond Spec Publ 533 pp 199–252

Ghienne J-F, Moussa A, Saad A, Barnabé D, Youssouf A (2023b) The Ordovician strata of the Ennedi, northeastern Chad (Erdi Basin). C R Géosci 355: 63–84

Kettler C, Wohlschlägl R, Russell C, Scharfenberg L, Ghienne J-F, Le Heron DP (2023) A world-class example of a Late Palaeozoic glaciated landscape in Chad. Sediment Geol 455: 106470. https://doi.org/10.1016/j.sedgeo.2023.106470

Klitzsch E (1983) Paleozoic formations and a Carboniferous glaciation from the Gilf Kebir-Abu Ras Area in southwestern Egypt. J Afr Earth Sci 1:17–19

Lang J, Yahaya M, El Hamet MO, Besombes JC, Cazoulat M (1991) Dépôts glaciaires du Carbonifère inférieur à l'Ouest de l'Aïr (Niger). Geol Runds 80:611–622

Le Heron DP (2016) The Hirnantian glacial landsystem of the Sahara: a meltwater-dominated system. In: Dowdeswell JA, Canals M, Jakobsson M, Todd BJ, Dowdeswell EK, Hogan KA (eds) Atlas of Submarine glacial landforms: modern, Quaternary and ancient. Geol Soc Lond Mem 46, pp 509–516

Le Heron DP (2018) An exhumed Paleozoic glacial landscape in Chad. Geology 46:91–94

Le Heron DP, Craig J (2008) First-order reconstructions of a Late Ordovician Saharan ice sheet. J Geol Soc London 165:19–29

Le Heron DP, Howard J (2010) Evidence for Late Ordovician glaciation of Al Kufrah Basin, Libya. J Afr Earth Sci 58:354–364

Le Heron DP, Sutcliffe O, Bourgig K, Craig J, Visentin C, Whittington R (2004) Sedimentary architecture of Upper Ordovician tunnel valleys, Gargaf Arch, Libya: implications for the genesis of a hydrocarbon reservoir. GeoArabia 9:137–160

Le Heron DP, Sutcliffe OE, Whittington RJ, Craig J (2005) The origins of glacially-related soft-sediment deformation structures in Upper Ordovician glaciogenic rocks: implication for ice sheet dynamics. Palaeogeogr Palaeoclimatol Palaeoecol 218:75–103

Le Heron DP, Craig J, Etienne JL (2009) Ancient glaciations and hydrocarbon accumulations in North Africa and the Middle East. Earth Sci Rev 93:47–76

Liégeois J-P (2019) A New Synthetic Geological Map of the Tuareg Shield: An Overview of Its Global Structure and Geological Evolution. In: Bendaoud A, Hamimi Z, Hamoudi M, Djemai S, Zoheir B (eds) The Geology of the Arab World—An Overview. Springer, Cham, pp 83–107

Lüning S, Craig J, Loydell DK, Storch P, Fitches W (2000) Lowermost Silurian "hot shales" in north Africa and Arabia: regional distribution and depositional model. Earth Sci Rev 49:121–200

McElhinny MW, Powell CMA, Pisarevsky SA (2003) Paleozoic terranes of eastern Australia and the drift history of Gondwana. Tectonophys 362:41–65
Monod T (1952) L'Adrar Mauritanien (Sahara occidental)/Esquisse géologique. Bull Dir Mines Afr occ fr, Dakar, vol 15
Moreau J (2011) The Late Ordovician deglaciation sequence of the SW Murzuq Basin (Libya). Basin Res 23:449–477
Moreau J, Ghienne J-F, Le Heron DP, Rubino J-L, Deynoux M (2005) 440 Ma ice stream in North Africa. Geology 33:753–756
Nutz A, Ghienne J-F, Štorch P (2013) Circular, cryogenic structures from the Hirnantian deglaciation sequence (Anti-Atlas, Morocco). J Sediment Res 83:115–131
Pachur HJ, Altmann N (2006) Die Ostsahara im Spätquartar: Ökosystemwandel in Grössten Hyperariden Raum des Erde. Springer-Verlag, Berlin, p 674
Shields-Zhou GA, Deynoux M, Och L (2011) The record of Neoproterozoic glaciation in the Taoudéni Basin, NW Africa. In: Arnaud E, Halverson GP, Shields-Zhou G (eds) The geological record of Neoproterozoic glaciations. Geol Soc Lond Mem 36, pp 163–171
Sougy J (1956) Nouvelles observations sur le "Cambro-Ordovicien" du Zemmour (Sahara occidental). Bul Soc Geol 6:99–113
Vaslet D (1990) Upper Ordovician glacial deposits in Saudi Arabia. Episodes 13:147–161
Wohlschlägl R, Kettler C, Le Heron D, Zboray A (2023) Chad's cabinet of icy curiosities: sharpening the interpretation of subglacial and ice marginal environments during the LPIA glaciation of the Ennedi Plateau. Sediment Geol 455: 106483. https://doi.org/10.1016/j.sedgeo.2023.106483

4 Sandstone Massifs

Barbara Sponholz

Abstract

Sandstones and sandstone massifs play an important role in the central Sahara's landscapes. These massifs represent geological units, tectonic movements and landmarks; they contain important aquifers and they belong to the most prominent geomorphological units in the central Sahara. General physical and mainly geomorphological features such as weathering crusts, sandstone karsts and wind-eroded shapes are discussed as well as the role of sandstone massifs for prehistoric and recent settlement.

Keywords

Sandstone · Geomorphology · Sandstone karst · Fulgurite · Oasis · Weathering crust · Saprolite · Ventifacts

4.1 Introduction

Within the central Sahara, sandstones are widespread, and isolated sandstone massifs are emerging from the sand seas and serir plains in large areas (Fig. 4.1). The sandstone massifs have attracted human population since the Paleolithic (e.g. Gabriel 1981; Tillet 1983; Tauveron et al. 2009) because of their outstanding morphological position and favourable settlement conditions, such as protection/good view over the surroundings, water availability, sandstone caves and shelters.

The oldest sandstones in the central Sahara date from the Paleozoic or even Precambrian ('Post-Pharusien' after Klitzsch 1970), but most of the central Sahara's massifs are developed from Mesozoic to Paleogene sandstones ('Continental Intercalaire', deposited mainly during the Jurassic/Cretaceous) and Post-Eocene 'Continental Terminal' (Greigert and Pougnet 1967). The central Sahara's main geological basins: Murzuk Basin (north-central Sahara) and Tchad (Chad) Basin (south-central Sahara, including the Bilma Basin) started subsidence since the end of the Mesozoic. Clastic sediments filled the basins from the late Mesozoic onwards with 2500–10,000 m of thickness in total (among them up to 7000 m of sandstone strata), slightly inclined towards basin centres (Savornin 1947; Klitzsch 1978; Greigert 1979). Most of the sandstones originate from continental or shallow marine environments. In many cases, they represent fluvial or delta deposits, but also ancient dune bedding occurs. Many of the sandstones are iron bearing and also feldspars are an important component. Apart from the fact that many of the basic geomorphological and pedological events in sandstone massif formation had already taken place long before the Quaternary, for more recent times the sandstone strata serve as a basis for geomorphodynamic processes and in a more general sense the formation of the present-day Saharan environment. Sandstone massifs are important sites for altitudinal change in climate and vegetation variations (Schulz 1988), and thus they contribute to the small-scale ecological compartmentation of the area.

4.2 Morphology

The evolution of the present sandstone relief generally dates back to the early Cenozoic (Hagedorn 1991; Besler 1992) and develops until today as a function of the various (palaeo) climate conditions. The sandstone deposits that fill up the geological basins have been sculpted mainly along tectonic faults, such as the Kawar Cuesta between Seggedim (north-central Sahara) and Bilma (south-central

B. Sponholz (✉)
Institute of Geography and Geology, Universität Würzburg, Am Hubland, 97074 Würzburg, Germany
e-mail: barbara.sponholz@uni-wuerzburg.de

J. Knight et al. (eds.), *Landscapes and Landforms of the Central Sahara*, World Geomorphological Landscapes, https://doi.org/10.1007/978-3-031-47160-5_4

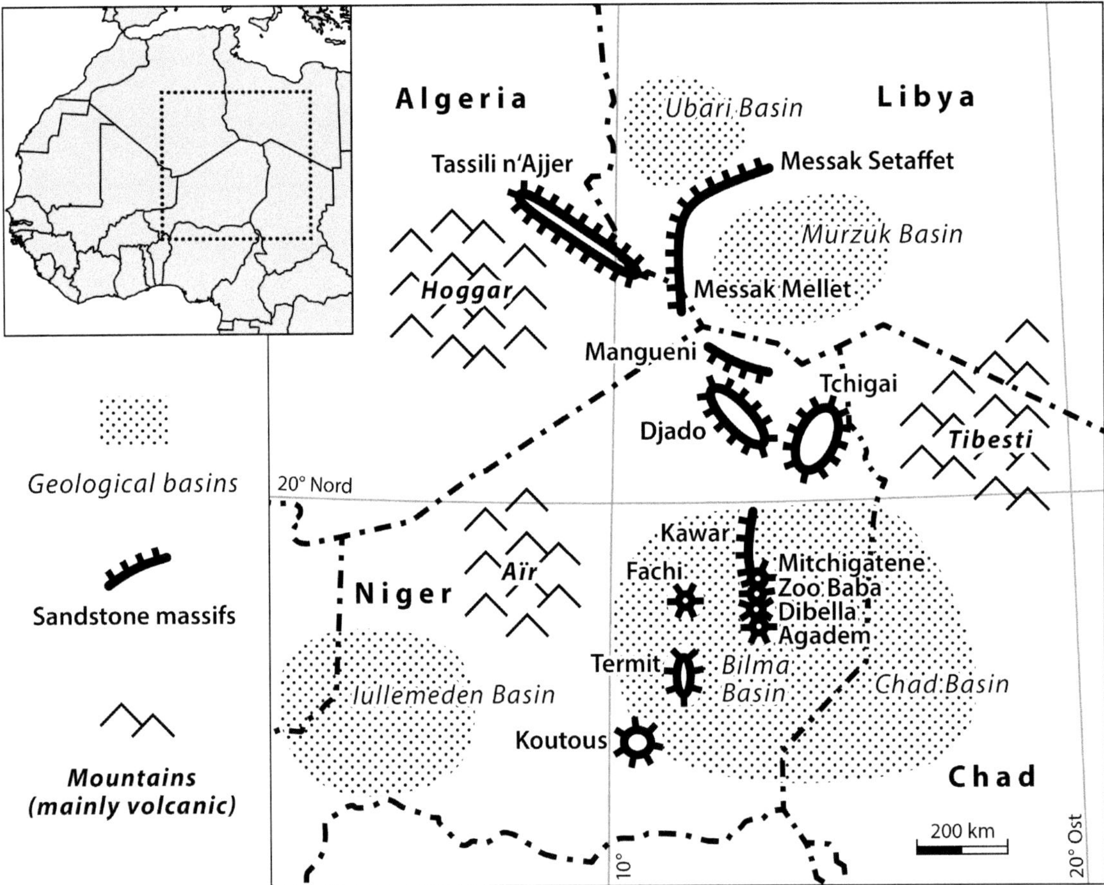

Fig. 4.1 Sandstone massifs in the central Sahara. They are related to bed inclination towards geological basins (Tassili, Messak cuestas, Djado, Mangueni, Tchigai) or to north–south-oriented tectonic faults (Kawar, Termit, Fachi and small isolated massifs)

Sahara). At its southward extent are the isolated massifs of Mitchigatène, Zoo Baba, Dibella and Agadem. The Termit Massif is also situated along a north–south-oriented fault and connected at its southern extremity to the Dilia structure, a neotectonic element (Faure 1963; Pouclet and Durand 1983). Rotational landslides that occur on almost all of the higher cuesta slopes of the central Saharan sandstone massifs are linked to about 200 mm minimum annual rainfall under semiarid climate conditions (Grunert and Busche 1980), and landslides have occurred in several periods during the Quaternary up to the mid-Holocene.

4.3 Weathering

During Cenozoic periods with wet and hot climate conditions, the sandstones were subject to intense chemical weathering. In the lower parts of the cuesta slope profiles, saprolites are observed where the original sandstone structure is preserved, but all the cement and partially even the grains have been chemically weathered. Feldspars and mica were transformed into clay minerals, and easily soluble components of the sandstone components except for iron and aluminium, as well as partially quartz, have been evacuated (Felix-Henningsen 2009). The saprolites have lower mechanical resistance than the original sandstones. Residual iron and aluminium oxides form weathering crusts and pedocrusts at the surface that may reach up to 2 m thickness. After induration, they form laterite and bauxite. Hardening of the sandstones by precipitation of dissolved silica in the sandstone pore space is common and forms quartzite-like silcrete (Fig. 4.2). Together with the saprolites, the surface crusts provide a 'stratification' of hard and soft materials. The crusts protect the underlying sandstone from erosion. In many cases, it is not the original sandstone bedding but the weathering crusts that are responsible for the preservation of typical cuesta profiles of the sandstone massifs (Busche 1983; Besler 1992).

Under present arid to hyperarid climatic conditions, the surface crusts inhibit infiltration into the deeper strata and enhance overland flow during the very rare rainfall events. They also inhibit pedogenesis, and in more humid areas (mountains or outermost areas of the Sahara), they are an important obstacle against farming because of the low soil

Fig. 4.2 Sandstone massif of Zoo Baba, south of the Kawar cuesta. Large rotational landslides are visible on the slopes. On top of the cuesta and on the landslide blocks blackish weathering crust fragments (mainly from iron and manganese) are visible. They protect the underlying light reddish and sometimes saprolised sandstones from erosion

quality. Therefore, many of the crust-covered sandstone plateaus are not used at all or just for extensive pasture and for hunting and gathering activities. Agriculture and horticulture remain restricted to oases near the groundwater table. Some specific studies on the weathering and pedogenesis of sandstone massifs in eastern Niger were undertaken by Gavaud (1977) on Saharan soils, by Völkel (1988) on Holocene pedogenesis on dunes and cuestas, and by Felix-Henningsen et al. (2009) on palaeoclimatic evidence of soils on ancient dunes and in foreland depressions.

4.4 Karst

Many sandstone outcrops underwent intense karst processes during humid climate periods mainly during the Neogene and Quaternary (Mainguet 1972; Sponholz 1989; Besler 1992; Busche 1993). This 'sandstone karst', or 'silicate karst' in a broader sense, also including karst phenomena in magmatic bedrock, for example the Aïr Mountains (Adrar Bous; Sponholz 1989), is very important on local and regional scales. This is because it characterises the aquifers and the groundwater flow in many sandstone areas (e.g. Bilma region) and provides natural cisterns for rainwater collection by local people (e.g. Koutous Massif). The process of silicate solution, however, is closely associated with intense chemical weathering (Sponholz 1989; Tietz 2007). In some areas, such as in the Termit Massif in eastern Niger, there is a close coincidence of sandstone karst/silica solution and piping (subterraneous mechanical evacuation of loose materials) (Busche and Sponholz 1988; Busche et al. 2005).

4.5 Hydrology

Since Savornin (1947) and Knetsch (1963), the importance of artesian aquifers in the Central Sahara in respect to freshwater needs and reserves has been discussed. Sandstone layers in geological basins are the most important groundwater bearing strata all over the Sahara. Sandstone massifs and the thickness and inclination of their sandstone strata are important for groundwater flow direction and therefore for the position of springs and wells, as well as for the constancy of their discharge. Most of the aquifers are under artesian pressure and springs, and even spring mounds, are common in the large and deep foreland depressions of the sandstone massifs. Drilling for groundwater also takes place under artesian conditions: in the Bilma oasis a groundwater drilling project was established in 1980 and delivered very high discharge over several years (see below) (Baumhauer and Hagedorn 1989). During Quaternary wet periods, continuous discharge together with seasonal rainfall was the basis for widespread lake formation (Gabriel 1986). Palaeolimnological studies on some of these lakes have been undertaken, such as by Servant-Vildary (1978), Baumhauer (1986, 1993), Joseph (1990) and Goschin (1992).

4.6 Fulgurites

Fulgurites (dune sands vitrified by lightning strikes) are a major indicator of Holocene climate change in the dune areas and sand sheets of the central and southern Sahara. Fulgarites are found concentrated in mid-slope positions of interdune and foreland depressions south of 18.5°N, and they are correlated with mid-Holocene archaeological sites. Their position indicates climate change taking place in the mid-Holocene from Mediterranean-style cyclone influence all year round, to monsoonal, summer rainfall (Sponholz et al. 1993). Rock fulgurites, however, occur on top of at least some of the sandstone massifs (e.g. Kawar, Djado, Mangueni) (Sponholz 2009). Their formation is not relevant for palaeoclimatic reconstruction, because lightning strikes to the highest point of the area are a general effect and not necessarily corresponding with climate variations. However, they are dependent on the presence of sandy substrates.

4.7 Human Impacts

Because of the presence of near-surface groundwater, the forelands of sandstone massifs are preferred settlement sites, while prehistoric fortresses were mainly situated on top of the massifs (e.g. Chemidour/Kawar cuesta). At some places, large sandstone caves have served as shelters for village populations during times of war (in Seggedim/Niger) (Carl and Petit 1954). Today, sandstone caves are used for storage of household as well as trade goods in the central Saharan nomadic living space. The sandstone massifs figure as important landmarks for trans-Saharan traffic. In historic times, caravans were lucky to see sandstone massifs from a distance and to find wells in their forelands. Today, this has the same effect for motor vehicles. In sandstone massifs, ore minerals may occur in placer deposits. The most recent exploitation of gold deposits has taken place since 2014 in the Djado Massif in northern Niger and in their reworked sands. The attraction of these deposits is very high for people not only from Niger but also from other countries of North and sub-Saharan Africa. Some of the most important sandstone massifs of the central Sahara are now discussed, from north to south.

4.8 Tassili N'Ajjer (Algeria) and the Messak Mellet and Messak Settafet (Libya)

The sandstone cuestas of the Tassili n'Ajjer (Algeria) and the Messak Mellet (south-central Sahara) and Messak Settafet (north-central Sahara) at the western rim of the Edeyen of Murzuk (Libya) are related to lowering of the Murzuk basin during the Paleogene, and to the uplift of the Hoggar Mountains, respectively, and the resulting inclination of their sandstone strata. Hence, the Tassilis are oriented NW–SE, while the Messak cuestas have a half-ring shape that also influences their hydrography and interactions with trade winds and therefore aeolian morphodynamics.

A variety of Nubian sandstones underlies the cuestas of these areas (Capot-Rey 1953). Capot-Rey describes the Messak and the Tadrart in western Libya as typical Saharan sandstone cuestas. Its part west of Ghat, the Akakous Massif, is a continuous wall-like cuesta and is more and more dissected towards the east. The fluvial dissection forms deep ravines with caves and rock shelters in their walls that are often decorated with rock paintings (Gallinaro 2014) as well as rock carvings in the Wadi Mathendous/Messak Settafet areas (Pachur and Altmann 2007). In the northern part, when these areas were exposed to climatic influence from the Mediterranean until the mid-Holocene, the Messak cuestas are mainly characterised by fluvial dissection. Their middle part shows large rotational landslides (Busche et al. 1979; Grunert 1980; Busche 1982; Besler 1992), and the southernmost part is now mainly exposed to wind action.

4.9 Djado and Tchigai Mountains and Plateau De Mangueni (Northeast Niger)

The important sandstone massifs of the Djado and the Tchigai Mountains that separate the Murzuk Basin from the Chad Basin are mainly comprised of Paleozoic sandstones, in some instances combined with claystone/siltstone and limestone strata. Both massifs dominate their forelands with 200–300 m relative relief, and they are intensely intersected by valleys. The western front of the Djado massif is a highly dissected cuesta, and in the foreland are found many inselbergs. The inselbergs show intense sandstone karst, and the caves are often used as shelter and for storage of goods (e.g. at Ehi Ouarek/Djado). The sandstone aquifers allow for hydrostatic groundwater flow and the karst systems contribute to a high discharge from the springs in the cuesta foreland. This easy water access was probably the basis for historic settlement such as the fortresses of Djado or Djaba.

While the Tchigai area mainly shows Carboniferous sediments, the Djado and the Plateau de Mangueni contain unspecified late Paleozoic Nubian Sandstones (McKee 1963; Greigert and Pougnet 1967). These sandstones show saprolitisation as well as silcrete formation on top. Iron crusts also occur, but their thickness is limited and surface hardening is less than it is in sandstone massifs farther south. Between the Djado and the Plateau of Mangueni are areas of important (palaeo-) fluvial dissection, the so-called Enneri Achelouma. This valley system is still occupied today by episodic overland flow, including the risk of landmine dislocation as remnants of the 1990s rebellion in Niger. The origin of the valley dates back to wet Neogene and Quaternary periods. During these wetter climate conditions, the slopes of the sandstone massifs were affected by large rotational landslides that surround the slopes in a festoon-like manner (Busche et al. 2005).

4.10 Cuesta of Kaouar and Isolated Small Sandstone Massifs in the Erg of Ténéré (Eastern Niger)

The Kawar Cuesta in eastern Niger is the most important sandstone massif in the Erg of Ténéré in size and in permanently inhabited oasis concentration. It is oriented north–south along a tectonic line, from the Djado in the north to the Lake Chad basin in the south. Over 200 km long, this cuesta is a more or less a continuous rim of 'Continental Intercalaire' sandstones, with an increase in height from only several metres up to 200 m over the foreland, and increases in steepness towards the south. South of the Rocher de Silemi (near the oasis of Bilma), the cuesta breaks into small and isolated sandstone massifs: Mitchigatène, Zoo Baba, Dibella, Agadem, each of which are about 200 m higher than the surrounding foreland. Nevertheless, they have all common characteristics together with the Kawar or even with the Djado massifs, where rotational landslides are common on the slopes, the sandstones are intensely weathered and even saprolised (Figs. 4.3 and 4.4). Sandstone karst is common (Sponholz 1989). The top of the cuesta shows lateritic to bauxitic weathering crusts, and in its western foreland, every cuesta has springs or wells with near-surface water levels because of artesian pressure. The oasis of Fachi, west of the Kawar, is characterised in the same way. In the Bilma area, spring mounds up to 15 m high are developed (Baumhauer 1986). In the 1980s, a failed water drilling project in Bilma with uncontrolled discharge up to 12,000 m^3/day caused disastrous flooding of parts of the oasis and loss of fossil water (Baumhauer and Hagedorn 1989). Since large amounts of irrigation water that is needed in the oasis today, the water level is continuously lowering, and groundwater quality has worsened significantly since the 1980s. There are no recent studies on water quality, but compared to the mid-1980s, elevated values of dissolved salts (NaCl and others) and iron have been measured in the groundwater (sampling by the author in 1986, 1989, 1990, 2005 and 2006). While the normal depth of the freshwater aquifer is at about 343 m asl in the Bilma basin, another aquifer near the surface (at about 360 m asl; Faure 1963) is heavily charged with salt and is exploited for salt production in the Kawar oasis between Seggedim and Bilma. These salt production sites are until today one of the bases of traditional caravan traffic (Fig. 4.5).

All the cuestas in this region are dissected by valleys and valley embayments that have formed by fluvial action and retrogressive erosion, and have sometimes eroded to form inselbergs (Fig. 4.5). Today, wind action is the most important geomorphological agent. Northeast trade winds transport large amounts of sand northeast–southwest and the north–south-oriented cuestas are an important obstacle in this transport. Because of wind flow dynamics, sand is either deposited at the back of the cuesta or, more often, on the western foreland as a dune ramp before it reorganises

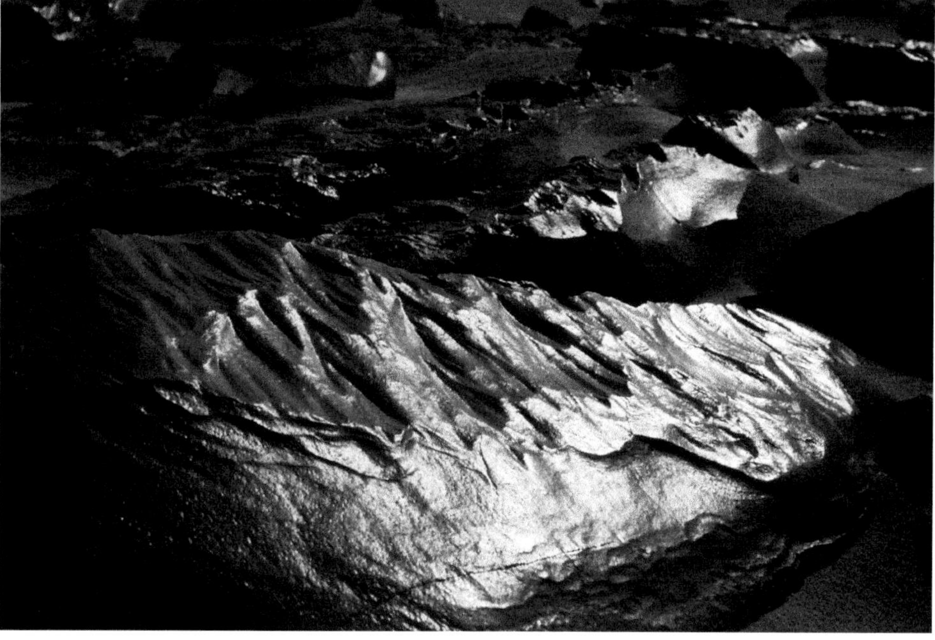

Fig. 4.3 Silcrete blocks on the top of the Dibella sandstone massif. The silcrete originates from silica precipitation in the pore spaces of sandstones, forming a continuous weathering crust. After breaking into blocks, the silcrete still protects the underlying sandstone from erosion. The silcrete itself is heavily polished by wind erosion

Fig. 4.4 Highly dissected sandstone massif/Kawar cuesta near Achenouma. The sandstone surface is covered by dark weathering crusts. Some whitish spots are outcrops of siltstone. In the centre of the photograph, traces of fluvial erosion and accumulation are visible

Fig. 4.5 View from the Kawar cuesta looking west. The oasis of Seggedim with its palm fringed sebkha is located at a freshwater well (artesian water level near the surface). The overlying shallow saltwater body is exploited for salt production

to barchans and finally to trade wind-oriented longitudinal dunes (Besler 1992). On top of the sandstone massifs, wind-polished surfaces with ventifacts cover most of the area (Fig. 4.3).

4.11 Termit Massif (Eastern Niger)

At the southern transition from the central Sahara to the northern Sahel, the Termit Massif is a very isolated massif in eastern Niger, formed of late Cretaceous to Paleogene 'Continental Terminal' sandstones. It follows the same characteristics as the Kawar farther northeast: intense weathering of the sandstones and surface covered by lateritic iron crusts. In the western foreland, are also spring mounds (up to 1.5 m high) and near-surface water in artesian wells. The Termit Massif seems to be oriented along a tectonic line leading from the Aïr Mountains to Lake Chad. The so-called Dilia, a graben-like structure in non-consolidated sands that leads from the Termit Massif to Lake Chad, indicates very recent tectonic movements (Pouclet and Durand 1983).

4.12 Koutous Massif (Eastern Niger)

Southwest of the Termit Massif, the Koutous Massif is a dissected sandstone tableland. Situated at the transition of the central Sahara to the northern Sahel, it receives regular rainfall of about 150 mm year^{-1}. This rainwater is stored in sandstone sinkholes: loose filling material is removed out of the sinkholes by local people, then the sinkhole walls are sealed with clay, and rainwater is directed into the sinkhole. The sinkhole 'cisterns' provide drinking water during the dry season, offering better water quality than wells in the foreland. This cistern management system was used by the local population until at least the 1990s, perhaps even today.

4.13 Summary

Today, all the presented sandstone massifs in the central Sahara are not accessible because of the security and political situation in the concerned countries. This makes it very difficult to realise new research about this area and also to consider its resources for economic and developmental aspects. Considering future research, it is important to keep established scientific contacts with local scientists and traditional knowledge in the massifs' populations. The sandstone massifs of the central Sahara have been important geographical and cultural landmarks over centuries, and they have to be preserved in order to keep this function in future.

References

Baumhauer R (1986) Zur jungquartären Seenentwicklung im Bereich der Stufe von Bilma (NE-Niger). Würzb Geogr Arb 65:235

Baumhauer R (1993) Zur Genese der Schichtstufenvorlandsenken in der südlichen zentralen Sahara. Würzb Geogr Arb 87:85–106

Baumhauer R, Hagedorn H (1989) Probleme der Grundwassererschließung im Kawar (Niger). Die Erde 120:11–20

Besler H (1992) Geomorphologie der ariden Gebiete. Erträge der Forschung 280, WBG, Darmstadt, p 189

Busche D (1982) Die geomorphologische Entwicklung des westlichen Murzuk-Beckens, des Djado-Plateaus und des nördlichen Kaouar (Zentrale Sahara). Habil, Würzburg, p 440

Busche D (1983) Silcrete in der Zentralen Sahara (Murzuk-Becken, Djado-Plateau und Kaouar, Süd-Libyen und Nord-Niger). Z Geomorphol 48:35–49

Busche D (1993) Das Sandsteinkarstrelief des Tchigai, Nordost-Niger. Würzb Geogr Arb 87:63–84

Busche D, Sponholz B (1988) Karsterscheinungen in nichtkarbonatischen Gesteinen der Republik Niger. Würzb Geogr Arb 69:9–44

Busche D, Grunert J, Hagedorn H (1979) Der westliche Schichtstufenrand des Murzuk-Beckens (Zentralsahara) als Beispiel für das Gefügemuster des ariden Formenschatzes. Fesctschrift Deutscher Geographentag Göttingen, pp 43–63

Busche D, Kempf J, Stengel I (2005) Landschaftsformen der Erde. Primus-Verlag, Berlin, p 360

Capot-Rey R (1953) Le Sahara français. Presses Universitaires, Paris, p 464

Carl L, Petit J (1954) La ville de sel, du Hoggar au Tibesti. Julliard, Paris, p 245

Faure H (1963) Inventaire des evaporites du Niger. Rapport du Bureau de Recherches Géologiques et Minières, Niamey, p 260

Felix-Henningsen P (2009) Jungquartäre Landschaftsentwicklung und aktuelle Morphodynaik in der zentralen Sahara (NE-Niger). Unpublished report, Gießen, p 120

Felix-Henningsen P, Kornatz P, Eberhardt E (2009) Palaeo-climatic evidence of soil development on Sahelian ancient dunes of different age in Niger, Chad and Mauritania. Palaeoecol Afr 29:91–106

Gabriel B (1981) Die östliche Zentralsahara im Holozän: Klima, Landschaft und Kulturen (mit besonderer Berücksichtigung der neolithischen Keramik). Préhistoire Africaine, Editions ADPF, Paris, pp 195–211

Gabriel B (1986) Die östliche Libysche Wüste im Jungquartär. Berliner Geographische Studien 19, Berlin, p 217

Gallinaro M (2014) Tadrart acacus rock art sites. In: Smith C (ed) Encyclopedia of global archaeology. Springer, New York, pp 7201–7208

Gavaud M (1977) Les grands traits de la pédogenèse au Niger Oriental. ORSTOM, Paris, p 102

Goschin M (1992) Geochemische Analysen des holozänen Sees El Atrun (Nordsudan; Nubien). Würzb Geogr Arb 84:45–58

Grunert J (1980) Geomorphologische Untersuchungen am West-und südrand des Murzuk-Beckens (Zentrale Sahara), unter Besonderer Berücksichtigung von Rutschungen an schichtstufenhängen. Habil, Würzburg, p 348

Greigert J (1979) Atlas des eaux souterraines du Niger, Etat des connaissances (1978). BRGM, Orleans, p 438

Grunert J, Busche D (1980) Large-scale fossil landslides in the Msāk Mallat and Ḥamadat Mānghīnī escarpment. In: Salem MJ, Busrewil MT (eds) The geology of Libya, vol III. Academic Press, London, pp 849–860

Greigert J, Pougnet R (1967) Essai de description des formations géologiques de la République du Niger. BRGM, Paris, p 239

Hagedorn H (1991) Bemerkungen zur Relief-und Klimageschichte der südlichen Zentralsahara. In: Plitzner K (ed) Im Bann der Wüsten dieser Erde. Verlag, Sigmaringendorf, pp 39–60

Joseph A (1990) Paléo-hydrologie de la région de Bilma, Niger. Rapport IAEA, p 15

Klitzsch E (1970) Die Strukturgeschichte Der Zentralsahara. Geol Rundsch 59:459–527

Klitzsch E (1978) Zur erdgeschichtlichen Entwicklung der Sahara. In: Sahara, 10.000 Jahre zwischen Weide und Wüste. Museen der Stadt Köln, Köln, pp 12–21

Knetsch G (1963) Geologische Überlegungen zu der Frage des artesischen Wassers in der westlichen ägyptischen Wüste. Geol Rundsch 5:640–650

Mainguet M (1972) Le modelé des grès, vol 1. IGN, Paris, p 227

McKee ED (1963) Origin of the Nubian and similar sandstones. Geol Rundsch 52:551–587

Pachur HJ, Altmann N (2007) Die Ostsahara im Spätquartär. Springer, Berlin, p 662

Pouclet A, Durand A (1983) Structures cassantes cénozoïques d'après les phénoménes volcaniques et néotectoniques au nord-ouest du lac Tchad (Niger oriental). Ann Soc Géol Nord 3:143–154

Savornin J (1947) Le plus grand appareil hydrolique du Sahara (nappe artésienne dite de l'Albien). Trav de l'Instit des recherches sahariennes 4, Alger, pp 26–66

Schulz E (1988) Der Südrand der Sahara. Würzb Geogr Arb 69:167–210

Servant-Vildary S (1978) Etude des diatomées et paléolimnologie du bassin tchadien au Cénozoique supérieur. ORSTOM, Paris, p 346

Sponholz B (1989) Karsterscheinungen in nichtkarbonatischen Gesteinen der östlichen Republik Niger. Würzb Geogr Arb 75, Würzburg, p 265

Sponholz B (2009) New discovery of rock fulgurites in the Central Sahara. Palaeoecol Afr 29:127–136

Sponholz B, Baumhauer R, Felix-Henningsen P (1993) Fulgurites in the southern Central Sahara, Republic of Niger, and their palaeoenvironmental significance. Holocene 3:97–104

Tauveron M, Striedter KH, Ferhat N (2009) Neolithic domestication and pastoralism in Central Sahara: The cattle necropolis of Mankhor (Tadrart Algérienne). Palaeoecol Afr 29:179–186

Tietz G (2007) Lösung und neuerliches Wachstum auf Quarzkörnern: ein Indikator chemischer Verwitterung unter tropischen Klimabedingungen. Zent Geol Paläontol 1:151–172

Tillet T (1983) Le paléolithique du bassin tchadien septentrional (Niger/Tchad). CNRS, Paris, p 379

Völkel J (1988) Zum jungquartären Klimawandel im saharischen und sahelischen Ost-Niger aus bodenkundlicher Sicht. Würzb Geogr Arb 69:255–276

Landforms and Landscapes of the Ennedi Plateau

5

Jasper Knight

Abstract

The Ennedi plateau region of north-east Chad is a spectacular sandstone erosional landscape comprising arches, pillars, mushroom rocks and other weathering forms. These landforms are found in associated with isolated sandstone hills (buttes and inselbergs), relict and sand-filled alluvial wadi channels, and in exceptional cases permanent water-filled pools (gueltas). Although little is known about these different landforms, with reference to similar sandstone landscapes elsewhere it is likely that rock structures and episodic surface availability have influenced spatial and temporal variations of erosional processes and thus landform development in the region. A model is proposed to explain Ennedi landscape evolution, set in a regional climatic and environmental context.

Keywords

Sandstone · Weathering · Chad · Arches · Gueltas · Pillars · Ennedi

5.1 Introduction

The Ennedi plateau is a significant topographic feature of north-east Chad. It is best known for its spectacular sandstone arches which give this region its distinctive geomorphic character (Master 2016) and, in turn, have given rise to surfaces that now host an important rock art heritage. As a result of these two elements in combination, the Ennedi plateau as a natural and cultural landscape was inscribed as a World Heritage Site (WHS) in 2016. The Ennedi plateau region itself covers ~ 60,000 km^2 and extends in an arc shape for 250 km west–east towards the Sudanese border in northeast Chad (Fig. 5.1). The northern margin of the plateau is relatively well defined and continuous, whereas the southern margin is more variable and broken up, with smaller upstanding sandstone plateau surfaces separated by wider sand-dominated areas and dry river valleys. The highest elevations on the plateau reach 1450 m asl, and the relative relief with respect to the surrounding landscape is up to 600 m. Annual rainfall varies between 50 and 150 mm (thus is hyperarid to marginally semiarid) with wetter regions found on the southern edge of the Ennedi. A general background to the formation of sandstones in the Sahara, and the weathering processes contributing to their breakdown, is given by Sponholz (this volume). Here, this chapter considers these weathering and erosion processes and their resulting landforms in more detail, focusing on the Ennedi as an example, and highlights the interconnections between the distinctive sandstone landforms and other landscape elements, including ecosystems and human activity.

The detailed geology of the Ennedi comprises a succession of flat-lying sandstone beds interbedded with minor conglomerates, siltstones and marls, deposited from the Ordovician to lower Carboniferous (Master 2016). The sandstones overlie a granitic Precambrian basement. This region was subsequently eroded by glaciers during the Viséan in the early Carboniferous (Le Heron 2018). Another significant geological feature within the Ennedi is the Mousso impact structure that is developed on, and therefore postdates, these sandstones (Buchner and Schmieder 2007). In detail, the sandstone beds themselves are highly variable, with a range of grain sizes, different mineral compositions, different sedimentary structures (e.g. cross beds) and a variable presence of trace fossils. This reflects their different environments of deposition, including fluvial and marine. These different properties in totality mean that there are bed-by-bed differences in rock hardness and weathering potential (Master 2016). This is particularly

J. Knight (✉)
School of Geography, Archaeology and Environmental Studies, University of the Witwatersrand, Johannesburg 2050, South Africa
e-mail: jasper.knight@wits.ac.za

J. Knight et al. (eds.), *Landscapes and Landforms of the Central Sahara*, World Geomorphological Landscapes, https://doi.org/10.1007/978-3-031-47160-5_5

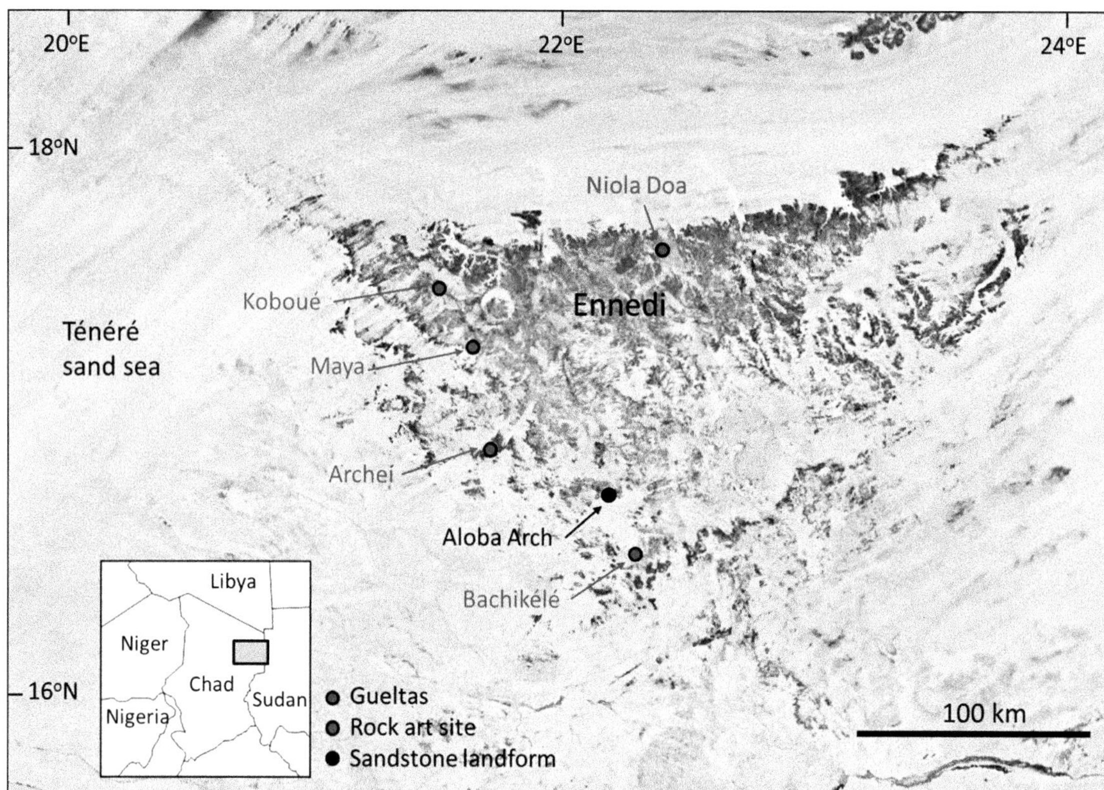

Fig. 5.1 Location of the Ennedi study area in the wider region (boxed in inset map) and sites named in the text (based on Google Earth background image, date 31 December 2016)

significant where climate-driven weathering rates are high and where wind and water erosion processes are also differentially active (e.g. Migoń et al. 2005), either across the Ennedi region as a whole (as a result of altitude) or on different sandstone layers. In turn, this stratigraphic variability in weathering and erosion potential provide the context for considering how and why spectacular sandstone landforms, such as the Ennedi arches, can result. This chapter first describes the characteristic sandstone landforms of the Ennedi plateau and then proposes an evolutionary model to explain their development.

5.2 Sandstone Features of the Ennedi Plateau

There are few studies that have examined and presented evidence for different styles of sandstone weathering and erosion, and thus the different landforms of the Ennedi. The most well-known features are sandstone arches, found in particular on the southern margins of the plateau where weathering and erosion are most extensive. In the Ennedi, a number of rock arches have been identified (IUCN 2016). The Aloba Arch (Arche d'Aloba) is 120 m high and 77 m wide, and is the second largest natural arch in the world (Fig. 5.2). Here, the uppermost third of the arch corresponds to harder rock layers that have helped maintain the top of the arch in place. Variations in sandstone hardness have also resulted in the variable widths and outline shapes of arches, pillars, hoodoos and mushroom-shaped rocks throughout the Ennedi, in which softer beds are eroded more quickly than harder beds (Fig. 5.3). It is notable that these features are mainly located around the south-western margins of the plateau, and as such likely reflect the areas that have been eroding for longest and thus where erosional landforms are most developed; in areas where sandstone hills are farthest apart; and near to the margins of the Ténéré sand sea where wind-blown sand may be most common. Another potential reason may be that this area is the most accessible and therefore best explored and documented of the Ennedi massif.

Interpretation of the sandstone landforms of the Ennedi can be informed by studies from other regions. In Jordan, sandstone landforms include isolated bedrock hills with steep slopes (inselbergs), sandstone pillars, columns and arches, and cavernous weathering (Migoń et al. 2005; Migoń and Goudie 2014). These mesoscale landforms result from extensive rockfall, slopewash during rainstorm events, and salt weathering. Based on the limited field evidence available, it does not appear that similar features are

Fig. 5.2 Photo of Aloba Arch. *Source* Wikimedia Commons

found in the Ennedi despite the similar hyperarid climate regime. This may be due to the prevalence of salt weathering in Jordan and the heightened role of wind abrasion in the Ennedi in keeping any rock surfaces clean (e.g. Loope et al. 2008). Other examples of sandstone landscapes worldwide emphasise the role of geological controls (bedrock fractures and joint sets) as a control on silicate karst (pseudokarst) development (e.g. de Melo and Giannini 2007; Yang et al. 2012); in particular where chemical weathering of impure sandstones results in kaolinite formation, which can dramatically weaken rock strength along joints and contribute to creation of rockfall blocks and steep inselberg slopes; and weaken intergranular bonds, thereby leading to granular disintegration of the sandstone body (Bruthans et al. 2012). Any development of small-scale solutional weathering forms (e.g. silicate or quartzitic pseudokarst; Wray 1997) is likely in this landscape but cannot be evaluated in detail based on the absence of field observations.

5.2.1 Controls on Sandstone Weathering Patterns

Several studies from other sandstone regions discuss the role of bedrock fractures and joints in influencing spatial patterns of weathering and, in turn, the landforms that

Fig. 5.3 Examples of sandstone landforms in the Ennedi. **a** Five Arch Rock; **b** hoodoos; **c** weathered wadi system with sand landforms within its base; **d** Wadi Archei with the water pool (guelta) at its base. All images reproduced under licence from Wikimedia Commons

result. The role of these structural features is twofold. First and most commonly, these can act as lines of weakness that can give rise to enhanced weathering, and can be exploited by fluvial channels. Models for the development of sandstone landscapes are usually based on the idea that such lines of weakness influence the size, shape and orientation of mesoscale bedrock landforms including inselbergs (Migoń et al. 2017, 2018; Duszyński et al. 2019). Second, these pre-existing bedrock fractures may be case-hardened as a result of subaerial weathering and chemical precipitation. In order to consider whether there is a preferred structural alignment of mesoscale sandstone landforms in the Ennedi, the orientation of valleys or sand-filled structures from across the Ennedi was measured in a relative sense, from Google Earth imagery. The plots suggest that there is a landform structural alignment in different sectors of the Ennedi (Fig. 5.4) imparted by geological control. Although this cannot be evaluated given the spatial scale examined, it is likely in turn that there is a structural control on individual landforms. For comparison, in the Arches National Park (Utah, USA), where the largest sandstone rock arches are found, multiple sandstone joint sets reflect the tectonic and structural evolution of the sandstone (Cruikshank and Aydin 1995) and thus can inform on the surface stress patterns that result (Rihosek et al. 2018). Bruthans et al. (2014) and Ostanin et al. (2017), adopting the same approach, modelled the development of sandstone landforms as a result of the interplay between stress fields within the sandstone body and erosion, and found that the gravity force exerted by tall structures such as rock cliffs reduces weathering and erosion rates by increasing fabric interlocking. Thus, tall rock structures such as arches and pillars may be able to survive for long periods. Building from this idea, Filippi et al. (2018) argued that cavernous weathering in sandstone is controlled by this gravity-induced stress, and Rihosek et al. (2016) showed how this can help explain the breakdown of sandstone monuments in Petra, Jordan. Evidence from the Ennedi in the form of arches, pillars and mushroom rocks bears further field investigation to test such hypotheses.

5.3 Other Landscape Features of the Ennedi Plateau

Several permanent ponds (gueltas) are located within incised sandstone valleys in the Ennedi. The largest gueltas are at Archei, Bachikélé, Maya and Koboué, all located on the south side of the plateau (Fig. 5.1). These gueltas are important not only as a water source for present-day herder farmers, but also in their relationship to ecosystems, and

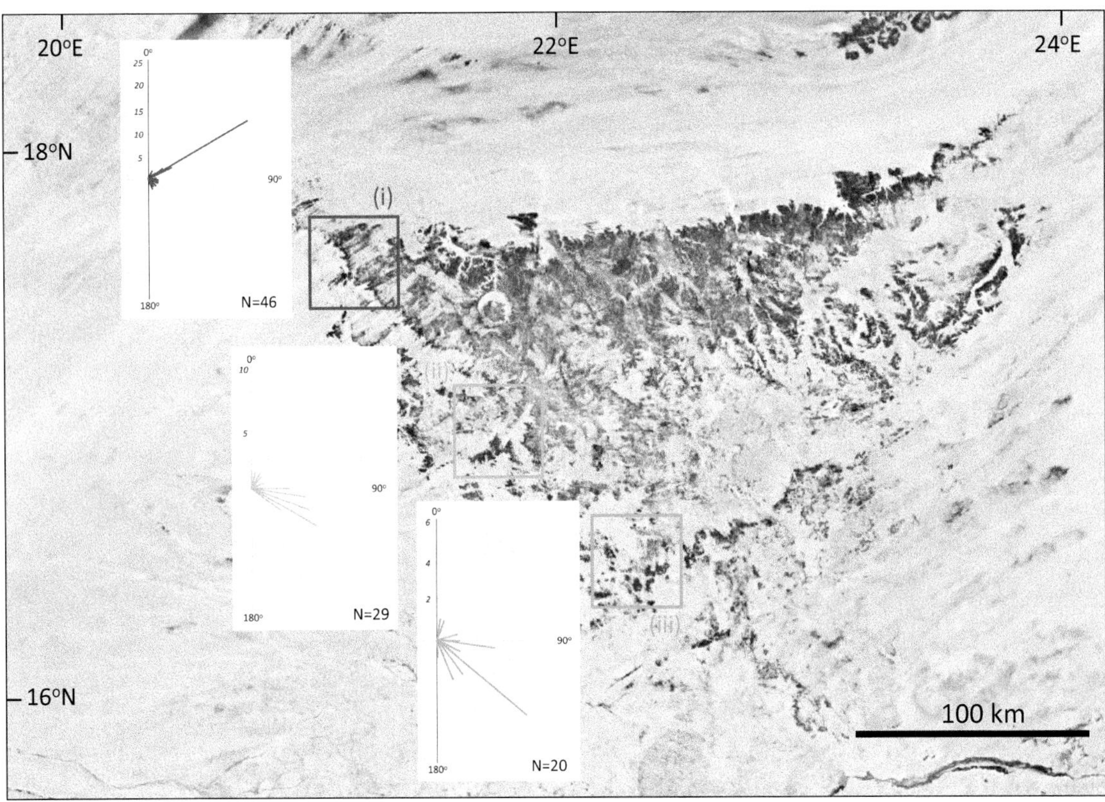

Fig. 5.4 Plot of the structural assignment of the orientation of different wadis and sand-filled valley bottoms in different sectors of the Ennedi, boxed (based on analysis of Google Earth data, date 31 December 2016)

the Ennedi has the highest species diversity of all Saharan rock massifs (at 526 species) and 44% of the tree species in Guelta Maya are relict (IUCN 2016). Guelta Bachikélé contains a relict population of quinine trees (*Rauvolfia caffra*) which is usually found in subtropical and equatorial Africa. Gueltas also host one of the few remaining refugial populations of the critically endangered Nile crocodile (*Crocodylus niloticus*). There are estimated to be only about 10 individuals left in the Guelta d'Archei (IUCN 2016). This species once had a more extensive distribution across the Sahara, during a mid-Holocene pluvial phase, but populations became restricted with the onset of a more arid climate (de Smet 1999). Relict populations of fish are also recorded in the area (Lévêque 1990). The past greater extent of many different now uncommon fauna is recorded in proxy in the range of rock art images present within the Ennedi.

5.3.1 Rock Art

Rock art is relatively common within the Ennedi, but as this region has not been systematically surveyed, the true extent and significance of the rock art is not clear. Much of the literature on the Ennedi has focused on the documentation and interpretation of rock art images, including phases of human occupation and relationships to past environmental regimes (Simonis et al. 2007; Lenssen-Erz 2012). One of the most important sites is at Niola Doa, on the northern side of the Ennedi (Fig. 5.1). Despite observations at such sites, the age and meaning of the Ennedi rock art are still unclear. Although undoubtedly reflecting elements of the surrounding landscape at their time of creation, rock art may also contain symbolic components that are difficult to decode (Fig. 5.5). Further, the anthropological context of rock art traditions emphasise the relationship of rock art to specific localities (di Lernia 2006, 2018), and hence the style and motifs of the rock art is likely related to the physical landscape and properties of the Ennedi. Likewise, sandstone outcrops across the central Sahara are known as significant hosts for both rock art and petroglyphs (Menardi Noguera 2017; Soukopova 2017), and thus there may be comparisons between the Ennedi and other sandstone cultural sites such as in the Tassili (Algeria), Termit (Niger) and Messak (Libya) (Simonis et al. 2007). This has not yet been fully explored.

Fig. 5.5 Example of rock art at Manda Guéli Cave, Ennedi. *Source* Wikimedia commons

5.4 Discussion

The presence of distinctive sandstone weathering and erosional forms in the Ennedi reflects the interplay between climatic and lithological controls, similar to studies of other sandstone regions (e.g. Migoń et al. 2005; de Melo and Giannini 2007; Yang et al. 2012; Migoń and Goudie 2014). Here, the hyperarid climate likely means that salt and thermal weathering, contributing to flaking/delamination, large-scale exfoliation and granular disintegration, work most effectively. Furthermore, bed-by-bed variability of the sandstone itself likely amplifies differential weathering, resulting in different landforms being developed over time. A significant process here, that can be identified from available field photos (Figs. 5.2 and 5.3), is case hardening of the bedrock surface. This can arise from both chemical (pore fluid migration) and biological processes (crust formation), identified by surface discolouration, smoothness, and patina on the rock surface. Furthermore, this can lead to undercutting into the softer rock beneath, and spalling of sheet-like rock surfaces. Bruthans et al. (2012) stated that sandstone rock surfaces in the Czech Republic, in a much wetter environment than is considered here, reach full tensile strength as a results of case hardening over as little as 6 years' exposure.

Several sandstone landscape evolutionary models have been proposed, based on the premise that an intact regional sandstone massif is broken down into smaller constituent units as weathering and erosion proceeds, with progressive escarpment/cliffline retreat and development of fluvial valleys/canyons over time (Migoń et al. 2018; Duszyński et al. 2019). These models thus consider that breakup of the sandstone massif takes place from its margins towards its centre, and that the degree of rock fragmentation can be used as a proxy for age, along the evolutionary pathway cliff → arch/hoodoo → pillar/mushroom rock. (Other geomorphic elements such as footslope pediments or landslides/rockfalls may also be developed, but are not considered here.) Although the principles of these types of evolutionary

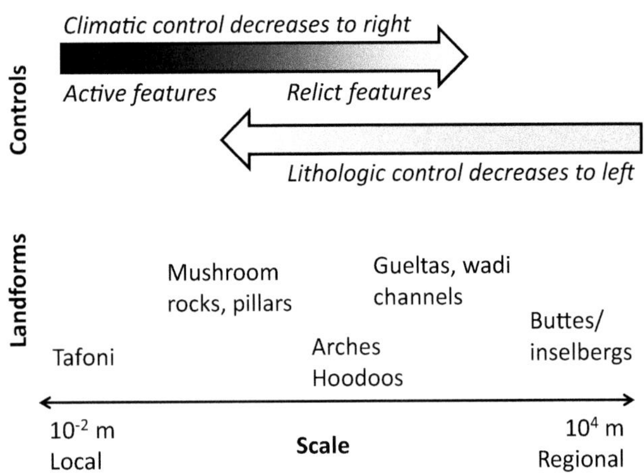

Fig. 5.6 Model showing the relationships between different sandstone erosional forms, and their likely controls on different scales

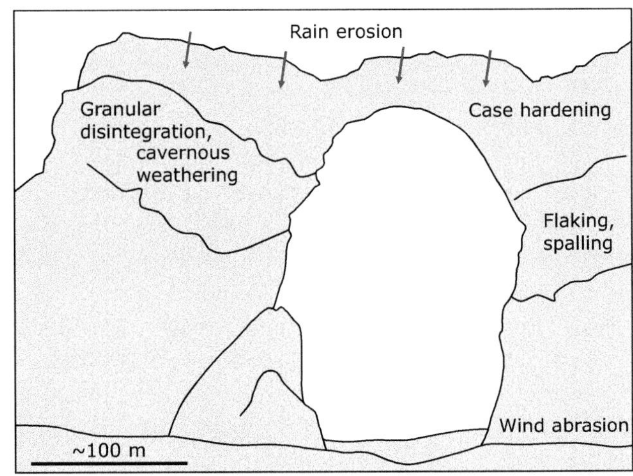

Fig. 5.7 Schematic diagram of the different weathering and erosion processes shaping the Aloba Arch (Arche d'Aloba), Ennedi plateau (sketched from a photo in Master 2016)

models are applicable to many different environments, the model can be used to explain the development of the Ennedi landscape. Here, the southernmost massif edge is broken into smaller blocks, and the size of these blocks generally decreases with distance outwards from the massif. As these blocks get smaller, the relative role of any regional lithological control decreases (Fig. 5.6) as more microscale controls (microclimatic or mineralogical) become more significant. In this model, isolated blocks (inselbergs, buttes) become smaller and the areas between them widen over time. Here, this is evidenced by the spread of sand from the adjacent Ténéré desert from the south-west. Further, the role of water erosion in developing and widening these valleys is much less in these environments than in other locations discussed in the literature (e.g. de Melo and Giannini 2007; Bruthans et al. 2012), where rainfall is greater. In the Ennedi, episodic overland flow and funnelling of this water across the massif surface and into small fluvial incisions likely takes place only over higher areas where orographic rainfall occurs. This water quickly dissipates away from the massif margins. Thus, the role of water erosion is considered to play a role only on the massif itself and associated with periods of initial sandstone surface incision.

Loope et al. (2008) discussed the role of wind abrasion on sandstone surfaces on the Colorado Plateau (USA), shaping large and curved troughs, scour pits and domes, and cavernous weathering features (tafoni). In the Ennedi, wind abrasion is potentially also very significant given the high sediment availability and it is likely that this is the most important process leading to erosion and undercutting of sandstone cliffs within the height of the saltation curtain (~ 1.5–2.5 m). Figure 5.7 schematically illustrates the potential weathering and erosion processes potentially affecting the Aloba Arch. It is evident that different processes are found in different sectors of the arch. Case hardened surfaces are geologically controlled but driven by water availability and therefore are best developed on the exposed top of the arch. Wind abrasion takes place nearest the ground surface. The midsection of the arch is farthest from both the top and bottom and here passive weathering and erosion by flaking, spalling and granular disintegration take place, giving rise to greatest microscale relief and cavernous weathering. Therefore, different sectors of the arch are sensitive to the effects of different processes, in this case, driven mainly by water availability. It is notable that models of arch evolution (Vařilová et al. 2011, 2015; Rihosek et al. 2018) consider all processes to be operating uniformly, which is clearly not the case. Further, these studies explicitly describe rock arches in terms of their 'maturity', but this Davisian concept is not helpful because it assumes that all rock arches evolve in a singular, deterministic way and that the same evolutionary path is followed in all cases. This is a false assumption.

Intraformational palaeosols within sand dunes in east Niger, including in the Bilma sand sea (south-east Ténéré) which is functionally and hydrologically connected to the Ennedi (IUCN 2016; Master 2018), indicate wetter phases and lake development during the late Pleistocene and early Holocene, then drying until 4.5 ka, wetter in the period 4.5–3.0 ka, and drier thereafter (Felix-Henningsen 2000; Baumhauer 2014). Therefore, the Ennedi region may have been influenced by different climatic phases and therefore different processes and rates of weathering and erosion during the late Quaternary. A significant area of uncertainty is the timescale over which sandstone erosional features develop, particularly in hyperarid to semiarid landscapes where there may be variations in rainfall. However, dating of rock art on sandstone surfaces (which has not been done in the Ennedi) can determine surface age and possible also weathering rates. Cosmogenic methods can also be used (which has also not been done in this area). The close association between sandstone landforms of the Ennedi and rock art and other cultural artefacts attests to the role of sandstone geology and topography in influencing human livelihood and cultural patterns. Further, rock art is well established on sandstone surfaces because these tend to be relatively smooth and porous, which allows for paint pigments to be absorbed into the rock surface and preserved relatively easily. Protection of the Ennedi as a WHS demonstrates this co-relationship between people and the environment and also highlights why sandstone landforms and landscapes in the Sahara are important.

5.5 Summary and Outlook

The Ennedi plateau shows a unique and spectacular combination of geomorphology and cultural heritage which reflects the interplay between geology, climate and human occupation throughout the late Quaternary. The spatial patterns and geomorphic types of sandstone landforms present in the Ennedi are not known in detail because of the lack of field investigations. This is clearly an important future research objective, subject to safe and reliable field access to the region. A better understanding of sandstone landforms can in turn help contextualise the rock art and ecosystems, thereby enabling a more integrated environmental management plan for the region. The latest Conservation Outlook Assessment report on the status of the Ennedi WHS (IUCN 2020) indicates potential threats from livestock grazing and wood harvesting, deriving from local community use of natural resources in the area. More significant threats are off-road driving and oil and gas exploration (mainly on the north side of the Ennedi) and incoherent management strategies put in place, despite its WHS inscription. This is critically important for maintaining the rock art heritage and the intactness of habitats for its endemic and endangered species.

Acknowledgements Jennifer Fitchett is thanked for commenting on a previous version of this chapter.

References

Baumhauer R (2014) Some new insights into palaeoenvironmental dynamics and Holocene landscape evolution in the Nigerian Central Sahara (Ténéré, Erg of Ténéré, Erg of Fachi-Bilma). Zbl Geol Paläont Teil I 2014:387–403

Bruthans J, Svetlik D, Soukap J, Schweigstillova J, Valek J, Sedlackova M, Mayo AL (2012) Fast evolving conduits in clay-bonded sandstone: characterization, erosion processes and significance for the origin of sandstone landforms. Geomorphology 177–78:178–193

Bruthans J, Soukup J, Vaculikova J, Filippi M, Schweigstillova J, Mayo AL, Masin D, Kletetschka G, Rihosek J (2014) Sandstone landforms shaped by negative feedback between stress and erosion. Nat Geosci 7:597–601

Buchner E, Schmieder M (2007) Mousso structure: a deeply eroded, medium-sized, complex impact crater in northern Chad? J Afr Earth Sci 49:71–78

Cruikshank KM, Aydin A (1995) Unweaving the joints in Entrada Sandstone, Arches National Park, Utah, U.S.A. J Struct Geol 17:409–421

de Smet K (1999) Status of the Nile crocodile in the Sahara desert. Hydrobiologia 391:81–86

de Melo MS, Giannini PCF (2007) Sandstone dissolution landforms in the Furnas Formation, Southern Brazil. Earth Surf Proc Landf 32:2149–2164

di Lernia S (2006) Cultural landscape and local knowledge: a new vision of Saharan archaeology. Libyan Stud 37:5–20

di Lernia S (2018) A (digital) future for Saharan rock art? Afr Archaeol Rev 35:299–319

Duszyński F, Migoń P, Strzelecki MC (2019) Escarpment retreat in sedimentary tablelands and cuesta landscapes: landforms, mechanisms and patterns. Earth Sci Rev 196:102890. https://doi.org/10.1016/j.earscirev.2019.102890

Felix-Henningsen P (2000) Paleosols on Pleistocene dunes as indicators of paleo-monsoon events in the Sahara of East Niger. Catena 41:43–60

Filippi M, Bruthans J, Řihošek J, Slavík M, Adamovič J, Mašín D (2018) Arcades: products of stress-controlled and discontinuity-related weathering. Earth Sci Rev 180:159–184

IUCN (2016) Ennedi Massif: Natural and Cultural Landscape (Chad)—ID No. 1475. World Heritage Nomination—IUCN Technical Evaluation Report, May 2016. IUCN, Geneva, p 10

IUCN (2020) Ennedi Massif: Natural and Cultural Landscape—2020 Conservation Outlook Assessment. IUCN, Geneva, p 13

Le Heron DP (2018) An exhumed Paleozoic glacial landscape in Chad. Geology 46:91–94

Lenssen-Erz T (2012) Adaptation or aesthetic alleviation: which kind of evolution do we see in Saharan herder rock art of northeast Chad? Cambr Archaeol J 22:89–114

Lévêque C (1990) Relict tropical fish fauna in Central Sahara. Freshwaters 1:39–48

Loope DB, Seiler WM, Mason JA, Chan MA (2008) Wind scour of Navajo Sandstone at the Wave (central Colorado Plateau, U.S.A.). J Geol 116:173–183

Master S (2016) The natural arches of the Ennedi Plateau in northeastern Chad. In: Anhaeusser CR, Viljoen MJ, Viljoen RP (eds) Africa's top geological sites. Struik Nature, Cape Town, pp 256–260

Menardi Noguera A (2017) The oval engravings of Nabara 2 (Ennedi, Chad). Arts 6:16. https://doi.org/10.3390/arts6040016

Migoń P, Goudie A (2014) Sandstone geomorphology of south-west Jordan, Middle East. Quaest Geogr 33:119–126

Migoń P, Goudie A, Allison R, Rosser N (2005) The origin and evolution of footslope ramps in the sandstone desert environment of south-west Jordan. J Arid Environ 60:303–320

Migoń P, Duszyński F, Goudie A (2017) Rock cities and ruiniform relief: forms—processes—terminology. Earth Sci Rev 171:78–104

Migoń P, Różycka M, Jancewicz K, Duszyński F (2018) Evolution of sandstone mesas—following landform decay until death. Progr Phys Geogr 42:588–606

Ostanin I, Safanov A, Oseladets I (2017) Natural erosion of sandstone as shape optimisation. Sci Rept 7:17301. https://doi.org/10.1038/s41598-017-17777-1

Rihosek J, Bruthans J, Masin D, Filippi M, Carling GT, Schweigstillova J (2016) Gravity-induced stress as a factor reducing decay of sandstone monuments in Petra, Jordan. J Cult Herit 19:415–425

Rihosek J, Slavík M, Bruthans J, Filippi M (2018) Evolution of natural rock arches: a realistic small-scale experiment. Geology 47:71–74

Simonis R, Scarpa Falce A, Calati D (2007) Tchad. Rock art of Sahara and North Africa, thematic study. ICOMOS World Heritage Convention, Paris, pp 71–81

Soukopova J (2017) Central Saharan rock art: considering the kettles and cupules. J Arid Environ 143:10–14

Vařilová Z, Přikrly R, Zvelebil J (2011) Pravčice rock arch (Bohemian Switzerland NationalPark, Czech Republic) deterioration due to natural and anthropogenic weathering. Environ Earth Sci 63:1861–1878

Vařilová Z, Přikrly R, Zvelebil J (2015) Factors and processes in deterioration of a sandstone rock form (Pravčická brána Arch, Bohemian Switzerland NP, Czech Republic). Z Geomorphol 59:81–101

Wray RAL (1997) A global review of solutional weathering forms on quartz sandstones. Earth Sci Rev 42:137–160

Yang G, Tian M, Zhang X, Chen Z, Wray RAL, Ge Z, Ping Y, Ni Z, Yang Z (2012) Quartz sandstone peak forest landforms of Zhangjiajie Geopark, northwest Hunan Province China: pattern, constraints and comparison. Environ Earth Sci 65:1877–1894

Caves and Rock Shelters of the Central Sahara and Their Records

6

Andrea Luca Balbo

Abstract

The central Sahara region has a range of caves and rock shelters located around the margins of bedrock massifs, and these are important because they reflect long-term bedrock weathering which is therefore climatically and geologically controlled, but they also provide evidence for past human occupation over millennial timescales. Caves and rock shelters are therefore important sites within the context of central Sahara. For example, the Tadrart Acacus displays one of the most important groupings of prehistoric cave art in the world and was inscribed as a UNESCO World Heritage Site in 1985. This chapter examines the physical processes leading to cave and rock shelter development and then discusses the palaeoenvironmental and archaeological evidence present at such sites. Examples of the Takarkori rock shelter and Uan Afuda cave, both from the Tadrart Acacus, show that human interactions with the physical environment can be reconstructed based on the study of the sediments trapped within such rock shelter and cave sites. Evidence from these sites can be used to inform on relationships between palaeoclimates, environmental resources and human activity in areas of the central Sahara that today are hyperarid.

Keywords

Archaeology · Caves · Rock shelters · Rock art · Sediment traps

A. L. Balbo (✉)
Prehistoric Archaeology and Anthropology Laboratory, University of Geneva, Boulevard Carl Vogt 66, CH-1211 Geneva, Switzerland
e-mail: andrea.balbo@planthro.org

A. L. Balbo
Platform anthropocene Inc., 160 Riverside Blvd, New York, NY 10069, US

6.1 Introduction

Caves and rock shelters found worldwide are sites that provide key archaeological and palaeoenvironmental information (Oestmo et al. 2014; Cremaschi et al. 2015; Parton and Bretzke 2020). These landforms develop in cliff face and cliff foot locations, and thus, there is a strong geologic and topographic control on cave and rock shelter distributions (Baumhauer 2010). However, the presence of caves and rock shelters is important because they can provide palaeoclimatic and palaeoenvironmental information in hyperarid areas where there are few other records available. There is also relatively high preservation potential of organic and isotopic evidence within caves and rock shelters because they provide protected environments and stable microclimatic conditions (Manning and Timpson 2014). In desert environments, diurnal and seasonal variation in moisture and temperature are significantly reduced in caves and rock cavities, and these landforms offer protection from wind and sandstorms. One limitation is that these very localised and site-scale records may not be representative of regional environmental and climatic patterns. Nonetheless, several archaeological studies have examined the evidence for human activity in these environments (Cremaschi and Zerboni 2010; Cremaschi et al. 2014). One of the most sensational discoveries in this sense has been that of the oldest known completely mummified human body. This body, found in the Uan Muhuggiag rock shelter in the Wadi Teshuinat of southwest Libya, was that of a 30-month-old boy (Cremaschi and di Lernia 1998). Two radiocarbon dates were obtained from the shroud covering this body (5405 ± 180 ^{14}C yr BP and 7438 ± 220 ^{14}C yr BP) (di Lernia 2002), calibrated to median ages of 6177 and 8244 yr BP, respectively. This discovery predates by several centuries the earliest cases of complete mummification known in the Nile Valley, located more than 2000 km to the east. Cave and rock shelter sites in the central Sahara are also known for their rock art where the Tassili n'Ajjer

J. Knight et al. (eds.), *Landscapes and Landforms of the Central Sahara*, World Geomorphological Landscapes, https://doi.org/10.1007/978-3-031-47160-5_6

massif of southern Algeria contains one of the most important groupings of prehistoric cave art in the world, and this region was inscribed as a UNESCO World Heritage Site in 1982.

This chapter considers the geomorphological processes contributing to cave and rock shelter development in the central Sahara and the unique palaeoenvironmental and archaeological records that they contain. This analysis provides a context for understanding human–environment relations in the region under late Quaternary climate change.

6.2 Formation Processes of Caves and Rock Shelters

Besides having large extents of aeolian, alluvial and colluvial deposits, the central Sahara is characterised by a number of mountain ranges with sedimentary and volcanic surface lithologies (Young and Young 1992) (Fig. 6.1). These massifs are variable in their geologic properties, and this has a first-order control on the nature of massif geomorphological development, including the development of caves and rock shelters. The Messak Mellet and Ennedi are developed in sandstone; the Tassili n'Ajjer contains sandstone overlying granite; Tibesti has basalt lava flows overlying sandstone; the Hoggar massif is developed in metamorphics with volcanic intrusions; and the Aïr massif is developed in granite. These different rock types have different physical and lithological properties and as such are affected by different weathering and erosion processes. The lithological distribution within the local mountain ranges affects the concentration of rock cavities. Caves and rock shelters prevail in sedimentary-dominated mountain ranges, such as the Hoggar and Tassili n'Ajjer of south Algeria or the Aïr of north Niger, while they are less common in volcanic dominated ranges such as the Tibesti of north Chad (Fig. 1). It is very likely that caves and rock shelters found in these different rock types develop different sizes and shapes, but this has not been investigated in detail (e.g. Piccini 2011; Al-Moadhen 2018). Broadly speaking, when all other parameters are equal, the larger the weathered cavity, the older the landform, but the geotechnical properties of the rock determine maximum cave size.

Rock weathering is the main factor contributing to the development of rock cavities, and different chemical and physical processes can affect their formation. Solutional cavity formation processes are perhaps the most common worldwide. They usually occur in limestone, but they may be observed in several other rocks, including chalk, dolomite, marble, salt beds and gypsum. They also take place on sandstone and igneous substrates, which are common in the central Sahara. Weathering on granitic terrain is dominated by chemical dissolution by which shallow circular depressions (gnammas) can be formed (Cooks and Pretorius 1987; Goudie and Migón 1997). Other weathering processes especially in desert regions include thermal heating/cooling which can result in sheet-like exfoliation of the rock surface and in granular disintegration of individual mineral grains or particles from the bedrock surface (Twidale and Bourne 1976; Eppes and Griffing 2010; Loope et al. 2020). These processes are effective on both igneous rocks (Pye 1985) and sandstones (Grab et al. 2011).

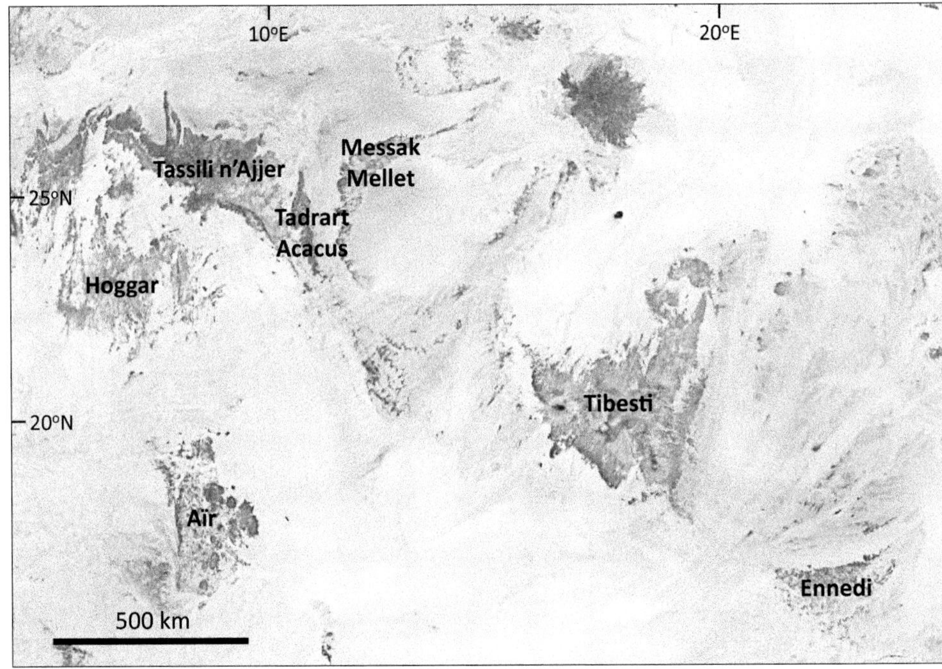

Fig. 6.1 Locations of central Sahara mountain ranges mentioned in the text (*Source* Google Earth)

While erosional processes are dominant in forming rock cavities and caves, solution by karstification during periods of wetter climates may have also taken place (Young and Young 1992; Parsons and Abrahams 2009). The longest and deepest solutional cave systems in Africa are found in the Atlas Mountains of north Algeria, only a few are found in the central Sahara owing to its dominantly desert climate. Busche (1980) described 'hanging embayments' which are incipient rock shelters developed as a result of chemical and physical weathering of sandstones in the Messak Mellet, These 'embayments' are located 10–20 m below the summit surface and form a laterally extensive bench <100 m wide that runs parallel to the summit edge. Soukopova (2017, 2018) described circular features on the floor at the mouths of sandstone rock shelters in the Tadrart Acacus and Tassili n'Ajjer. These were interpreted as human-made engravings that have functional and symbolic values to cave dwellers. However, based on their morphology and position, they may be better interpreted as natural potholes formed by weathering and erosion, in particular by dripping water falling from the bedrock overhang at the cave or shelter entrance.

Water and wind erosion also affect footslope locations and can be instrumental in excavating the base of bedrock cliffs (Baumhauer 2010). Orographic rainfall over summits can lead to rapid overland flow over bedrock surfaces, and this may be channelised over the cliff face (Fig. 6.2). Sand may be blown from dune or wadi surfaces on adjacent lowlands. Honeycomb or tafone weathering is common and has been observed in the Hoggar massif of southern Algeria (Côte 1957; Smith 1978). Tafoni are developed by a range of weathering processes, and the tafone cavities enlarge over time (Mellor et al. 1997; Mol and Viles 2012). Tafoni are also developed on both exposed bedrock surfaces across bedrock summits and within the margins of caves and bedrock cavities and are thus not found exclusively in association with these landforms. Hagedorn (1980, his Fig. 8) showed large (m-scale) tafone developed in sandstone in Tibesti, which represent incipient circular-shaped caves. Busche (1998, pp. 46–68) provides specific examples of weathering landforms from various central Saharan massifs. Bedrock fractures are zones of enhanced weathering, especially associated with bedding planes within sandstone (Loope et al. 2020), and these features can give rise to cave development (Al-Moadhen 2018). Bedding plane surfaces are often case hardened because of chemical and hydrological changes within the host rock. Case hardening is noted as a key control on the development of sandstone landforms in the Ennedi (Knight, this volume).

6.3 Caves and Rock Shelters as Sediment Traps

Dry climates in desert environments, with low precipitation, strong winds and sparse vegetation cover, generally hamper sediment stabilisation and soil formation processes (Parsons and Abrahams 2009). In such environments, where stratified archaeological and palaeoenvironmental sedimentary archives are scarce, caves and rock shelters constitute exceptional sediment traps, allowing the accumulation and stabilisation of sediments, favoured by wind and sun protection as well as a more stable temperature and moisture microclimate. These conditions, mixed with the hyperarid surroundings, allow a high degree of preservation of organic and inorganic materials. Cave sediment processes and stratigraphies have been described from some sites in Africa and in other dryland environments. For example, in Spain cave deposits contain soils that mark wetter late Pleistocene phases, and these correspond chronologically with North Atlantic Dansgaard-Oeschger cycles and Heinrich events (Courty and Vallverdu 2001; Vallverdú i Poch 2017). In the peri-Sahara region, cave and rock shelter stratigraphies have also been recorded. In the Ethiopian Highlands, a 2-m-long record from Sodicho rock shelter reflects climate changes over the last 27,000 years (Hensel et al. 2021). Anthropogenic activity within this rock shelter appears strongly controlled by climate, and the prominent African Humid period (AHP, ~15–5 kyr BP) is archaeologically sterile, comprising massive silt with low organic and

Fig. 6.2 Mori rock shelter carved out in sandstone in context with the nearby wadi (a group of visitors on the jeep track provides the scale)

high iron content, suggesting an environment dominated by weathering. At Mochena Borago rock shelter, in the same region, basal sediments are >50 ka in age and vary across the floor of the rock shelter, due to both geogenic and anthropogenic processes (Lanzarone et al. 2019). At Rhafas cave in eastern Morocco, 4 m of sediment has a basal luminescence age of 107 ± 12 ka suggests stratigraphically that Mousterian occupation of the site took place around 90–80 ka (Mercier et al. 2007; Doerschner et al. 2016). Such age control can help constrain both environmental and human phases within caves and rock shelters (Cherkinsky and di Lernia 2013).

It is from such preserved materials that most evidence of past human presence and behaviour in the central Sahara has been inferred, and sediments are used to define trajectories of human–environment interaction (Cremaschi 2002; Cremaschi et al. 2014). Specifically, work in the mountain ranges of the Tadrart Acacus and Tassili n'Ajjer has provided key evidence of climate change in the late Pleistocene and Holocene periods, necessary to understand climate-related trajectories of human populations in the central Sahara (di Lernia 1999).

6.3.1 Example of the Takarkori Rock Shelter

The Takarkori rock shelter, positioned on a structural terrace of the Tadrart Acacus sandstone massif, some 100 m above the wadi floor, developed in an alcove-type morphology through interlayer undersapping of the local sandstone (Young and Young 1992; Cremaschi et al. 2014). The sediment sequence preserved inside the Takarkori rock shelter has provided a stratigraphic reference for the Holocene period in the central Sahara region (Biagetti and di Lernia 2013). Archaeological, palynological and micromorphological methods have thus been combined to infer relationships between human activity and their adaptive strategies during periods of severe climatic and environmental change. In the Takarkori rock shelter, hunter-gatherer groups were present in the early Holocene around 10.2 ka BP, when the area was characterised by high availability of water, freshwater habitats and sparsely wooded savannah vegetation. The last hunter-gatherers to occupy the rock shelter, left around 8.2 ka BP, in a period marked by increased aridity, decreasing grassland, the spread of cattails and with generally reduced waterbodies and freshwater habitats in the vicinity of the rock shelter. Between ~8.2 and 5.6 ka BP pastoral groups occupied the Takarkori rock shelter, in a period characterised by increasing water availability and the general spread of water bodies. The top of the stratigraphic sequence at ~5.7–4.6 ka BP points to increasing environmental instability and the establishment of dryer climatic conditions, at the end of the AHP (Gasse 2000; Hoelzmann et al. 2004; Lézine et al. 2011). Regionally, dry savannah was expanding thereafter, and freshwater habitats became sparser, setting the stage for the hyperarid conditions that characterise the region in the present day (deMenocal et al. 2000; Mamtimin et al. 2011).

6.3.2 Example of the Uan Afuda Cave

The Uan Afuda cave is situated in the Wadi Kessan of the Tadrart Acacus mountain range, the easternmost stretch of the Tassili n'Ajjer mountains in Libya. Compared to the Takarkori rock shelter, Uan Afuda provides insight into human population and environmental conditions during an earlier chronological phase. Specifically, the archaeological record preserved in Uan Afuda can shed lights on hunter-gatherer groups occupying the Sahara as early as ~90 ka BP, and the cave was consistently settled in the early Holocene ~0–8 ka BP, being abandoned thereafter (di Lernia 1999). Rock paintings attributed to Holocene hunter-gatherer groups have also been found inside the Uan Afuda cave. The paintings are thus associated to one of the earliest phases of cave art in central Sahara, known as 'Round Heads', named after its characteristic round-headed human silhouettes, usually filled in solid red pigments (Mori 1965, p. 101). Red pigments have been broadly used throughout prehistory (Attard Montalto 2010), mostly obtained from hematite iron oxides (Henshilwood et al. 2011), and these are widely distributed in the central Sahara. Inside Uan Afuda, for example, a large body of in situ hematite iron oxide was found inside the cave, which points to dissolution and precipitation processes in a silicate karst environment, probably during the Neogene. Different hues of yellow and white pigments could have been easily obtained from exposed levels within rock shelters incised at the base of sandstone cliffs located along the main wadis (Fig. 6.3).

6.4 Caves, Rock Shelters and Rock Art

The highest concentrations of rock art in the central Sahara are found in association with caves and rock shelters, where smooth sedimentary rock surfaces provide a canvas that can be carved using relatively soft tools and easily impregnated with coloured powders and pastes (Gallinaro 2013 and this volume). In the central Sahara, a broad array of coloured materials would have been readily available to the local populations within or in the near proximity of rock cavities. This includes iron oxide giving red pigments, charcoal or manganese oxides for black pigments and gypsum for white pigments (Iriarte et al. 2018). There may also be a relation between the distribution of rock art locations and climate (Soukopova 2011). The earliest phases of rock art

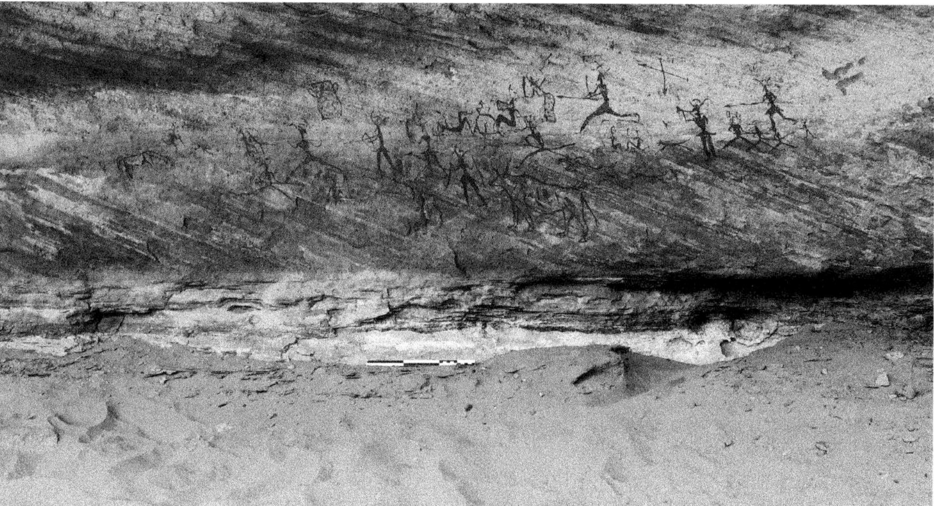

Fig. 6.3 Painted scene in correspondence with pigmented mineral levels from the Acacus mountain range (scale 30 cm).

in the central Sahara (~10–8 ka BP) appears to immediately postdate or be at the end of the wetter AHP of the late Pleistocene/early Holocene periods, when there may have been active development of rock cavities (di Lernia 1999; Damnati 2000; Cremaschi et al. 2010).

6.5 Summary and Outlook

Caves and rock shelters reflect the operation of weathering and erosion processes, controlled by rock type and climate, over potentially long periods of the Quaternary. One key element in the study of caves and rock shelters is to set them in a wider landscape context (e.g. Oestmo et al. 2014; Parton and Bretzke 2020) which can inform on the role of these sites as a place of human habitation amid the provision of environmental resources (e.g. food, water, minerals). Even when climatic conditions became drier from the mid Holocene, mountain regions where caves and rock shelters are mainly located remained relatively more hospitable than surrounding lowlands, offering lower temperatures, shaded environments and seasonal mountain water reserves (gueltas). This then facilitated the use of caves and rock shelters as refugia for human occupation.

Although many studies have examined the archaeological and palaeoecological evidence from caves and rock shelters (e.g. Moeyersons et al. 2002; Pirson et al. 2006; Collins et al. 2017; Watson 2017), these are not always set into a landscape context. The discovery of new African cave and rock shelter sites (Sahle et al. 2019) provides an opportunity to develop such a research approach. The profusion of rock art in caves and rock shelters within such climatic and physiographical contexts has made the central Sahara one of the highest density regions of preserved rock art in the world (McLaren and Reynolds 2009). The 1972 UNESCO World Heritage Convention recognises outstanding places of global hybrid, natural and cultural universal value. In this sense, the natural and cultural significance of caves and rock shelters in the central Sahara constitutes a potential asset for the region. The Tadrart Acacus was inscribed as a UNESCO World Heritage Site in 1985 on the basis of its rock art heritage, but since 2016 it has been on the Sites in Danger List because of the lack of active management by the Government of Libya and issues such as site vandalism. Specific management issues at this site were discussed by di Lernia and Gallinaro (2011). The presence of similar sites elsewhere in the region requires integrated and sustainable conservation strategies and limiting risks from, among others, local conflicts, unregulated tourism and oil exploitation, as well as ongoing climate change.

References

Al-Moadhen AAH (2018) The origin and morphology of Al-Tar caves, West Karbala, Iraq. Arab J Geosci 11:696. https://doi.org/10.1007/s12517-018-4044-y

Attard Montalto N (2010) The Characterisation and Provenancing of Ancient Ochres. Ph.D. thesis. Cranfield University

Baumbauer R (2010) Scarpfoot depressions in south-central Sahara—State of knowledge and critical review. Z Geomorphol 54:377–392

Biagetti S, di Lernia S (2013) Holocene deposits of Saharan rock shelters: the case of Takarkori and other sites from the Tadrart Acacus Mountains (Southwest Libya). Afr Archaeol Rev 30:305–338

Busche D (1980) On the origin of the Msāk Mallat and Hamādat Mānghīnī escarpment. In: Salem MJ, Busrewil MT (eds) The geology of Libya, vol III. Academic Press, London, pp 837–848

Busche D (1998) Die Zentrale Sahara. Justus Perthes Verlag, Gotha, p 284p

Cherkinsky A, di Lernia S (2013) Bayesian approach to ^{14}C dates for estimation of long-term archaeological sequences in arid environments: the Holocene site of Takarkori Rockshelter, southwest Libya. Radiocarbon 55:771–782

Collins JA, Carr AS, Schefuß E, Boom A, Sealy J (2017) Investigation of organic matter and biomarkers from Diepkloof Rock Shelter,

South Africa: insights into Middle Stone Age site usage and palaeoclimate. J Arch Sci 85:51–65
Cooks J, Pretorius JR (1987) Weathering basins in the Clarens Formation sandstone, South Africa. S Afr J Geol 90:147–154
Côte H (1957) Quelques aspects de la morphologie de l'Ahaggar. Rev Géogr Lyon 23:321–332
Courty N-A, Vallverdu J (2001) The microstratigraphic record of abrupt climate changes in cave sediments of the Western Mediterranean. Geoarchaeology 16:467–500
Cremaschi M, di Lernia S (eds) (1998) Wadi Teshuinat: palaeoclimate and prehistory in south-western Fezzan (Libyan Sahara). Centro Interuniversitario di Recerca per le Civilta e l'Ambiente del Sahara Antico. Milan, CNR Quaderni di Geodinamica Alpina e Quaternaria
Cremaschi M, Zerboni A (2010) Human communities in a drying landscape: Holocene climate change and cultural response in the central Sahara. In: Martini IP, Chesworth W (eds) Landscapes and societies: selected cases. Springer, Dordrecht, pp 67–89
Cremaschi M, Zerboni A, Spötl C, Felletti F (2010) The calcareous tufa in the Tadrart Acacus Mt. (SW Fezzan, Libya). an early Holocene palaeoclimate archive in the central Sahara. Palaeogeogr Palaeoclimatol Palaeoecol 287:81–94
Cremaschi M, Zerboni A, Mercuri AM, Olmi L, Biagetti S, di Lernia S (2014) Takarkori rock shelter (SW Libya): an archive of Holocene climate and environmental changes in the central Sahara. Quat Sci Rev 101:36–60
Cremaschi M, Zerboni A, Charpentier V, Crassard R, Isola I, Regattieri E, Zanchetta G (2015) Early-Middle Holocene environmental changes and pre-Neolithic human occupations as recorded in the cavities of Jebel Qara (Dhofar, southern Sultanate of Oman). Quat Int 382:264–276
Cremaschi M (2002) Late Pleistocene and Holocene climatic changes in the Central Sahara: the case study of the Southwestern Fezzan, Libya. In: Hassan FA (ed) Droughts, food and culture. Ecological change and food security in Africa's later prehistory. Kluwer, New York, pp 65–81
Damnati B (2000) Holocene lake records in the Northern Hemisphere of Africa. J Afr Earth Sci 31:253–262
deMenocal P, Ortiz J, Guilderson T, Adkins J, Sarnthein M, Baker L, Yarusinsky M (2000) Abrupt onset and termination of the African Humid Period: rapid climate responses to gradual insolation forcing. Quat Sci Rev 19:347–361
di Lernia S, Gallinaro M (2011) Working in a UNESCO WH Site. Problems and practices on the rock art of Tadrart Akakus (SW Libya, Central Sahara). J Afr Arch 9:159–175
Doerschner N, Fitzsimmons KE, Ditchfield P, McLaren SJ, Steele TE, Zielhofer C, McPherron SP, Bouzouggar A, Hublin J-J (2016) A new chronology for Rhafas, Northeast Morocco, spanning the North African Middle Stone Age through to the Neolithic. PLoS ONE 11:e0162280. https://doi.org/10.1371/journal.pone.0162280
Eppes MC, Griffing D (2010) Granular disintegration of marble in nature: a thermal-mechanical origin for a grus and corestone landscape. Geomorphology 117:170–180
Gallinaro M (2013) Saharan rock art: local dynamics and wider perspectives. Arts 2:350–382
Gasse F (2000) Hydrological changes in the African tropics since the Last Glacial Maximum. Quat Sci Rev 19:189–211
Goudie AS, Migón P (1997) Weathering pits in the Spitzkoppe area, Central Namib Desert. Z Geomorphol NF 41:417–444
Grab SW, Goudie AS, Viles HA, Webb N (2011) Sandstone geomorphology of the Golden Gate Highlands National Park, South Africa, in a global context. Koedoe 53:985. https://doi.org/10.4102/koedoe.v53i1.985
Hagedorn H (1980) Geological and geomorphological observation on the northern slope of the Tībistī Mountains, central Sahara. In: Salem MJ, Busrewil MT (eds) The geology of Libya, vol III. Academic Press, London, pp 823–835
Hensel EA, Vogelsang R, Noack T, Bubenzer O (2021) Stratigraphy and chronology of Sodicho Rockshelter—A new sedimentological record of past environmental changes and human Settlement Phases in Southwestern Ethiopia. Front Earth Sci 8:611700. https://doi.org/10.3389/feart.2020.611700
Henshilwood CS, d'Errico F, van Niekerk KL, Coquinot Y, Jacobs Z, Lauritzen S-E, Menu M, García-Moreno R (2011) A 100,000-year-old ochre-processing workshop at Blombos Cave, South Africa. Science 334:219–222
Hoelzmann P, Gasse F, Dupont LM, Salzmann U, Staubwasser M, Leuschner DC, Sirocko F (2004) Palaeoenvironmental changes in the arid and sub arid belt (Sahara-Sahel-Arabian Peninsula) from 150 kyr to present. In: Battarbee RW, Gasse F, Stickley CE (eds) Past climate variability through Europe and Africa. Springer, Berlin, pp 219–256
Iriarte M, Hernanz A, Gavira-Vallejo JM, Sáenz de Buruaga A, Martín S (2018) Micro-Raman spectroscopy of rock paintings from the Galb Budarga and Tuama Budarga rock shelters, Western Sahara. Microchem J 137:250–257
Lanzarone P, Seidel M, Brandt S, Garrison E, Fisher EC (2019) Ground-penetrating radar and electrical resistivity tomography reveal a deep stratigraphic sequence at Mochena Borago Rockshelter, southwestern Ethiopia. J Arch Sci: Rept 26:101915. https://doi.org/10.1016/j.jasrep.2019.101915
di Lernia S (ed) (1999) The Uan Afuda cave. Hunter-Gatherer Societies of Central Sahara. AZA Monograph 1, Edizioni All'Insegna del Giglio, Firenze, 272 p
di Lernia S (2002) Dry climatic events and cultural trajectories: adjusting Middle Holocene pastoral economy of the Libyan Saharan. In: Hassan FA (ed), Droughts, food and culture. Ecological change and food security in Africa's later prehistory. Kluwer, New York, pp 225–250
Lézine A-M, Hély C, Grenier C, Braconnot P, Krinner G (2011) Sahara and Sahel vulnerability to climate changes, lessons from Holocene hydrological data. Quat Sci Rev 30:3001–3012
Loope DB, Loope GR, Burberry CM, Rowe CM, Bryant GC (2020) Surficial fractures in the Navajo sandstone, south-western USA: the roles of thermal cycles, rainstorms, granular disintegration, and iterative cracking. Earth Surf Proc Landf 45:2063–2077
Mamtimin B, Et-Tantawi AMM, Schaefer D, Meixner FX, Domroes M (2011) Recent trends of temperature change under hot and cold desert climates: comparing the Sahara (Libya) and Central Asia (Xinjiang, China). J Arid Env 75:1105–1113
Manning K, Timpson A (2014) The demographic response to Holocene climate change in the Sahara. Quat Sci Rev 101:28–35
McLaren SJ, Reynolds T (2009) Early humans in dryland environments: a geoarchaeological perspective. In: Parsons AJ, Abrahams AD (eds) Geomorphology of desert environments. Springer, Berlin, pp 773–798
Mellor A, Short J, Kirkby SJ (1997) Tafoni in the El Chorro area, Andalucia, southern Spain. Earth Surf Proc Landf 22:817–833
Mercier N, Wengler L, Valladas H, Joron J-L, Froget L, Reyss J-L (2007) The Rhafas Cave (Morocco): chronology of the mousterian and aterian archaeological occupations and their implications for quaternary geochronology based on luminescence (TL/OSL) age determinations. Quat Geochron 2:309–313
Moeyersons J, Vermeersch PM, Van Peer P (2002) Dry cave deposits and their palaeoenvironmental significance during the last 115 ka, Sodmein Cave, Red Sea Mountains, Egypt. Quat Sci Rev 21:837–851

Mol L, Viles HA (2012) The role of rock surface hardness and internal moisture in tafoni development in sandstone. Earth Surf Proc Landf 37:301–314

Mori F (1965) Tadrart Acacus. Arte rupestre e culture del Sahara preistorico. Einaudi, Torino, 257 p

Oestmo S, Schoville BJ, Wilkins J, Marean CW (2014) A Middle Stone Age Paleoscape near the Pinnacle Point caves, Vleesbaai, South Africa. Quat Int 350:147–168

Parsons AJ, Abrahams AD (eds) (2009) Geomorphology of desert environments. Springer, Berlin, p 831

Parton A, Bretzke K (2020) The PalaeoEnvironments and ARchaeological Landscapes (PEARL) project: recent findings from Neolithic sites in Northern Oman. Arab Arch Epig 31:194–201

Piccini L (2011) Recent developments on morphometric analysis of karst caves. Acta Carsol 40:43–52

Pirson S, Haesaerts P, Court-Picon M, Damblon F, Toussaint M, Debenham N, Draily C (2006) Belgian cave entrance and rock-shelter sequences as palaeoenvironmental data recorders: the example of Walou Cave. Geol Belg 9:275–286

Pye K (1985) Granular disintegration of gneiss and migmatites. Catena 12:191–199

Sahle Y, Giusti D, Gossa T, Ashkenazy H (2019) Exploring karst landscapes: new prehistoric sites in south-central Ethiopia. Antiquity 93:e21. https://doi.org/10.15184/aqy.2019.94

Smith BJ (1978) The origin and geomorphic implications of cliff foot recesses and tafoni on limestone hamadas in the Northwest Sahara. Z Geomorphol 22:21–43

Soukopova J (2011) The earliest rock paintings of the central Sahara: approaching interpretation. Time Mind 4:193–216

Soukopova J (2017) Central Saharan rock art: considering the kettles and cupules. J Arid Env 143:10–14

Soukopova J (2018) Decorated boulders and other neglected features of the Central Saharan rock art. J Arid Env 156:96–105

Twidale CR, Bourne JA (1976) Origin and significance of pitting on granitic rocks. Z Geomorphol NF 20:403–416

Vallverdú i Poch J (2017) Soil-stratigraphy in the cave entrance deposits of Middle Pleistocene age at the Trinchera del Ferrocarril sites (Sierra de Atapuerca, Spain). Quat Int 433:199–210

Watson DJ (2017) Bosumpra revisited: 12,500 years on the Kwahu Plateau, Ghana, as viewed from 'On top of the hill.' Azania 52:437–517

Young R, Young A (1992) Sandstone landforms. Springer, Berlin, p 163

Landslides and Alluvial Fans

7

Jasper Knight

Abstract

Landslides and alluvial fans have not been commonly reported from the central Sahara, but are found around the margins of bedrock massifs which act as sediment source areas for these landforms. Rotational landslides may correspond to groundwater sapping or slope undercutting of weaker rock units. Alluvial fans are usually located at the outlet of wadi systems and may be associated with terrace fragments or alluvial-filled bedrock valleys. Alluvial fan activity therefore corresponds to episodic rainstorm events when wadi systems are reactivated. Both landslides and alluvial fans in the central Sahara are considered to be largely relict features, with low activity under present climatic conditions. Their precise timing of activity, however, is largely unknown, although previous radiocarbon dating studies of river terrace sediments suggest several aggradational phases during the Holocene, but there is very little preserved evidence for earlier activity. More systematic mapping and field investigations (including sedimentology and dating) are needed in order to understand the spatial and temporal patterns of landslides and alluvial fans and their controls.

Keywords

Climate change · Groundwater sapping · Holocene · Land surface instability · Mass movements · Wadi systems

J. Knight (✉)
School of Geography, Archaeology & Environmental Studies, University of the Witwatersrand, Johannesburg 2050, South Africa
e-mail: jasper.knight@wits.ac.za

7.1 Introduction

Landslides and alluvial fans can be considered as some of the most common signatures of land surface instability on and adjacent to weathered bedrock slopes which act as sediment sources to these landforms. Other landforms associated with land surface instability on bedrock slopes include rockfall/exfoliation sheets, gullies caused by soil erosion and water runoff and fluvial terraces and river channels contained within bedrock valleys. Thus, (1) landslides and alluvial fans are but two elements of wider slope instability systems, and (2) all of these elements are common in many areas of low latitude Africa that have high weathering rates and heavy, episodic rainstorms. Long-term epeirogenic uplift and the presence of coarse crystalline sedimentary and metamorphic rocks also facilitate land surface weathering and erosion that contribute to downslope sediment supply. In these environments, the land surface can become unstable as a result of erosional undercutting, seismic activity, subaerial weathering or by the internal loss of geotechnical strength in subsurface rocks. (Landslides in most areas of Africa including the Sahara are associated with climate and land cover change rather than seismic activity, however; Ngecu et al. 2004.) Landslides and alluvial fans have somewhat different triggering factors, geomorphic processes and physical expressions in the landscape, but they both reflect the erosion of massif margins and are thus found in the same areas geographically. This means, therefore, that landslides and alluvial fans are not located uniformly across the central Sahara region but are found associated with mountain massifs, including the Tibesti, Messak Mallet and Ahaggar massifs (Fig. 7.1). It is notable however that there are very few studies that focus on these landforms, and as such very little is known, including their distribution, age and controls. Lee et al. (2013) discussed the fact that shifts between humid and arid regimes in the central Sahara can give rise to different geohazard types, including landslides around escarpment edges and episodic fluvial activity in wadis.

J. Knight et al. (eds.), *Landscapes and Landforms of the Central Sahara*, World Geomorphological Landscapes, https://doi.org/10.1007/978-3-031-47160-5_7

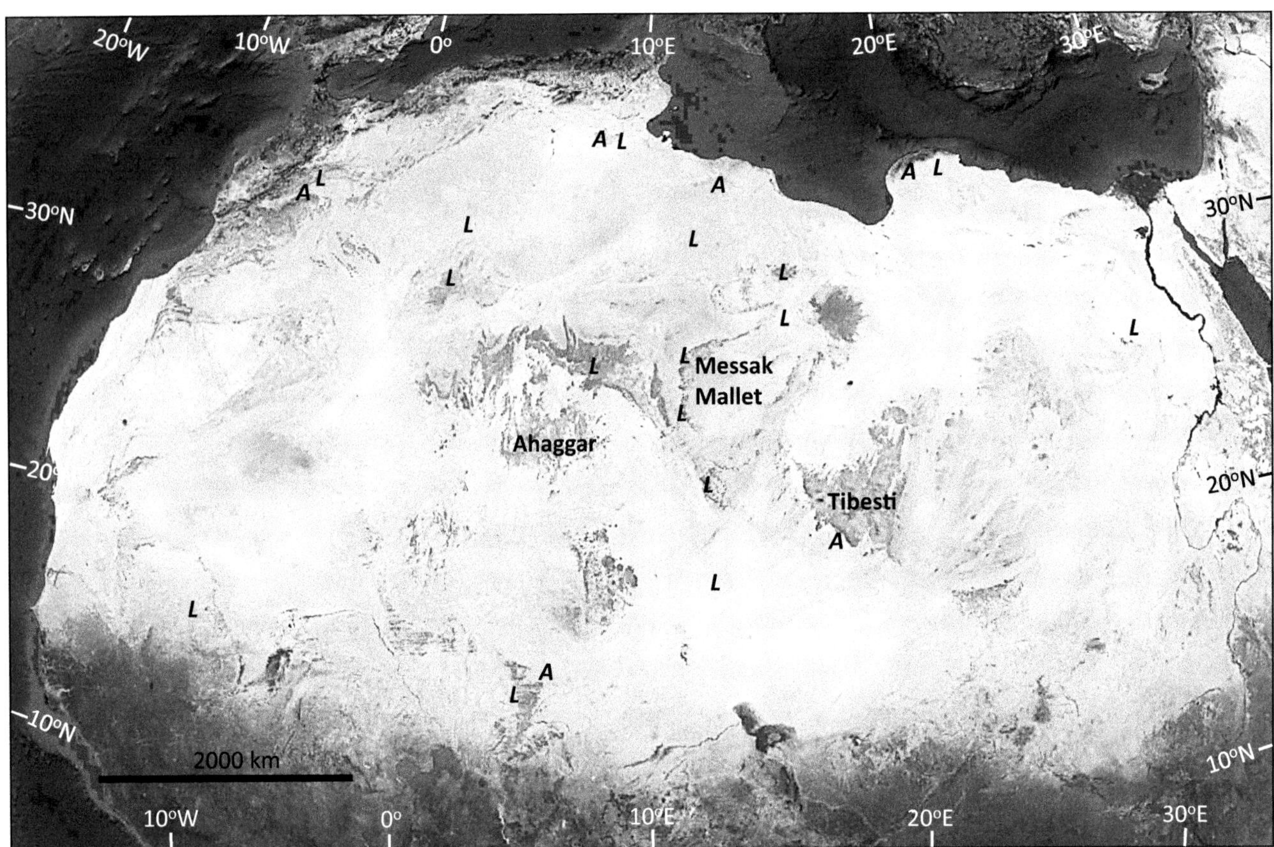

Fig. 7.1 Locations of relict landslides (L) and alluvial fans (A) in the central Sahara region that are mentioned in previous studies and named massifs. Background image from Google Earth (date: 31 December 2016)

This chapter describes and discusses examples of landslides and alluvial fans from the central Sahara, based on evidence from the literature and new examples discovered from the region that illustrate more modern ways of thinking of these landforms in terms of their dynamics and controls. The chapter presents and discusses evidence for (1) landslides and (2) alluvial fans, and then (3) these landforms are discussed from a palaeoenvironmental perspective. The wider context of these landforms is then examined.

7.2 Examples from the Central Sahara

7.2.1 Landslides

Fossil (relict) landslides have been identified in several locations in the central Sahara, around the margins of bedrock massifs. Their wide distribution suggests a regional-scale control such as climate rather than local triggering mechanisms such as earthquakes (Busche 2001). Grunert and Busche (1980) showed that different landslide styles on the Messak Mallet reflect their geological context. Nested rotational landslides have sharp and extensive (1 km long) crestlines with escarpments 150–400 m high and are developed in sandstone substrates (Messak Sandstone) underlain by Triassic clays (Tilemsin Formation) within which basal slip takes place (Fig. 7.2). Where these clays dominate, mass movement is more uncontrolled and disaggregated bedrock blocks break down in a downslope direction to yield mass and debris flows. These mass movements are less laterally extensive but are of greater downslope extent (< 2 km).

The rotational landslides comprise 3–5 successive slide elements that are nested in a downslope direction (Fig. 7.2). The uppermost surfaces of each slide are usually upslope-dipping, creating accommodation space on these surfaces that is sometimes infilled with sand or desiccated clay pans. These infills therefore postdate slide movement and stabilisation and thus can be used in a dating or stratigraphic context to evaluate slide age. Cited examples of these landslides are found around Passe de Salvador and Col d'Anai, Messak Mallet (Grunert and Busche 1980). Examples of clay-dominated landslides are found in the Zouzoudinga

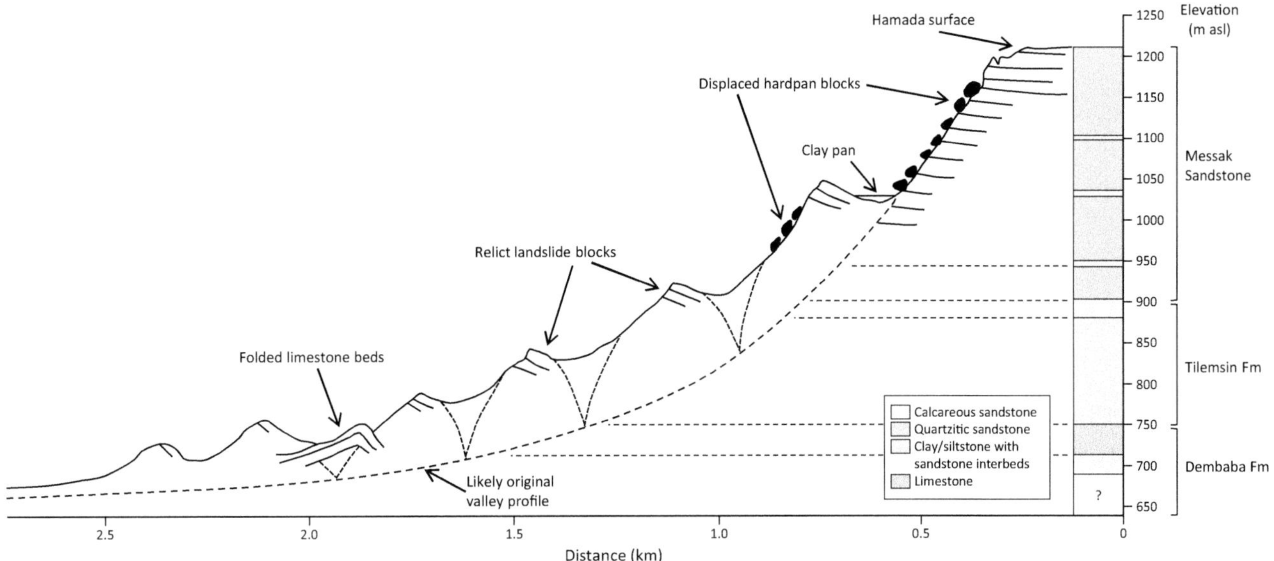

Fig. 7.2 Schematic cross-section of a rotational landslide system, southwest Libya, redrawn from Grunert and Busche (1980) with additional notes. The specific location described is 20 km southwest of Col d'Anai (24° 19′ N, 11° 27′ E)

area where clays and marls dominate and where landslide blocks tilt forwards (antithetic) rather than backwards. This mechanism is typical of groundwater sapping at the base of a slope, along the spring line, where higher water availability gives rise to enhanced basal slip. This is known to be a key process of landslide generation in the Western Desert of Egypt (Luo et al. 1997). Such a process, however, is strongly geologically controlled, dependent on both lithology and structure and position of the groundwater table. Variations in the position of the latter in response to climate and/or changes in base level are therefore a first-order control on such rotational slides.

Lee et al. (2013) described other processes of escarpment erosion, related to landslides. These include displacement of protective caprock slopes (weathered) by undercutting and subsurface collapse by the development of soil pipes and gullies in badland environments. These processes can also variously trigger or act in response to larger-scale landslides or can operate to further degrade slopes that have been affected by landsliding, thereby reducing slope angle and relief over time (Smith 1978). The presence of preserved landslide blocks in the landscape, however, suggests that subaerial weathering and erosion rates following landsliding are very low (Busche 2001). This is also confirmed by the preservation of Eocene coastal landslides on the Mediterranean coast of Libya (Ammar 1993).

The landslide studies reported above refer to single examples and from single, isolated locations. It is therefore likely that these studies significantly under-represent the distribution and number of landslides in the region. Remote sensing methods may be used to identify large-scale rotational landslides, where scarp faces can be identified (e.g. Mwaniki et al. 2017), but smaller and complex landslides may be difficult to image. Thus, landslide mapping is still in its infancy across the Sahara.

7.2.2 Alluvial Fans

Alluvial fans can be considered as just one type of depositional landform found in semiarid and arid footslope locations. Others include slope sediment mantles (related to colluvial processes), fluvial terraces and fluvial channel/floodplain elements and bajadas (coalescent alluvial fans). The geometry and physical properties of alluvial fans are relatively well known globally (Tooth 2008), and the structure of alluvial fans typically found in the Sahara region is presented by White (1991). The fans are incised by a fan-head trench at the fan apex, with distributary or anabranching channels feeding outwards and in a downslope direction over the fan surface. These distributary channels are variously activated or abandoned (as wadis) during flash flood events. Alluvial fans have not been previously described in any detail from the central Sahara region although they have been identified around the margins of the Sahara in Tunisia (White 1991) and Morocco (Stokes and Mather 2015) and within dune systems in Niger (Talbot and Williams 1979). In Tunisia, alluvial fans are located at footslopes and comprise largely relict surfaces that reflect past fluvial activity in the mountain hinterland (White 1991; White et al. 1996). The phasing relationships of fluvial, debris torrent and mudflow activity on channel incision and infilling and channel abandonment can be inferred based on weathering rind depth and desert varnish on surface boulders (e.g. Al-Farraj and Harveu 2000).

Compared to arcuate alluvial fans, alluvial spreads or valley-fill elements dissected into terraces have been more commonly identified in the Sahara, within and at the mouths of massif valleys. For example, Rognon (1980) described paired and unpaired terraces on the margins of the Atakor and Tibesti massifs. Based on correlation across the region constrained by radiocarbon dates, these terraces were formed in distinct late Pleistocene phases, at 15–7.5 and 5–2 ka BP. A reduction in stream capacity over time was noted, shown by an upward decrease in sediment size and likely related to decreased valley accommodation space and channel narrowing. Sediment transport distance was relatively short, with sediments largely contained within these valleys and not widely transported outside of this zone onto the massif hinterland. A similar pattern of Holocene alluvial terrace and floodplain deposition and incision is also noted in Tunisia, where climate and also human disturbance are given as forcing factors (Zerai 2009). Different floodplain elements are distinguished mainly on the basis of their internal sediments, including carbonates and buried soils. The oldest terrace fragments vary in thickness from 3 to 15 m and are found in particular on south-facing wadi mouths. This oldest terrace was formed prior to ~ 35 ka based on poorly constrained radiocarbon dates on carbonates and on the presence of Acheulian lithics that were eroded from terrace sediments. Middle terrace fragments were formed following a period of extensive downcutting and are characterised by the deposition of organic silts and grass pollen indicative of wetter conditions and a temperate/Mediterranean climate (Maley 1977). Calcrete and travertine deposition also took place at this time. Radiocarbon dating suggests that this terrace formation period took place between ~ 15 and 6 ka BP. The final terrace stage took place during the mid-Holocene following an early Holocene arid phase and was characterised variously by sediment aggradation and reworking indicative of spatially varying climatic conditions. A very similar stratigraphy of different fan evolution phases is also recorded in northwest Libya where fan gravels are intercalated with aeolian sands and floodplain silts (Anketell and Ghellali 1988).

Some river terrace fragments—which may potentially also include alluvial fan elements—are reported on the northern slope of the Tibesti Mountains (Hagedorn 1980). However, these are not widespread in their distribution, and they have been strongly affected by dissection, likely as a result of base level uplift. These terrace fragments are relatively thin (5–10 m thick), comprised interbedded sands and gravels and were deposited in late Pleistocene as well as late Holocene phases. Chemical weathering (rubification) of older terrace fragments attests to their potential age. Around river systems of the Tibesti, including the Enneris Aozou, Tidedi and Bardagué-Zoumri river systems, different upper, middle and lower terraces are identified on the basis of their elevation (Hagedorn 1980). This record of terrace formation phases reflects humid climatic periods when water flow within wadi systems took place more frequently. This is consistent with regional evidence for both increased precipitation and also its variability during the late Quaternary, leading to the reinvigoration of surface drainage patterns (e.g. Pachur and Kröpelin 1987; Cremaschi 2001; Reade et al. 2018).

Some river terrace fragments are wholly contained within massif wadi systems, whereas some others are connected functionally to fronting alluvial fans. For example, along the southwest Tibesti, Chad, at least three generations of alluvial fans and land surface incision can be distinguished based on their cross-cutting relationships (Fig. 7.3). These alluvial fans issue through narrow and bedrock-controlled wadi systems, with sediment issuing outwards away from the fan apex. No clear fan head trench is observed; this is likely due to cut-and-fill during episodic flood events in which sediment is deposited during waning flow stages. It is notable that these alluvial fans are coalescent, forming bajadas. These phasing relationships are not constrained by dating or stratigraphic evidence, but may correspond to variations in late Pleistocene humidity. Charif et al. (2014) described three generations of reworked and overlapping alluvial fans and debris cones in the southern Atlas Mountains in Morocco, and these may have evolved over different time periods since the Plio-Pleistocene. Dated alluvial fan sediments from coastal Libya show aggradational phases that correspond with more humid stadials during the Quaternary (Rowan et al. 2000).

7.3 Palaeoenvironmental Interpretations of Mass Transport Events and Processes

Both landslides and alluvial fans are an expression of land surface instability. The major control on landslide activity is moisture availability; Grunert and Busche (1980) considered that precipitation of 200 mm yr^{-1} is required in order to activate basal slip on underlying clays, contrasting with present precipitation in the central Sahara of ~ 20 mm yr^{-1}. In addition, the small size of most catchments, drawing water by precipitation only, means that landslides likely reflect triggering by single rainstorm events. Based on palaeoclimate records, it also confirms that these landslides are relict features, corresponding to previously wetter climate regimes. Landslides were also active during the late Pleistocene or early Holocene, based on radiocarbon dated freshwater fauna within displaced river terraces (Grunert and Busche 1980). Smith (1978) described how past more

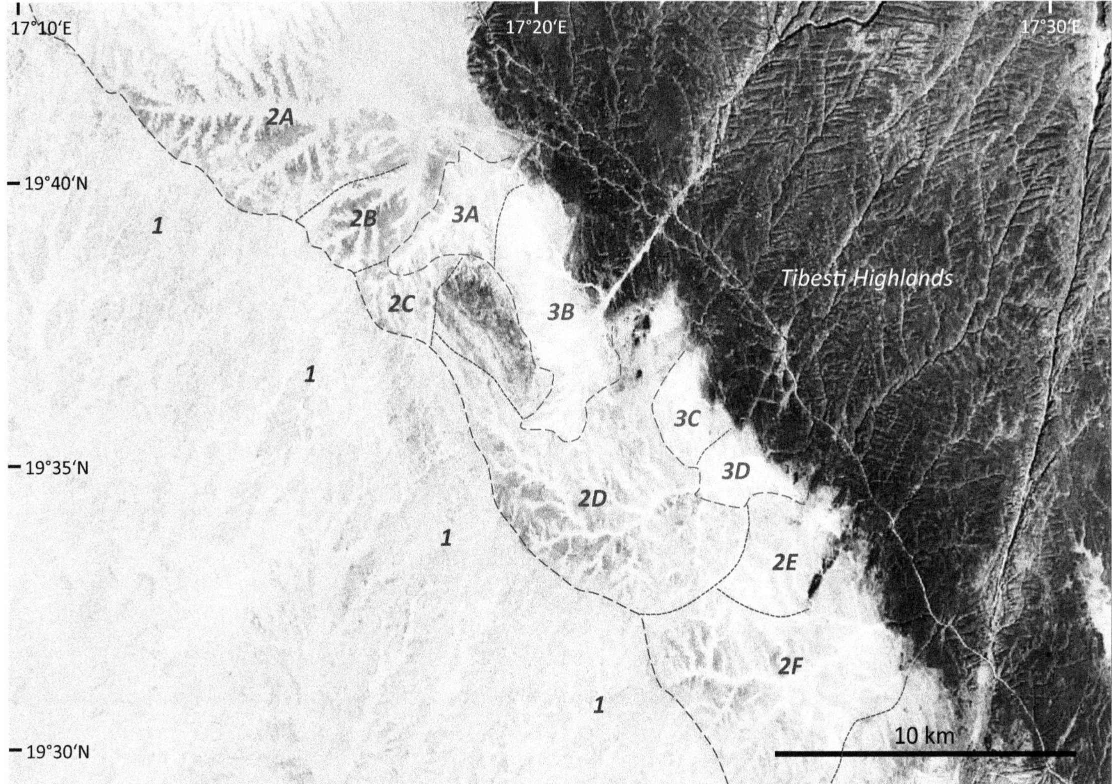

Fig. 7.3 Example of different alluvial fan phases (1 oldest–3 youngest) in the southern Tibesti Highlands, Chad. Background image from Google Earth (date: 28 September 2012)

humid climates and the effects of slope aspect have influenced bedrock weathering processes that impact on the susceptibility of the land surface to be affected by mass movements. Steeper ENE and WSW-facing slopes are attributed to the effects of greater diurnal thermal weathering. In the Western Desert of Egypt, Luo et al. (1997) also described groundwater sapping at the scarp foot, evidenced by spring tufa formation. Groundwater presence at the boundary of shale interbeds can in turn trigger instabilities that can lead to landslides, as well as form soil pipes that can lead to land surface gullying and collapse.

The widespread distribution of fossil landslides throughout the Sahara region, developed in relict etchplain surfaces of Neogene age, suggests a regional-scale control on landslide activity, most likely by a regional humid climate (Ammar 1993; Busche 2001; Lee et al. 2013). The most significant property of these landslides is that they are relict throughout and have not been mobilised since their initial development. This suggests that the geotechnical threshold for sliding has not been crossed since initial movement. A significant problem however is the size of contributing catchments located upslope of the landslide zone which may not be large enough to trigger landslides by water infiltration alone (Busche 2001). An alternative mechanism may be (1) water ponding and infiltration in the back-tilted depressions of rotational landslides (e.g. Fig. 7.2) and/or (2) rising groundwater table at the scarp foot, triggering instability above (Luo et al. 1997).

The fact that slide toes and alluvial fan elements may be buried by later blown sand (Fig. 7.3) indicates a switch from more humid to more arid conditions. It is conceivable that alluvial fan margins may be interbedded with dune sands, stratigraphically recording alternate humid–arid phases (El-Baz et al. 2000; Drake et al. 2009), with palaeo-fan or palaeosol surfaces marked by a desert pavement lag. This type of stratigraphy has been recorded on some slope surfaces in Libya (Zerboni et al. 2015). Likewise, some alluvial fans, palaeosols or palaeolake sediment may be completely buried by later dune migration and have no present surface expression at all (e.g. Maxwell and Haynes 2001). Megafans are large-scale features (10^3–10^5 km^2; Wilkinson et al. 2006) typical of low-angle, distributary and avulsion river systems, and Quaternary examples have been reported from Amazonia (Rossetti et al. 2018), central Brazil (Cuiabá megafan; Pupim et al. 2017), Namibia (Cubango megafan; Lindenmaier et al. 2014), Botswana (Okavango megafan; Podgorski et al. 2013) and Iran (Arzani 2012). It is very likely that larger-scale alluvial fans underlie some of the Saharan sand seas, corresponding to previously more humid environments. Geophysical methods may be able to inform on this possibility (e.g. Drake et al. 2009; Lindenmaier et al. 2014).

7.4 Discussion

Landslides and alluvial fans are records of land instability and likely triggered by climate changes across the Sahara region, but very little is known about the distribution of these landforms, their properties, age and significance. It is notable that landslides and alluvial fans that have been described in the literature are usually found in the same settings and locations (Fig. 7.1). It is also likely that these landforms can inform on landscape dynamics during pluvial phases of the Pleistocene when adjacent dune systems were largely inactive. As a result, some future research directions can be indicated. Development of alluvial fans during pluvial phases may have been associated with sheetwash facies interbedded with aeolian facies, especially at fan edges. It may be that prolonged arid phases resulted in fluvial facies becoming buried or extensively reworked, resulting in laterally extensive and planar/slightly dipping lagged surfaces. These can potentially be used as stratigraphic markers. Seismic data can be used to confirm deep stratigraphy; satellite radar data (e.g. SRTM) can be used to map topographic variations, including alluvial fans. Radar and borehole data suggest buried palaeovalleys of Miocene to Plio-Pleistocene age, that are now subsequently infilled with clastic sediments, are present across the central Sahara (e.g. Ghellali and Anketell 1991; Paillou et al. 2009). It is not clear whether exposed or buried terraces or fans are associated with these palaeovalleys. The relationship between fluvial sediments and aeolian sediments is also unclear—it is likely that present aeolian sediment was derived by weathering of underlying rocks, but no geochemical studies have examined sediment source areas or spatial/temporal patterns of transport in such systems, unlike for river channels. Likewise, aeolian transport of water-deposited sediment may have also taken place along wadi systems, but these relationships have not been examined. Such questions are exciting areas of future research.

7.5 Summary and Outlook

Landslides and alluvial fans in the Sahara region provide evidence for land surface responses to climate forcing, likely from the Miocene onwards, despite a lack of dating control. Although often found around the margins of bedrock massifs, neither landform type has been systematically mapped, and thus, their distributions and properties are largely unknown. Remote sensing and geophysical methods offer the best options for this type of research. The fact that these landforms are largely relict today suggests that (1) they were active during more humid climate phases, likely corresponding to Quaternary stadials; (2) there is high preservation potential for such landforms, consistent with low weathering rates and relatively stable land surfaces or beneath a surface sediment cover. The Sahara region is tectonically subdued despite the existence of mantle swells, and regionally extensive laterites and bauxites suggest land surface stability since the late Cretaceous (Burke and Gunnell 2008). Thus, landslides and alluvial fans could potentially be millions of years old and provide insight into Neogene landscapes in the Sahara. This hypothesis could be tested with cosmogenic exposure dating.

References

Al-Farraj A, Harveu AM (2000) Desert pavement characteristics on wadi terrace and alluvial fan surfaces: Wadi Al-Bih, U.A.E. and Oman. Geomorphology 35:279–297

Ammar AA (1993) An analysis of Eocene mass movements in the Wadi Athrun, Cyrenaica. Libyan Stud 24:19–26

Anketell JM, Ghellali SM (1988) Stratigraphy of the Qasr Al Haj Formation, fan-gravel deposits of Quaternary age, Tripolitania, S.P.L.A.J. Libyan Stud 19:123–131

Arzani N (2012) Catchment lithology as a major control on alluvial megafan development, Kohrud Mountain range, central Iran. Earth Surf Proc Landf 37:726–740

Burke K, Gunnell Y (2008) The African erosion surface: a continental-scale synthesis of geomorphology, tectonics, and environmental change over the past 180 million years. GSA Memoir 201, Geological Society of America, Boulder CO, p 66

Busche D (2001) Early Quaternary landslides of the Sahara and their significance for geomorphic and climatic significance. J Arid Environ 49:429–448

Charif A, Aït Malek H, Chaïbi M, Tannouch-Bennani S (2014) Evolution géomorphologique quaternaire de la bordure anti-atlasique de la partie centrale de la plaine du Souss (Maroc). Quaternaire 25:49–66

Cremaschi M (2001) Holocene climatic changes in an archaeological landscape: the case study of Wadi Tanezzuft and its drainage basin (SW Fezzan, Libyan Sahara). Libyan Stud 32:3–27

Drake N, White K, Salem M, Armitage S, El-Hawat A, Francke J, Hounslow M, Parker A (2009) DMP VIII: palaeohydrology and palaeoenvironment. Libyan Stud 40:171–178

El-Baz F, Maingue M, Robinson C (2000) Fluvio-aeolian dynamics in the north-eastern Sahara: the relationship between fluvial/aeolian systems and ground-water concentration. J Arid Environ 44:173–183

Ghellali SM, Anketell JM (1991) The Suq al Jum'ah palaeowadi, and example of a Plio-Quaternary palaeo-valley from the Jabal Nafusah, northwest Libya. Libyan Stud 22:1–6

Grunert J, Busche D (1980) Large-scale fossil landslides in the Msāk Mallat and Ḥamadat Mānghīnī escarpment. In: Salem MJ, Busrewil MT (eds) The Geology of Libya, vol III. Academic Press, London, pp 849–860

Hagedorn H (1980) Geological and geomorphological observation on the northern slope of the Tībistī Mountains, central Sahara. In: Salem MJ, Busrewil MT (eds) The Geology of Libya, vol III. Academic Press, London, pp 823–835

Lee EM, Fookes PG, Hart AB (2013) Observations on the impact of climate change on landform behaviour and geohazards in the Algeria Sahara. Q J Eng Geol Hydrogeol 46:107–116

Lindenmaier F, Miller R, Fenner J, Christelis G, Dill HG, Himmelsbach T, Kaufhold S, Lohe C, Quinger M, Schildknecht S, Symons G, Walzer A, van Wyk B (2014) Structure and genesis of the Cubango Megafan in northern Namibia: implications for its hydrogeology. Hydrogeol J 22:1307–1328

Luo W, Arvidson RE, Sultan M, Becker R, Crombie MK, Sturchio N, El Alfy Z (1997) Ground-water sapping processes, Western Desert, Egypt. GSA Bull 109:43–62

Maley J (1977) Palaeoclimates of Central Sahara during the early Holocene. Nature 269:573–577

Maxwell TA, Haynes CV (2001) Sand sheet dynamics and Quaternary landscape evolution of the Selima Sand Sheet, southern Egypt. Quat Sci Rev 20:1623–1647

Mwaniki MW, Kuria DN, Boitt MK, Ngigi TG (2017) Image enhancements of Landsat 8 (OLI) and SAR data for preliminary landslide identification and mapping applied to the central region of Kenya. Geomorphology 282:162–175

Ngecu WM, Nyamai CM, Erima G (2004) The extent and significance of mass-movements in Eastern Africa: case studies of some major landslides in Uganda and Kenya. Env Geol 46:1123–1133

Pachur H-J, Kröpelin S (1987) Wadi Howar: Paleoclimatic evidence from an extinct river system in the southeastern Sahara. Science 237:298–300

Paillou P, Schuster M, Tooth S, Farr T, Rosenqvist A, Lopez S, Malezieux J-M (2009) Mapping of a major paleodrainage system in eastern Libya using orbital imaging radar: the Kufrah River. Earth Planet Sci Lett 277:327–333

Podgorski JE, Green AG, Kgotlhang L, Kinzelbach WKH, Kalscheuer T, Auken E, Ngwisanyi T (2013) Paleo-megalake and paleo-megafan in southern Africa. Geology 41:1155–1158

Pupim FN, Assine ML, Sawakuchi AO (2017) Late Quaternary Cuiabá megafan, Brazilian Pantanal: channel patterns and paleoenvironmental changes. Quat Int 438:108–125

Reade H, O'Connell TC, Barker G, Stevens RE (2018) Increased climate seasonality during the late glacial in the Gebel Akhdar, Libya. Quat Sci Rev 192:225–235

Rognon P (1980) Comparison between the Late Quaternary Terrces around Atakor and Tībistī. In: Salem MJ, Busrewil MT (eds) The Geology of Libya, vol III. Academic Press, London, pp 815–821

Rossetti DF, Gribel R, Tuomisto H, Cordeiro CLO, Tatumi SH (2018) The influence of late Quaternary sedimentation on vegetation in an Amazonian lowland megafan. Earth Surf Proc Landf 43:1259–1279

Rowan JS, Black S, Macklin MG, Tabner BJ, Dore J (2000) Quaternary environmental change in Cyrenaica evidenced by U-Th, ESR and OSL dating of coastal alluvial fan sequences. Libyan Stud 31:5–16

Smith DJ (1978) Aspect-related variations in slope angle near Béni Abbès, western Algeria. Geogr Ann 60A:175–180

Stokes M, Mather AE (2015) Controls on modern tributary-junction alluvial fan occurrence and morphology: High Atlas Mountains, Morocco. Geomorphology 248:344–362

Talbot MR, Williams MAJ (1979) Cyclic alluvial fan sedimentation on the flanks of fixed dunes, Janjari, central Niger. Catena 6:43–62

Tooth S (2008) Arid geomorphology: recent progress from an Earth System Science perspective. Progr Phys Geogr 32:81–101

White K (1991) Geomorphological analysis of piedmont landforms in the Tunisian Southern Atlas using ground data and satellite imagery. Geogr J 157:279–294

White K, Drake N, Millington A, Stokes S (1996) Constraining the timing of alluvial fan response to late Quaternary climatic changes, southern Tunisia. Geomorphology 17:295–304

Wilkinson MJ, Marshall LG, Lundberg JG (2006) River behavior on megafans and potential influences on diversification and distribution of aquatic organisms. J South Am Earth Sci 21:151–172

Zerai K (2009) Chronostratigraphy of Holocene alluvial archives in the Wadi Sbeïtla basin (central Tunisia). Géomorphol: relief, proc, environ 2009:271–286

Zerboni A, Perego A, Cremaschi M (2015) Geomorphological map of the Tadrart Acacus massif and the Erg Uan Kasa (Libyan central Sahara). J Maps 11:772–787

Hamadas and Desert Pavements

8

Jasper Knight

Abstract

Landforms associated with the presence of bedrock surfaces are common throughout the central Sahara, in regions where bedrock surfaces are readily exposed, which is mainly on upland massifs. Weathering over unknown but potentially long timescales (10^4–10^6 yr) leads to development of hamada, serir and desert pavement surfaces. This chapter reviews the processes and controls responsible for the formation of such surfaces, their locations in the central Sahara region and the wider implications of such evidence for land surface age, stability and human occupation. Despite being widely distributed and important geomorphic elements of many desert environments, hamadas, serir and desert pavements are critically under-researched.

Keywords

Clasts · Deflation · Desert pavement · Hamada · Soil development · Vesicles · Wind abrasion

8.1 Introduction

Land surfaces of the central Sahara are geomorphically complex because they reflect the operation of physical and chemical weathering processes caused by climatic changes throughout the Cenozoic (last 66 million years) and in particular during the last ~ 7 million years of the late Miocene, the Pliocene and Quaternary (Swezey 2009). Using different analytical and dating methods, including biological data, arid conditions throughout Africa were developed in the middle Miocene around 17–16 million years ago (Senut et al. 2009), and today's Africa deserts likely started to develop around this time, based on evidence of dust export recorded in Atlantic marine cores (Goudie 2013). Cosmogenic dating of desert dune sands from southern Africa suggests significant sand accumulation by at least 1 million years ago (Vermeesch et al. 2010). Palaeomagnetic and geochemical evidence from the Qattara depression, western Egypt, suggests land surface stability from at least the early Miocene (Abdeldayem 1996). Several studies have found that some elements of today's Saharan landscape, in particular now-dry drainage networks identified using remote sensing imagery, date from the Miocene and Pliocene (e.g. Drake et al. 2008; Hounslow et al. 2017). An important inference from this evidence is that weathering of bedrock outcrops, generating the loose sediment that forms today's surrounding sand seas, has likely persisted since the middle Miocene. Based on regional-scale remote sensing classification from different spectral bands from MODIS satellite data, the land surface across the north of Africa as a whole comprises 22% sand dunes, 21% desert pavements, 15% sand sheets, 14% alluvial fans, 13% vegetation and 9% mountains (Ballantine et al. 2005) (Fig. 8.1). However, these values correspond to the entire region, and smaller areas where dune sand and rock surfaces are found together may show values that vary due to seasonal to interannual changes in climate. Thus, there is a variety of substrate types and topographic contexts across this region, and that desert pavements are a common landscape element. The weathering and erosion forms found on mountain massifs such as Ahaggar and Tibesti in the central Sahara may reflect both a long time period of evolution, and changes in climate that can affect the combination of different weathering processes and their rates of operation on different bedrock surfaces. Despite this role as potential evidence for climate and weathering over long time periods, the physical properties and evolution of bedrock surfaces are very poorly known.

J. Knight (✉)
School of Geography, Archaeology & Environmental Studies, University of the Witwatersrand, Johannesburg 2050, South Africa
e-mail: jasper.knight@wits.ac.za

J. Knight et al. (eds.), *Landscapes and Landforms of the Central Sahara*, World Geomorphological Landscapes, https://doi.org/10.1007/978-3-031-47160-5_8

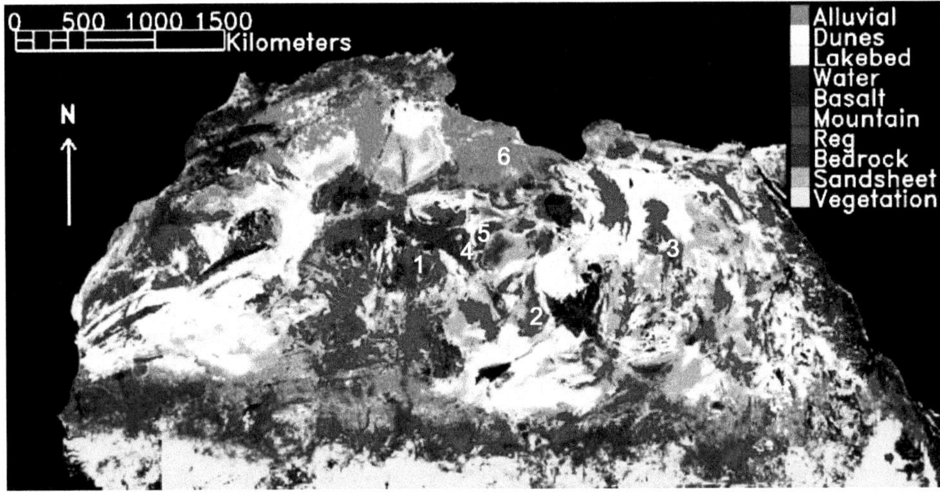

Fig. 8.1 Landforms of north Africa, based on autoclassification from MODIS data (Ballantine et al. 2005, their Fig. 7), with places named in the text. 1 = Ahaggar highlands, 2 = Tibesti highlands, 3 = Gilf Kebir, 4 = Tadrart Acacus, 5 = Messak Settafet and 6 = Jabal as Sawdā'

This chapter examines the different weathering and erosion forms found on bedrock outcrops, with some specific examples from the central Sahara region, focusing on the key landforms of hamadas and desert pavements. Globally, there is a lack of evidence on the physical properties, dynamics and evolution of hamadas and desert pavements. This chapter thus provides new insight on these enigmatic landforms and their development, with reference to the geomorphic evolution of Saharan bedrock massifs.

8.1.1 Definition of Terms

The terms *hamada* and *desert pavement* refer to genetically related but somewhat different landforms, typical of semiarid and arid environments and particularly found in hot deserts (Knight 2019). There are very few evolutionary models that describe the development of these landforms over time or space. As such, these landforms and thus arid bedrock environments more generally are very poorly known.

The term *hamada* is derived from an Arabic descriptor of a flat bedrock surface or summit (حمادة) and refers to eroded bedrock surfaces that are usually covered by a litter of loose pebbles or boulders. These clasts sit directly on bedrock; the bedrock irrespective of lithology generally has a hard, planar surface. Any loose and small weathering products have been either transported away from the bedrock surface by wind and water or initially retained on the land surface in the form of a thick soil or protolith and then eroded away by wind or flash floods. Any past soil or protolith is not retained on the bedrock surface. Hamadas can thus be considered as net erosional landscapes. As such, hamadas are strongly controlled by bedrock properties and are usually found on relatively flat and planar sedimentary bedrock surfaces where granular disintegration processes dominate, which are controlled in hot deserts mainly by thermal weathering (cataclastism). The clasts on hamada surfaces are usually randomly scattered and do not show a consistent spatial pattern, but are commonly quite large (few tens of cm to ~ 1 m in diameter) and vary from subangular to well-rounded. Apart from rock surface weathering and erosion, hamadas can also develop by larger-scale stripping of a deep-weathered sediment mantle, developed by mainly chemical weathering under warm, moist conditions, such that the underlying bedrock surface is revealed by net erosion (Laity 2008, p. 159). Under this model, hamadas are therefore erosional landforms developed where duricrust-covered surfaces (etchplains) are incised and stripped away over time, moving loose sediments downslope and revealing a bedrock plateau surface that is slightly elevated above the surrounding landscape (Busche 1980; Kroonenberg and Melitz 1983; Dauteuil et al. 2015). Guillocheau et al. (2018) identified and discussed both mantled and stripped etchplain types in central Africa, determined by the extent and style of sediment stripping. Thus, hamadas and inselbergs may be progressively revealed over time, and pediments and terraces formed as sediment accumulate in footslope locations. Here, the term *hamada* is used in a geomorphological (bedrock surface) and not a geographical sense (summit plateau).

The term *desert pavement* (or stone pavement) is a residual accumulation of clasts on the land surface, commonly of one clast thickness and overlying unconsolidated sediments or soil (Cooke 1970; Laity 2008, pp. 159–167; Knight 2019). These clasts are usually tightly packed, and thus, they can help protect underlying sediments from erosion. The clasts comprising desert pavements are usually only a few cm in diameter and are therefore much smaller than those found on hamadas. Desert pavements are commonly found throughout desert environments globally, and different regional names have been used to describe them. In the western Sahara, they are commonly termed *reg*, in central Asia they are termed *saï* or *gobi*, and in Australia they are termed *gibber*. All of these terms generally refer to an extensive planar surface that is made up of surface clasts that are subangular to subrounded and overlying a sedimentary cover. Desert pavements comprised of well-rounded clasts are called *serir* (gravel plain), a term common in the eastern Sahara. Here, the term *desert pavement* is used generically to describe all such pavement types, irrespective of location or clast roundness. Laity (2008, p.162) describes two main types of pavement compositions—those pavements composed of mixed lithologies and clast sizes and commonly developed on the surface of alluvial fans and terraces and formed mainly by water erosion and to a lesser extent deflation; and those composed of local bedrock and broken duricrust fragments and which thus correspond to a residual, in situ weathering origin. Both of these types exist in the central Sahara region. These varied terms for desert pavements are not used consistently, and are often used interchangeably, in the literature. In addition, hamadas and desert pavements may also be confused with one another, according to the degree of loose debris (blown sand or weathered rock) that covers or exposes a bedrock surface.

Goudie (2013, pp. 99–103) described the processes contributing to the formation and stability of a desert pavement, including deflation by wind, water sorting and upward clast migration by frost and salt heave. These processes tend to remove fine particles from the pavement surface and concentrate coarse particles (Cooke 1970; Williams and Zimbelman 1994). The presence of a clast lag on pavement surfaces can act to stabilise this surface, resulting in pavements having high potential longevity in desert landscapes. Several studies have used cosmogenic radionuclide dating to establish the age of desert pavements. These fall in the age range 30–332 ka in the Sonoran Desert, Arizona (Seong et al. 2016) and < 2 Ma in the Dead Sea area, Israel (Matmon et al. 2009; Guralnik et al. 2010). As yet, no similar dating studies have been undertaken in the central Sahara. Key environmental factors contributing to preservation of desert pavements are a relatively flat topography in a tectonically quiescent area and hyperarid conditions (Fujioka and Chappell 2011). These criteria are found in the central Sahara, and thus it can be inferred that desert pavements and associated landforms such as hamadas may be also geologically old and relict (inactive) features.

8.2 Hamadas and Desert Pavements: Distinctive Features of Desert Landscapes

Many desert landscapes worldwide show evidence for the development of hamadas and desert pavements, and these can be considered as part of a suite of net erosional landforms developed where the capacity for sediment transport is greater than sediment generation by weathering (Fujioka and Chappell 2011; Knight 2019). This means that weathered products are efficiently removed from the surface, leaving bare rock exposed. However, in detail, models of long-term landscape evolution in deserts show that surface boulders or coarse weathered products can accumulate on the land surface over time, reducing the rate of bedrock weathering (Adelsberger and Smith 2009; Knight and Zerboni 2018). The reason for this is that a surficial weathered cover reduces ground-level wind speeds and therefore deflation potential and protects underlying rock surfaces from extreme heating–cooling regimes. For example, Dorn (2018) argued that bedrock surfaces in southwest USA deserts can be mantled by weathered products over a timescale of 89–600 ka, which is within the range of ^{10}Be cosmogenic dating evidence for the age of rock surfaces in the same region (Seong et al. 2016). Surficial boulders or coarse sediment patches can also deflect sheetwash. The net result is that bedrock denudation decreases over time once a critical threshold of surface debris cover is reached. Mapping studies have also identified spatial relationships between debris-covered summits or rock slopes and footslopes where weathered debris can accumulate in the form of fans or aprons (e.g. Perego et al. 2011). There may therefore be a genetic relationship between rock surface weathering and wider patterns of sediment dynamics, although this has not been fully explored in most desert studies.

8.3 Examples from the Central Sahara

8.3.1 Hamadas

Throughout much of the central Sahara, the term hamada is often used as a place name of a summit plateau, but this chapter uses this term in a geomorphic sense only. In the central Sahara, hamadas such as the Gilf Kebir are usually developed on sedimentary bedrock and where thermal weathering and granular disintegration dominate on the rock surface. In the Tibesti mountains, granite inselbergs are affected by exfoliation and tafoni weathering. These contribute to sediment accumulation in footslope locations (Hagedorn 1980). In the Messak plateau region, denudation surfaces form bare bedrock hamadas developed mainly in sandstone, and these are incised by river valleys that are now dry (Busche 1980). This suggests that hamada surfaces have developed over long—but largely unknown—timeframes (Zerboni et al. 2011) and in concert with development of downslope landforms such as alluvial fans that correspond to variations in upslope sediment supply (see Knight, this volume). Hamada surfaces have been described from highest elevation areas of the Tadrart Acacus and the Messak Settafet in SW Libya (Perego et al. 2011; Zerboni et al. 2015) (Fig. 8.2). Indeed, the dark colour of the varnish-covered Messak Settafet hamada is how this place derives its name (Cremaschi 1996). These areas comprise relict etchplains affected by net erosion in which residual patches of palaeosols are preserved in isolated locations on the etchplain surface. This evidence favours hamada development, by exhumation of the underlying bedrock surface, during Quaternary arid phases when erosion of a pre-existing sediment cover was most vigorous (Zerboni et al. 2011). Surface boulders on hamada surfaces vary in angularity and concentration, and thus such hamadas are transitional to desert pavements. The precise nature of these surface boulders has not been described in detail, but the apparent absence of pedestals beneath the boulders suggests either slow weathering rates of rock surfaces or that these boulders are mobile when episodic flash floods flow across the bedrock surface. If boulders are geomorphically stable, then this can increase the preservation potential of any underlying residual patches of palaeosols (Perego et al. 2011; Knight and Zerboni 2018). In the Messak Settafet, desert varnish may also be present on hamada rock surfaces (Zerboni 2008).

Griffin (2011) describes some examples of hamada surfaces from SE Libya that likely formed in the late Miocene/early Pliocene. These surfaces are laterally extensive etchplains and part of the Yangara Palaeosurface, incised by late Quaternary river systems. Based on remote sensing data, the hamadas show bedrock very clearly, are free from debris and are interpreted to be strongly abraded by wind action. Lateritic palaeosols of a similar Mio-Pliocene age are present on plateau surfaces at the southern Sahara limits in Nigeria (Zeese 1991).

8.3.2 Desert Pavements

Desert pavements are commonly described from different arid and semiarid environments globally, including from the central Sahara region. The specific physical properties of desert pavements are not described in any way in the literature when based on field evidence, and these properties cannot be evaluated with any certainty from remote sensing imagery. Thus, desert pavements are not well known. Desert pavement surfaces from upland massifs in Libya have been identified in several remote sensing studies based on their spectral reflectance (Perego et al. 2011; Zerboni et al. 2015). Adelsberger and Smith (2009) described different pavement surfaces from the eastern Sahara in central Egypt, with different properties of clast size, lithology and

Fig. 8.2 Examples of hamada and desert pavement surfaces from the Fazzan region, SW Libya (photographs: Stefania Merlo)

density/spacing. These properties appear unrelated to slope angle, from which it is inferred that slopewash is not a significant formation mechanism. Busche (1998, pp.142–147) presented photographs of different desert pavement surfaces (Wüstenpflaster) from across the Sahara. What is notable is that most surfaces in detail have clasts that are roughly the same size and have roughly the same spacing and are located on uniform substrates. In some instances, however, clast size is notably bimodal with around 5% of all clasts of the much larger size class. In the Fazzan region, SW Libya, desert pavement surfaces vary spatial due to variations in clast size, angularity and surficial blown sand cover (Fig. 8.2). Clast properties here are likely controlled by the fracturing and weathering styles of the igneous bedrock.

Cremaschi (1996) described the common occurrence of manganese rock varnishes over hamada and desert pavement surfaces. The layered microstructure of this varnish and its variable presence over rock engravings (petroglyphs) of different ages were used to identify the relative timing of varnish-forming events, corresponding to late Holocene arid phases. Several studies have also examined the geochemistry of rock varnishes, and these show a layered stratigraphy (μm scale to ~ 1 mm thickness) through the weathered profile of different rock surfaces. These often show an upper case-hardened Fe–Mn crust below which there is a thicker zone of weaker weathered material (Aftabi and Atapour 2018; Macholdt et al. 2018; Xu et al. 2019). Development of the surface crust likely reflects the interplay of water vapour, atmospheric dust and biological processes; however, these controls appear to be site-specific and indirectly influenced by rock mineralogy.

In the Sahara region, artefactual evidence has also been used to infer the age and stability of desert pavement surfaces, including the processes by which lithics of different ages can be moved around and concentrated on the land surface (Adelsberger et al. 2013; Foley and Lahr 2015).

8.4 Discussion

8.4.1 Processes of Hamada and Desert Pavement Development

Hamadas and desert pavements are genetically related and can both be considered as self-limiting systems in which their dynamical function in the landscape decreases and effectively stops when they have attained surface stability. For hamadas, this is where the underlying rock surface is fully exposed and subject to geological-background weathering rates. For desert pavements, this is where clasts form a closed pavement surface, protecting any underlying sediments from erosion. In both these cases, the land surface is stable and may persist relatively unchanged for possibly millions of years (e.g. Matmon et al. 2009).

Hamadas have been traditionally attributed to deflation and wind abrasion, where sand is blown from the bedrock surface (Garner 1974, p.351). However, more recent work focuses on the role of flash floods (Williams and Zimbelman 1994; Goudie 2013, p.102). These processes together are important in cleaning bedrock surfaces and contributing sediment to footslopes (Jäkel 1980) but not in shaping these surfaces in the first place. The size, area and characteristics of hamadas likely vary considerably over space (less so over time), but there is very little analysis of any patterns of hamadas either within the central Sahara or globally. This is a limitation of any discussion on the development of hamadas. Jäkel (1980) described hamada and desert pavement development at 800–900 m asl around Jabal as Sawdā', central Libya. Granular disintegration and thermal weathering are the main processes in operation here. As varnishes on hamada and desert pavement surfaces develop and thicken over time due to dust impacts and biological processes, they can form a crust over the rock surface, protecting it from weathering and abrasion (Xu et al. 2019). This case-hardening effect increases rock surface hardness but can lead to flaking of thin rock layers, in particular on hamada surfaces, and thus contribute to a different process of hamada weathering. This flaking process operates on a different scale to the processes of unroofing and exfoliation of inselbergs and domes—which can also be described as hamada surfaces—from the Tibesti mountains (Hagedorn 1980) or arising from thermal weathering (McFadden et al. 2005).

The processes contributing to desert pavement development are likewise not well known (Knight 2019). In the southern Sahara in western Niger, an experimental plot in a lowland plain showed that deflation was the main process leading to formation of a desert pavement (Symmons and Hemming 1968). Busche (1998, his foto 299, p.145) also shows cm-scale pedestals below flat cobbles that correspond to enhanced erosion between these surface clasts (e.g. Symmons and Hemming 1968). Deflation may be limited however where surface clasts are closely spaced or where they are resting on a hard duricrust (Zeese 1991). In many areas of desert pavements, the lower surface of pavement clasts and the upper few cm of the soil or sediment immediately below these clasts often have a vesicular appearance or show an immature soil A horizon (Laity 2008; Zerboni et al. 2011). This Av (vesicular) horizon is characterised by a high clay content, attributed to both atmospheric dust and clay movement within the soil horizon in response to evaporation after rainfall, giving rise to columnar soil structures caused by clay shrinkage and swelling (Jäkel 1980; Bockheim 2010). The presence of vesicles on clast undersides reflects chemical weathering under higher humidity conditions, where water is unable to evaporate freely from the soil after rainfall, and can thus be considered as a microclimate effect on pavements (Ugolini et al. 2008).

A common model proposed for the development of desert pavements is that surface deflation and/or sheetwash removes fine sediments and leaves larger clasts behind (Knight and Stratford 2020). Over time, the land surface thus decreases in elevation and the concentration (density) of clasts increases along with average clast size. At some critical threshold, clast density is such that the underlying fine sediments are protected from deflation and the pavement surface becomes stabilised. It is likely that development of varnishes as well as the in situ accumulation of cosmogenic isotopes indicative of land surface stability is also enhanced at this time, when clasts become immobile (Matmon et al. 2009; Seong et al. 2016). This model of desert pavement development is well established in the literature (Goudie 2013, p. 100) and results in a condensed stratigraphy in which older buried sediment layers or soils can be brought closer to the surface (Knight and Zerboni 2018; Knight and Stratford 2020) and have been argued to be a principal factor contributing to mixed lithic assemblages within pavements (Foley and Lahr 2015). This model therefore assumes that a closed clast surface is the ultimate end-member of pavement development. However, no field studies have examined whether this is indeed the case. Busche (1998) showed photographs of different clast densities on pavement surfaces. Assuming that the model described above is correct, pavement surfaces with low clast densities may be considered to be immature when compared to other pavements with more dense clast distributions. Thus, within any one landscape, pavements may reflect different evolutionary stages of pavement development. This is an untested hypothesis. Likewise, differences in clast density may also reflect clast size, surface slope, exposure to sheetwash, etc. This is likewise untested.

8.4.2 The Role of Other Weathering and Erosion Processes on Hamada and Desert Pavement Development

Apart from wind and water erosion, other processes are now known to be significant in the development of hamada and desert pavement surfaces. Thermal (insolation) weathering is driven by differential volumetric changes of rock constituents caused by increased shortwave insolation and air temperatures during the daytime and by rapid cooling during the night (Molaro and McKay 2010). This results in expansion and contraction of individual mineral grains, respectively, detaching them from the bedrock surface. These grains are then available to be blown or washed away. Thermal weathering is known to fracture clasts on desert pavements (McFadden et al. 2005; Moores et al. 2008; Eppes et al. 2010). Clasts and bare rock surfaces, in particular on hilltops, can also be fractured by lightning strikes (Longinelli et al. 2012). Jäkel (1980) argued that active frost shattering (congelifraction) takes place in the Jabal as Sawdā'area, based on the presence of fractured clasts, but this is an incorrect interpretation because there is not enough available water here to cause ice crystal formation.

Clasts on desert pavement surfaces may be affected by wind abrasion (Goudie 1989), and this has been described from the Sahara (McCauley et al. 1980; Soleilhavoup 2011). As clasts on desert pavements become more closely spaced, the potential for deflation and wind abrasion decreases. Knight and Zerboni (2018), based on photographs of sample quadrats of desert pavement surfaces in the Messak Settafet taken by Foley and Lahr (2015), estimated that 44% of clasts show evidence of wind abrasion (ventifaction), whereas 18% of clasts show primary fractures indicative of thermal weathering.

8.5 Summary and Outlook

Desert land surfaces can be considered as geomorphological palimpsests, because they reflect multiple phases of land surface evolution and are developed under different climatic regimes, in particular during the Quaternary. Landscape palimpsests are now being increasingly being identified across Africa, reflecting the age and evolution of the land surface (Dauteuil et al. 2015; Knight and Grab 2016; Guillocheau et al. 2018; Knight and Stratford 2020). Land surfaces in the central Sahara vary according to the presence or absence of loose sediments and sediment thickness. In lowland depressions and basins, sand seas are present which have a considerable thickness of land surface sediment and are characterised by high sediment mobility. By contrast, on upland massifs, bedrock is most commonly exposed on the land surface and available for subaerial weathering (Ballantine et al. 2005). These bedrock massifs may exhibit different responses to climate forcing when compared to adjacent sand seas.

These considerations mean that landforms such as hamadas and desert pavements found on bedrock massifs may potentially be used as palaeoenvironmental indicators and thus inform on late Quaternary climate changes (Cremaschi 1998; Zerboni 2008) or over longer time periods than desert dunes. This study shows that despite this potential, hamadas and desert pavements remain poorly studied in terms of their detailed properties, age, dynamics and evolution. New remote sensing platforms and analytical methods have potential for mapping of remote desert landscapes in more detail and to allow for better and more meaningful data analysis, including land surface properties (e.g. Mu et al. 2018).

Acknowledgements I thank Jennifer Fitchett for commenting on a previous version of this chapter.

References

Abdeldayem AL (1996) Palaeomagnetism of some Miocene rocks, Qattara depression, Western Desert, Egypt. J Afr Earth Sci 22:525–533

Adelsberger KA, Smith JR (2009) Desert pavement development and landscape stability on the Eastern Libyan Plateau, Egypt. Geomorphology 107:178–194

Adelsberger KA, Smith JR, McPherron SP, Dibble HL, Olszewski DI, Schurmans UA, Chiotti L (2013) Desert pavement disturbance and artifact taphonomy: a case study from the Eastern Libyan Plateau, Egypt. Geoarchaeology 28:112–130

Aftabi A, Atapour H (2018) A new record of silica-rich coating on carbonate substrates in southeast-central Iran: constraints on geochemical signatures. Sediment Geol 372:64–74

Ballantine J-AC, Okin GS, Prentiss DE, Roberts DA (2005) Mapping North African landforms using continental scale unmixing of MODIS imagery. Remote Sens Environ 97:470–483

Bockheim JG (2010) Evolution of desert pavements and the vesicular layer in soils of the Transantarctic Mountains. Geomorphology 118:433–443

Busche D (1980) On the origin of the Msāk Mallat and Hamādat Mānghīnī escarpment. In: Salem MJ, Busrewil MT (eds) The Geology of Libya, vol III. Academic Press, London, pp 837–848

Busche D (1998) Die Zentrale Sahara. Justus Perthes Verlag, Gotha, p 284

Cooke RU (1970) Stone pavements in deserts. Ann Assoc Am Geogr 60:560–577

Cremaschi M (1996) The rock varnish in the Messak Settafet (Fezza, Libyan Sahara), age, archaeological context, and paleo-environmental implication. Geoarchaeology 11:393–421

Cremaschi M (1998) Late Quaternary geological evidence for environmental changes in south-western Fezzan (Libyan Sahara). In: Cremaschi M, di Lernia S (eds) Wadi Teshuinat—Palaeoenvironment and Prehistory in south-western Fezzan (Libyan Sahara). CNR, Roma-Milano, pp 13–48

Dauteuil O, Bessin P, Guillocheau F (2015) Topographic growth around the Orange River valley, southern Africa: a Cenozoic record of crustal deformation and climatic change. Geomorphology 233:5–19

Dorn RI (2018) Necrogeomorphology and the life expectancy of desert bedrock landforms. Progr Phys Geogr 42:566–587

Drake NA, El-Hawat AS, Turner P, Armitage SJ, Salem MJ, White KH, McLaren S (2008) Paleohydrology of the Fazzan basin and surrounding regions: the last 7 million years. Palaeogeogr Palaeoclimatol Palaeoecol 263:131–145

Eppes MC, McFadden LD, Wegmann KW, Scuderi LA (2010) Cracks in desert pavement rocks: further insights into mechanical weathering by directional insolation. Geomorphology 123:97–108

Foley RA, Lahr MM (2015) Lithic landscapes: early human impact from stone tool production on the central Saharan environment. PLoS ONE 10:e0116482. https://doi.org/10.1371/journal.pone.0116482

Fujioka T, Chappell J (2011) Desert landscape processes on a timescale of millions of years, probed by cosmogenic nuclides. Aeolian Res 3:157–164

Garner HF (1974) The origins of landscapes. A synthesis of geomorphology. Oxford University Press, New York, p 734

Goudie AS (1989) Wind erosion in deserts. Proc Geol Assoc 100:83–92

Goudie AS (2013) Arid and semi-arid geomorphology. Cambridge University Press, Cambridge, p 454

Griffin DL (2011) The late Neogene Sahabi rivers of the Sahara and the hamadas of the eastern Libya-Chad border area. Palaeogeogr Palaeoclimatol Palaeoecol 309:176–185

Guillocheau F, Simon B, Baby G, Bessin P, Robin C, Dauteuil O (2018) Planation surfaces as a record of mantle dynamics: the case example of Africa. Gondwana Res 53:82–98

Guralnik B, Matmon A, Avni Y, Fink D (2010) ^{10}Be exposure ages of ancient desert pavements reveal Quaternary evolution of the Dead Sea drainage basin and rift margin tilting. Earth Planet Sci Lett 290:132–141

Hagedorn H (1980) Geological and geomorphological observation on the northern slope of the Tibistī Mountains, central Sahara. In: Salem MJ, Busrewil MT (eds) The Geology of Libya, vol III. Academic Press, London, pp 823–835

Hounslow MW, White HE, Drake NA, Salem MJ, El-Hawat A, McLaren SJ, Karloukovski V, Noble SR, Hlal O (2017) Miocene humid intervals and establishment of drainage networks by 23 Ma in the central Sahara, southern Libya. Gondwana Res 45:118–137

Jäkel D (1980) Current weathering and fluvio-geomorphological processes in the area of Jabal as Sawdā'. In: Salem MJ, Busrewil MT (eds) The Geology of Libya, vol III. Academic Press, London, pp 861–875

Knight J (2019) Wind erosion. In: Livingstone I, Warren A (eds) Aeolian Geomorphology: a new introduction. Wiley-Blackwell, Oxford, pp 61–80

Knight J, Grab SW (2016) A continental-scale perspective on landscape evolution in southern Africa during the Cenozoic. In: Knight J, Grab SW (eds) Quaternary environmental change in southern Africa: physical and human dimensions. Cambridge University Press, pp 30–46

Knight J, Stratford D (2020) Investigating lithic scatters in arid environments: the Early and Middle Stone Age in Namibia. Proc Geol Assoc 131:778–783

Knight J, Zerboni A (2018) Formation of desert pavements and the interpretation of lithic-strewn landscapes of the central Sahara. J Arid Environ 153:39–51

Kroonenberg SB, Melitz PJ (1983) Summit levels, bedrock control and the etchplain concept in the basement of Suriname. Geol en Mijnb 62:389–399

Laity J (2008) Deserts and desert environments. Wiley-Blackwell, Chichester, p 342

Longinelli A, Serra R, Sighinolfi G, Selmo E, Sgavetti M (2012) Oxygen isotopic composition of fulgurites from the Egyptian Sahara and other locations. Rapid Comms Mass Spectrom 26:1980–1984

Macholdt DS, Al-Amri AM, Tuffaha HT, Jochum KP, Andreae MO (2018) Growth of desert varnish on petroglyphs from Jubbah and Shuwaymis, Ha'il region, Saudi Arabia. Holocene 28:1495–1511

Matmon A, Simhai O, Amit R, Haviv I, Porat N, McDonald E, Benedetti L, Finkel R (2009) Desert pavement-coated surfaces in extreme deserts present the longest-lived landforms on Earth. GSA Bull 121:688–697

McCauley JF, Breed CS, Grolier MJ, El-Baz F (1980) VIII. Pitted rocks and other ventifacts in the western desert. Geogr J 146:84–85

McFadden LD, Eppes MC, Gillespie AR, Hallet B (2005) Physical weathering in arid landscapes due to diurnal variation in the direction of solar heating. GSA Bull 117:161–173

Molaro JL, McKay CP (2010) Processes controlling rapid temperature variations on rock surfaces. Earth Surf Proc Landf 35:501–507

Moores JE, Pelletier JD, Smith PH (2008) Crack propagation by differential insolation on desert surface clasts. Geomorphology 102:472–481
Mu Y, Wang F, Zheng B, Guo W, Feng Y (2018) McGET: a rapid image-based method to determine the morphological characteristics of gravels on the Gobi desert surface. Geomorphology 304:89–98
Perego A, Zerboni A, Cremaschi M (2011) Geomorphological Map of the Messak Settafet and Mellet (Central Sahara, SW Libya). J Maps 2011:464–475
Senut B, Pickford M, Ségalen L (2009) Neogene desertification of Africa. C R Geosci 341:591–602
Seong YB, Dorn RI, Yu BY (2016) Evaluating the life expectancy of a desert pavement. Earth Sci Rev 162:129–154
Soleilhavoup F (2011) Accumulation and erosion microforms on desert rock surfaces of Sahara. Géomorphologie: relief, proc, environn 2011:173–186
Swezey CS (2009) Cenozoic stratigraphy of the Sahara, Northern Africa. J Afr Earth Sci 53:89–121
Symmons PM, Hemming CF (1968) A note on wind-stable stone-mantles in the southern Sahara. Geogr J 134:60–64
Ugolini FC, Hillier S, Certini G, Wilson MJ (2008) The contribution of aeolian material to an Aridisol from southern Jordan as revealed by mineralogical analysis. J Arid Environ 72:1431–1447
Vermeesch P, Fenton CR, Kober F, Wiggs GFS, Bristow CS, Xu S (2010) Sand residence times of one million years in the Namib Sand Sea from cosmogenic nuclides. Nature Geosci 3:862–865
Williams SH, Zimbelman JR (1994) Desert pavement evolution—an example of the role of sheetflood. J Geol 102:243–248
Xu X, Li Y, Li Y, Lu A, Qiao R, Liu K, Ding H, Wang C (2019) Characteristics of desert varnish from nanometer to micrometer scale: A photo-oxidation model on its formation. Chem Geol 522:55–70
Zeese R (1991) Paleosols of different age in central and northeast Nigeria. J Afr Earth Sci 12:311–318
Zerboni A (2008) Holocene rock varnish on the Messak plateau (Libyan Sahara): chronology of weathering processes. Geomorphology 102:640–651
Zerboni A, Trombino L, Cremaschi M (2011) Micromorphological approach to polycyclic pedogenesis on the Messak Settafet plateau (central Sahara): formative processes and palaeoenvironmental significance. Geomorphology 125:319–335
Zerboni A, Perego A, Cremaschi M (2015) Geomorphological Map of the Tadrart Acacus Massif and the Erg Uan Kasa (Libyan Central Sahara). J Maps 11:772–787

9 The Development and Characteristics of Sand Seas in the Central Sahara

Jasper Knight and Stefania Merlo

Abstract

The central Sahara, although geomorphically diverse, is best known for its sand seas and their component sand dunes. This study describes the geomorphic setting of the different sand seas (ergs) of the central Sahara region, including their distribution, history of development and their principal geomorphic properties. This background then sets the scene for considering the detailed geomorphology, including the dunes, of the Grand Oriental Sand Sea and the Fachi-Bilma Sand Sea (Ténéré Desert), based on the existing literature and new analysis of remote sensing imagery. These results show that the sand seas contain different dune types in different areas, and that these dunes have highly variable properties which suggests they have different evolutionary histories. The outcome of this study is that, unlike its popular conception, Saharan sand seas and sand dunes are not monotonous and uniform, but have highly diverse properties which likely reflect a punctuated evolutionary history, in particular during the Quaternary, as well as different contemporary dynamic behaviours in response to present-day wind regimes.

Keywords

Draa · Dune migration · Ergs · Sand dune dynamics · Sand seas · Remote sensing · Seif dunes · Zibar dunes · Star dunes

J. Knight (✉)
School of Geography, Archaeology and Environmental Studies, University of the Witwatersrand, Johannesburg 2050, South Africa
e-mail: jasper.knight@wits.ac.za

S. Merlo
McDonald Institute for Archaeological Research, University of Cambridge, Downing Street, Cambridge CB2 3ER, UK
e-mail: sm399@cam.ac.uk

9.1 Introduction

Sand dunes can be considered as the most distinctive and iconic landform of the Sahara region, but sand seas (or *ergs*; edeyen/Idehan/Idhan in the Tamasheq language) of the central Sahara represent only 28% of its land area. In addition, of the 13 sand seas identified in the Sahara as a whole, only seven fall within the central Sahara region (Lancaster 1996) (Fig. 9.1). The term erg is derived from the Arabic word عرق which means a dune field, and although there is some discussion in the literature on the different meanings of the terms erg and sand sea because of the regional context of the former term (Wilson 1973; Mainguet and Callot 1978; McKee 1979; Goudie 2013), they are used synonymously herein. A sand sea can be defined as a large-scale topographic and sedimentary basin containing unconsolidated sands, where individual dunes merge into one another to form an undulating and relatively continuous sand landscape (Capot-Rey 1970; Wilson 1971, 1973). Because of this, the boundaries (and therefore the calculated area) of sand seas are relatively subjective and vary considerably from one source to another. Their generalised regional distributions are shown in Fig. 9.1. A recent discussion of the evolution and significance of ergs within desert landscapes is given by Embabi (2018) and Hesse (2019). Globally, active sand seas are found exclusively in hyperarid areas within the limits of the 150 mm isohyet (Wilson 1973), and hyperarid environments cover 22.6% of land area in Africa including the central Sahara (UNEP 1992). Bubenzer et al. (2020) defined a sand sea where it has a total area of 5000 km^2 or more and with a sand coverage > 50% of its area, whereas Wilson (1973) did not give a minimum size but stated that sand must cover more than 20% of the area and be large enough to contain large and continuous dunes (*draa*). Hesse (2019) gave a minimum sand sea area of 30,000 km^2. Globally, the most important study on sand seas remains the report by McKee (1979), and there have been relatively few recent case studies on

J. Knight et al. (eds.), *Landscapes and Landforms of the Central Sahara*, World Geomorphological Landscapes, https://doi.org/10.1007/978-3-031-47160-5_9

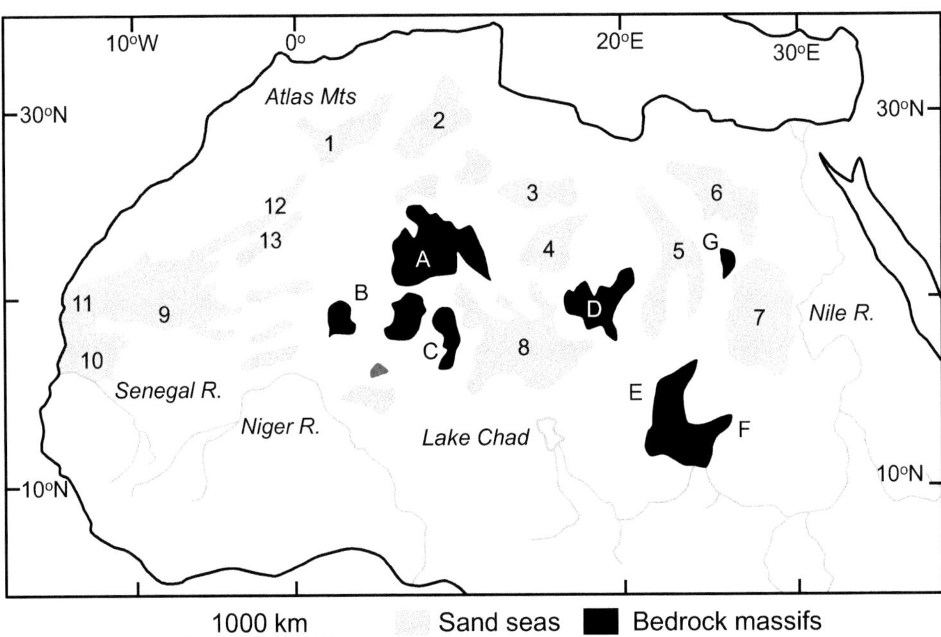

Fig. 9.1 Physiographic regions of the Sahara (redrawn from Lancaster 1996). Rivers and lakes are marked in blue. Sand seas are: (1) Grand Erg Occidental, (2) Grand Erg Oriental, (3) Ubārī, (4) Murzuk, (5) Calanscio, (6) Great Sand Sea, (7) Selima, (8) Fachi-Bilma and Ténéré, (9) Majabat al Koubra, (10) Aouker, (11) Akchar, (12) Iguidi and (13) Chech. Massifs are: (A) Hoggar, (B) Tanezrouft Iforas, (C) Aïr, (D) Tibesti, (E) Ennedi, (F) Jebel Marra and (G) Gif Kebir Uweinat

the geomorphology of specific sand seas, despite the significant capability of remote sensing data to now inform on sand sea properties and dune dynamics (Bubenzer and Bolten 2008; Abdelkareem et al. 2020; Bubenzer et al. 2020). Free (mobile) dunes can be considered as a key element of sand sea landscapes although dune typologies may vary significantly within them (Wilson 1973). Mainguet and Callot (1978) identified three types of erg, classified according to whether dunes within them were active, partly active or inactive.

Although several studies have examined the landscape-scale geomorphology of the central Sahara, including the presence of sand dunes and their characteristics (e.g. Busche 1998; El-Baz et al. 2000; Ballantine et al. 2005; Bubenzer and Bolten 2008; Hereher 2011; Telbisz and Keszler 2018; Hu et al. 2021), there has been hitherto no systematic descriptions of the different sand seas, including an evaluation of their different properties and present dynamics. This evaluation is critical, however, because the formation and dynamics of sand dunes can inform on long-term climatic controls on sediment supply (weathering) and sediment dynamics (synoptic circulation patterns driving aeolian sediment transport). Thus, sand seas and their dunes can be set within a wider climatic and geologic context. This is particularly the case with respect to global dust production (Bullard and Baddock 2019). The most common methodology used to investigate Saharan sand seas and their component dunes, on a regional scale, is remote sensing using air photographs and, over the last 50 years, satellite imagery (e.g. Breed and Grow 1979; Callot et al. 1994; Bubenzer and Bolten 2008; Al-Masrahy and Mountney 2013; Telfer et al. 2015; Herzog et al. 2021; Hu et al. 2021). Subsequently, hyperspectral data have been used to distinguish sand grain size and geochemical patterns across sand dune surfaces (White et al. 2001; Ewing et al. 2015). Elevational data from the Shuttle Radar Topography Mission (SRTM) and Advanced Spaceborne Thermal Emission and Reflection Radiometer (ASTER) sensors have been used for dune mapping and construction of digital elevation models (DEMs) (Bubenzer and Bolten 2008; Telfer et al. 2015; Telbisz and Keszler 2018; Abdelkareem et al. 2020). Several studies have also used these different satellite data in order to map and then spatially analyse Saharan desert dunes (Bubenzer and Bolten 2008; Taniguchi et al. 2012; Telfer et al. 2015; Telbisz and Keszler 2018). Particular emphasis has been paid to barchan dunes, in part because they have an outline shape that can be imaged and mapped clearly, where the position and orientation of barchan horns are sensitive to prevailing wind direction (Khatelli and Gabriels 1998; Elbelrhiti 2012; Boulghobra and Dridi 2016). This highlights the potential of such varied remote sensing datasets for informing on sand sea and sand dune properties, but these technologies have not been applied to all locations, and the detailed analytical methods and time series used have also not been applied consistently or uniformly. This limits the extent to which individual case studies can be compared, especially if they are using datasets that have different spectral or spatial resolutions.

The aim of this chapter is to describe the major properties of the central Saharan sand seas, set in a geological and geomorphic context. The chapter then presents evidence from two of these sand seas (Grand Oriental, Fachi-Bilma), considering sand sea properties and dune patterns based on first-pass remote sensing imagery. This comparison between different sand seas has not been done from the central Sahara before and highlights the variability of these sand seas and their component dunes. The implications of these different properties for sand sea evolution and contemporary dune activity are then discussed.

9.2 Geological Context for the Development of Saharan Sand Seas

A trend towards aridity across Africa as a whole started at the onset of the Quaternary around 2.6 Ma (deMenocal 1995). However, there is evidence for sand dunes developing in the Sahara region from the late Miocene onwards, at around 7 Ma (Schuster et al. 2006; Zhang et al. 2014), as a result of aridification forced by global tectonics and contraction of the Tethys Ocean (Senut et al. 2009). This was later than development of deserts elsewhere in Africa, which was during the mid-Miocene (~ 17 Ma) (Senut et al. 2009). Changes in tectonic and palaeogeographic setting increased the sensitivity of African climate to orbital forcing and monsoon seasonality (Sepulchre et al. 2009; Zhang et al. 2014), and this may have been a trigger for increased aridity and sand mobility. However, there is very little evidence for the development of desert landforms from this early period, and dune sands overlie palaeodrainage channels in the central Sahara that are of Neogene age (Ghellali and Anketell 1991; Maxwell and Haynes 2001; Drake et al. 2008), suggesting that large-scale aridity and dune activity generally postdate this period. Within the Quaternary in the Sahara, however, more arid and humid phases existed, but these were not spatially or temporally uniform (Swezey 2001). Evidence for phases more humid than present exists in the form of palaeosols, lake deposits, organic materials within caves and palaeontological evidence also exists for widespread animal presence in areas that are now dry (Cremaschi 1998; Mauz and Felix-Henningsen 2005; Kröpelin et al. 2008). More arid phases, based on modern analogues, are usually characterised by the reactivation or increased mobility of sand dunes, but this evidence has a very low preservation potential (Bubenzer et al. 2007). This means that the timing, length and spatial extent of arid phases (unlike humid phases) are not well known.

9.2.1 Bedrock Weathering and the Formation of Loose Sand Grains

Key to the availability of loose sediments for development of desert dunes is weathering of sandstones and coarse-grained volcanics and metamorphics, producing sand-sized quartz grains (Walker 1979). The bedrock geology of the central Sahara includes all of these rock types (Schlüter 2006). Precambrian metavolcanics and metasediments dominate in the central Sahara region of southern Libya/Algeria and northern Mali/Niger/Chad. Marine sediments (sandstones and limestones) and low-grade metasediments of mainly Carboniferous age are present in Mali, Algeria and southern Libya. Cenozoic marine sediments and volcanics are found in Mali and Libya. Long-term weathering of these rocks has resulted in the generation of loose sand grains and detached bedrock boulders on the land surface. The nature of weathering and erosion over long (millions of year) timescales in the Sahara is not well known, but several studies have described the effects of more recent weathering on land surface geomorphology, in particular relationships between climate and bedrock slopes (e.g. Smith 1978; Mainguet and Chemin 1983; Busche 2001). Spatial mapping of different landforms has also been undertaken on some bedrock massifs using a combination of remote sensing and ground truthing (e.g. White 1991; Perego et al. 2011; Zerboni et al. 2015). By contrast, there have been very few studies that have shown spatial patterns of dunes or relationships of dunes to other landforms, with exceptions being Callot and Oulehri (1996) who described these landforms at the northern edge of the Grand Erg Occidental, Hu et al. (2021) in the western Sahara and Glennie and Singhvi (2002), Al-Masrahy and Mountney (2013) and Abdelkareem et al. (2020) from the Saudi Arabian dune fields. Physical and chemical weathering processes dominant in the Sahara during the Neogene and Quaternary were described by Busche (1998, his ch1), and total sediment thickness within tectonic basins where sand seas are now located is on the order of 300–400 m (McKee and Tibbitts 1964). Mainguet and Chemin (1983) discussed that, based on a simple sediment budget approach, Saharan sand seas are dominantly relict and often experience a negative mass balance as a result of deflation and dune migration. This highlights that the geomorphic patterns observed today, of both dunes and other sand sea elements such as bedrock massifs, may reflect past rather than present environmental conditions. Thus, a key factor in the development of sand seas is high sediment availability (Laity 2008, pp. 197–199). This arises in this case from long-term weathering and sediment transport by wind and to a lesser extent by water. Capot-Rey (1970) argued that sand seas were developed from the reactivation by wind of pre-existing fluvial sediments within river basins.

The detailed processes of weathering and erosion are not well known in desert environments, in part because of the lack of in situ monitoring and measurement. Thermal weathering (cataclastism), however, is a dominant process of rock fracture in deserts, due to high absolute temperatures and often high diurnal temperature range (McFadden et al. 2005; Moores et al. 2008). Molaro and McKay (2010) also showed that the rate of temperature change over intervals of seconds is controlled by wind speed, a factor that is not accounted for in weathering models. On sandstones, thermal, chemical and honeycomb weathering contribute to granular disintegration and exfoliation of the rock surface, and these processes are significant in the Sahara region (Robinson and Williams 1992). Quartz grains also have a high coefficient of linear thermal expansion (Halsey et al. 1998), meaning that quartz-rich sandstones are particularly susceptible to thermal weathering. In addition, chemical and biological weathering also operate in desert landscapes, forming desert varnish that can variously protect rock surfaces or lead to enhanced selective weathering by case hardening, as in the Ennedi (see Knight, this volume). Bland et al. (1998), based on the degree of weathering experienced by meteorites in deserts, suggested that weathering is initially rapid then decreases over time because, as rock porosity decreases, the rock sample becomes less susceptible to moisture penetration, even in hyperarid environments. An important note is that weathering and formation of loose sandy sediments may substantially predate the accumulation of these sediments into sand sea basins, and sediment capture in these basins may also extend over long time periods.

The slow weathering rates, lack of rainfall and low vegetation cover in deserts mean that desert landscapes and quartz sand grains tend to be long-lived, confirmed by cosmogenic dating (Fujioka and Chappell 2011). Radionuclide dating of surficial sediment grains from the Kalahari and Namib sand seas, southwest Africa, suggests that these contemporary deserts started to form around 1 million years ago (Vermeesch et al. 2010). It is very likely that the central Sahara has a similar evolutionary history, although there is a general lack of data on its long-term evolution. In the southern Sahara, available luminescence dates from stable sand dunes are very patchy but show several clusters that have different temporal expressions, during the periods 21–15 kyr BP (in Nigeria, Niger and Mauritania), 14–10 kyr BP (Niger and Mauritania), 11–6 kyr BP (Sudan), 7–4 kyr BP (Nigeria and Chad) and 2–0 kyr BP (Sudan and Mali) (Bristow and Armitage 2016). It is notable that the dates obtained thus far come mainly from dunes in the semiarid periphery of the Sahara (Lancaster et al. 2016), rather than from individual sand sea basins within the Sahara. Thus, the age and evolution of these sand seas remain undetermined. Bubenzer et al. (2020) for the Great Sand Sea of western Egypt gave luminescence ages of 20–24 ka for dunes on the west side and 10–12 ka for the east side of the sand sea. Such dune studies, however, are concerned with near-surface samples only, and the age of accumulation of deeper sediments is completely unknown.

Throughout, sand is examined here. *Sand* refers to particles 0.0625–2.000 mm (62.5–2000 μm) in *b*-axis length, and quartz is the most dominant mineral that makes up sand grains in Saharan sand seas. Geochemical analysis of Algerian dune sand samples shows that quartz (SiO_2) makes up 98% of all sand grains, with a very minor component of calcite ($CaCO_3$, derived from weathering and erosion of carbonate source rocks) and some oxides (e.g. Al_2O_3, Fe_2O_3, MgO), derived from either impure source rocks or from subsequent subaerial chemical weathering) (Abdelhak et al. 2014; Meftah and Mahboub 2020). This compares to 70–99% quartz from sand dunes in Saudi Arabia (Benaafi and Abdullatif 2015). In samples from the Ouargla sand dunes, in the northern part of the Grand Oriental Sand Sea, Algeria, the proportion of quartz decreases with increased grain size, from 82% for very fine sand, to 67% (fine sand), 67% (medium sand), 78% (coarse sand) and 18% (very coarse sand) (Beddiaf et al. 2017). However, this analysis was based on a very limited number of samples and very small sample sizes.

9.3 Sand Seas and Their Component Dunes

Sand sea outline shapes are usually highly irregular and commonly bounded by bedrock outcrops. In North Africa, 27 sand seas larger than 12,000 km^2 in area were listed by Wilson (1973, his Table I). In the central Sahara, sand seas vary in size between ~ 3900 and 70,000 km^2 (Lancaster 1996), but these estimates are highly uncertain given that boundaries of sand seas are transitional and thus difficult to identify and may change over time. The sand seas are generally separated from each other by bedrock highlands (either massifs or areas of low-relief bedrock surfaces that have little or no sediment cover) (e.g. Zerboni et al. 2015; Bubenzer et al. 2020). As such, the sand seas can be considered as compartmentalised and functionally distinct in terms of their sedimentary, geomorphic and dynamical properties. These sand seas are dominated by active or free dunes (draa) with a high sand cover, rather than by fixed dunes (Wilson 1973). Apart from active dunes, sand seas may also contain bare rock, interdune slacks, soils, sand sheets and evidence for previous fluvial and lacustrine sediments interbedded within dune sands (Al-Masrahy and Mountney 2013). Most sand seas, however, have not been fully surveyed. Wilson (1973, his Table II) lists the characteristics of the Algerian sand seas Grand Erg Oriental, Grand Erg Occidental and Issaouane-N-Irarraren. These have a mean sediment thickness of 30 m, mean dune height of 115 m and mean dune spacing of 1.9 km. Irrespective of sand sea size and shape, the greatest bare sand cover, thickest sand layer, greatest mean dune height and longest dune wavelength are found towards the middle of these sand seas, supporting the idea that sand is captured within basins. However, Telbisz and Keszler (2018) argued that sand accumulation and dune development take place preferentially

Table 9.1 Composition of dune types (% area) in sand seas in different sectors of the Sahara (values from Fryberger and Goudie 1981)

Dune type	Locations in the Sahara			
	North (%)	South (%)	Northeast (%)	West (%)
Crescentic/barchan	33.33	28.38	14.52	19.17
Linear (longitudinal and transverse)	22.83	24.08	17.01	35.49
Star	7.92	0	23.92	0
Dome	0	0	0.80	0
Sand sheets/streaks	35.92	47.54	39.25	45.34
(Undifferentiated)	0	0	4.50	0
Total (%)	100	100	100	100

around the topographic 'rim' of sand sea basins. Dunes within sand seas, however, may show consistent and self-organised spatial patterns that reflect high sand availability within a consistent wind field (Bishop 2010; Ewing et al. 2015; and cf. Besler 2008). Busche (1998, pp. 209–229) provides a general overview of some Saharan dune forms. Fryberger and Goudie (1981) list some of the primary morphometric properties of dunes in different unnamed sectors of the Sahara (Table 9.1). This overview shows that different sectors are characterised by different dune morphologies and, by inference, different dynamical behaviours. It also shows that high variability exists across the Sahara in terms of dune type and thus relationships to factors such as sediment availability, surrounding rock type and synoptic circulation.

The properties of six sand seas in Egypt, thus east of the central Sahara, were described by Bubenzer et al. (2020). The area of these sand seas vary individually from 10,400 to 114,400 km^2 with sand making up 51–92% of the sand sea area. Bubenzer et al. (2020) also identified that the Egyptian sand seas are very flat (slopes of 1.36–1.80‰) and contain large draa that have wavelengths of 2.5–3.0 km, widths of 0.7–1.9 km and heights of 20–50 m. Recent work by Goudie (2020; Goudie et al., 2021a, b) has described the morphologies of different dune types globally, including examples from the central Sahara.

9.4 Examples of Sand Sea and Dune Characteristics

Sand sea properties and dune morphologies from the Grand Oriental Sand Sea and Fachi-Bilma Sand Sea are now described. These two areas are chosen because there have been some previous studies reporting dune types and morphometric patterns, and these are located in different climatic zones which may inform on the relationship of sand sea and dune properties to contemporary climate. This study then extends what is known on these two sand seas by looking in detail at dune morphologies and their patterns as shown on Google Earth imagery. The 10 × 10 km squares shown as examples below follow the broad methodology for analysing dune morphometric properties adopted by Goudie et al. (2021a, b).

9.4.1 Grand Oriental Sand Sea

The Grand Oriental Sand Sea (or Grand Erg Oriental) was described by Wilson (1973) who attributed an age of ~ 1.35 Ma to the sand sea. It has an area of 192,000 km^2, with a mean sand thickness of 26 m (Mainguet and Jacqueminet 1984) (Telbisz and Keszler (2018) reported it as 13.8 m) and sand cover of 70%. Draa bedforms have a mean height of 117 m (maximum height of 320 m) and spacing of 1.6 km. Telbisz and Keszler (2018) presented topographic cross-sections (based on SRTM and ASTER data) showing linear dune relief of > 200 m in the south-western part of the sand sea, compared to ~ 50 m relief of network dunes (crescentic transverse dunes) in the north-east. Dome dunes in this sand sea have a mean height of 50–94 m and width of 920–1540 m (Goudie et al. 2021b). Sediment in the sand sea generally coarsens from north to south, and coarser sediments are found within draa compared to finer sediments found in interdune areas. Mapped patterns of some of these attributes are presented in Fig. 9.2. The complex outline shape of the sand sea reflects geological control from underlying Cretaceous rocks and the presence of adjacent bedrock uplands. Reported mean sediment fluxes in the north of the sand sea are 3 m^3/m/year^{-1} (Dubief 1952), and Wilson (1973, p. 92) argued that sand deposition in the Grand Oriental Sand Sea is more recent in the north than the south. This follows the distribution of overall sand thickness in the sand sea (Fig. 9.2b). Sand is more continuous in the centre than at the margins of the sand sea, but Wilson (1973, p. 92) argues that, because bedrock outcrops are more common towards the margins, the underlying bedrock surface could be reconstructed.

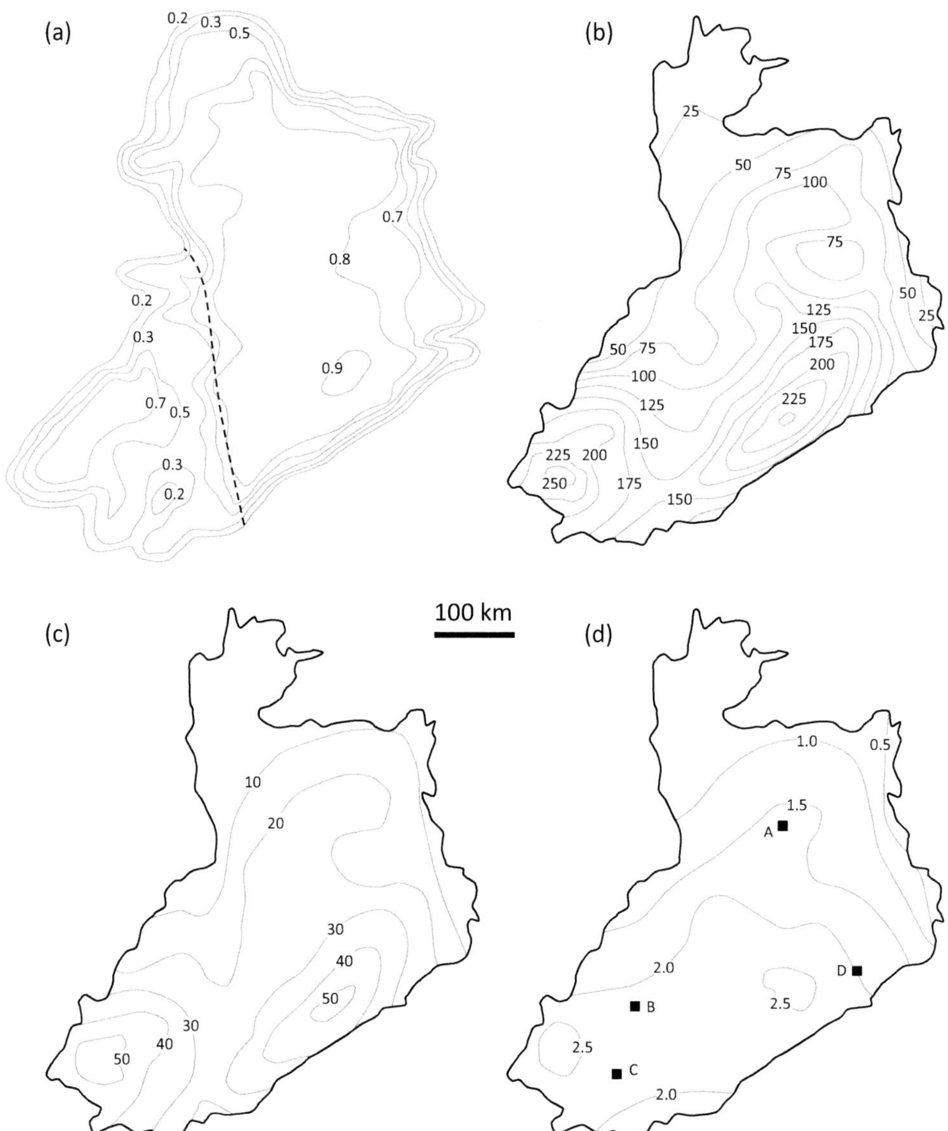

Fig. 9.2 Properties of the Grand Oriental Sand Sea, redrawn from Wilson (1973). **a** Proportion of sand cover, **b** mean sediment thickness (m), **c** mean draa dune height (m) and **d** mean draa dune wavelength (km). The dotted line in panel **a** is the route of a previously unidentified palaeodrainage system. Locations A–D in panel **d** are the locations of insets in Fig. 9.3

This suggests that the overall sand cover within the sand sea is quite shallow, and that the greatest volume of sand is contained within the dunes themselves (Mainguet and Jacqueminet 1984). It is notable that running through the sand sea is a distinctive but hitherto unidentified palaeodrainage system (along the route of the road between Hassi Bel Guenbour to Ouargla, marked by a dotted line in Fig. 9.2) that separates the sand sea into distinctive geomorphic zones.

Telbisz and Keszler (2018) presented patterns of dunes, dune morphometry and calculated sand volumes in the Grand Oriental Sand Sea, using SRTM- and ASTER-based DEMs. Linear and star dunes are the most common bedforms in the region, and these vary in size and spacing depending on wind regime and sediment thickness. Linear dunes are found dominantly in the southwest, star dunes in the south and southeast and network dunes in the north and northeast (Breed et al. 1979). Dome dunes are found in the north and east of the Grand Oriental Sand Sea where they have a mean diameter of 1.51 km (Breed and Grow 1979; Breed et al. 1979) and extend from northwest to southeast over a distance of 350 km. Smaller dome dunes (a few m or so in height) are noted southwest of Orgala (Algeria) (Goudie et al. 2021a, b). Star dunes are also present in the Grand Oriental Sand Sea where they cover the greatest proportion of the sand sea area than any other sand sea globally (Goudie et al. 2021a, b). These dune types were first described and mapped by Aufrère (1934) who termed them 'collines de sable', and Mainguet and Jacqueminet (1984) mapped their distributions in the Grand Oriental Sand Sea and estimated that they occupied two thirds of its area.

With respect to drift potential (DP) from the most geomorphically effective winds, star dunes reflect variable wind directions, whereas the linear dunes lie slightly oblique to present winds. Goudie et al. (2021a) reported average star dune width of 1723 m, height of 135 m and density of 25 per 100 km^2 in the Grand Oriental Sand Sea. However, they noted different values in different sectors of the sand sea. A single study in the far north of the Grand Oriental Sand Sea shows a mix of fixed and free dunes, with deflated interdune areas and wadis (Callot and Oulehri 1996). Several studies in this area also confirm that quartz sand grains dominate, with minor amounts of gypsum, basanite and calcite (Beddiaf et al. 2017; Meftah and Mahboub 2020; Hadjadj and Chihi 2022). Evidence for mid-Holocene palaeolakes and terraces around the eastern margin of the sand sea (Petitmaire et al. 1991) may suggest that dunes in this area have expanded over the dry lake bed over the last few thousand years.

Examples of different dune types in the Grand Oriental Sand Sea are shown in Fig. 9.3. These should be considered as indicative rather than representative of all dune morphologies in this sand sea. The dome dunes shown in Fig. 9.3a are good examples of this type and have widths consistent with those recorded from the Grand Oriental Sand Sea by Goudie et al. (2021b). These authors also reported a density of 26–53 dome dunes per 100 km^2 in this sand sea; those shown in Fig. 9.2a have a density of 39 per 100 km^2. Gao et al. (2018) argued that dome dunes are transitional to barchan dunes, although this is not clear from the examples shown here. The linear dunes shown in Fig. 9.3b, c represent the northern and southern ends of the southwestern part of the sand sea, respectively. Their properties can therefore be compared because they are part of the same system. In detail, the linear dunes extend across this portion of the sand sea, with individual ridges being of composite form, being made up of several superimposed ridges of different sizes. The ridges as a whole join and split along their composite length (in total ~ 200 km long) and are generally aligned in a NNE–SSW direction, terminating at a clear line at its southern margin and corresponding to the rising ground of the bedrock Hoggar massif (occasionally at the front of alluvial fans). The northern edge of this sector of the sand sea is less well defined where the ridge crests become broken up into linear mounds of star dunes (Fig. 9.3b). This may be indicative of higher sediment availability (Fig. 9.2). In the south (Fig. 9.3c), the composite ridges are wider, and the flat areas separating the ridges are also wider and show subdued hardground surfaces with dry palaeochannels. This morphological variability has been previously identified in the linear dunes of the Grand Oriental Sand Sea (Telbisz and Keszler 2018). Star dunes in this sand sea generally represent areas of high sediment availability but low migration rates, as they are affected by multidirectional rather than unidirectional winds (e.g. Dong et al. 2013). The range of values for star dune width and spacing reported here (Fig. 9.3d) is somewhat greater than values presented by Lancaster (1989) and Goudie et al. (2021a) for the same area, but the calculated mean width is robust at 1400 m. The morphology of both dome and star dunes, however, is sometimes not clear within sediment-rich draa fields, and within these environments, dune peaks within the general draa landscape are termed *rhourds* (Wilson 1973) or *ghourds* (Mainguet and Callot 1978). These are therefore likely to be larger compound features that can be reworked by wind into other dune forms, particularly where the ghourds have loose sand moving down slip faces. Wilson (1973, p.99) provided an example of these compound landforms from the Grand Oriental Sand Sea.

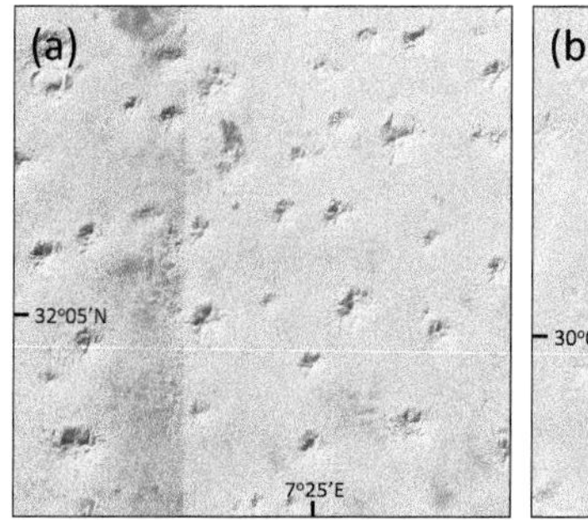

(a) Dome dunes
- Width 230-1150 m
- Nearest neighbour spacing 290-1000 m

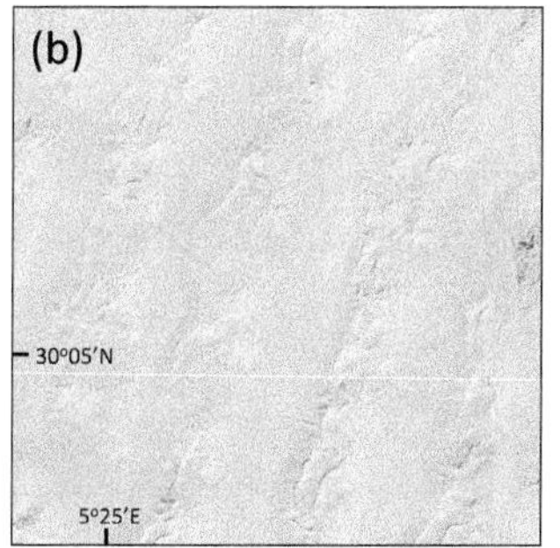

(b) Linear dunes
- Width 1380-1660 m
- Dune spacing 520-1090 m

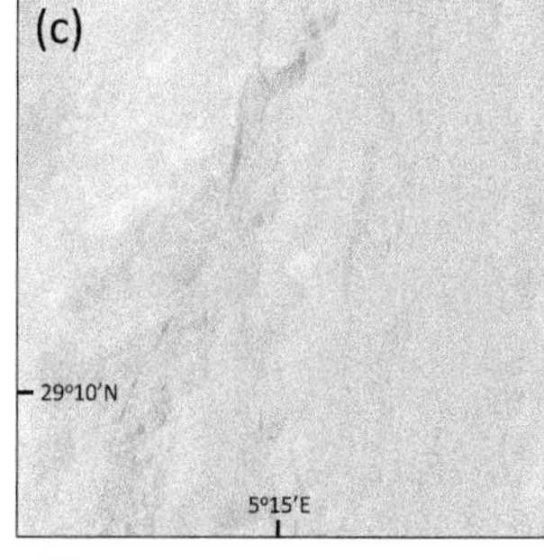

(c) Linear dunes
- Width 4420-5400 m
- Dune spacing 2870-4260 m

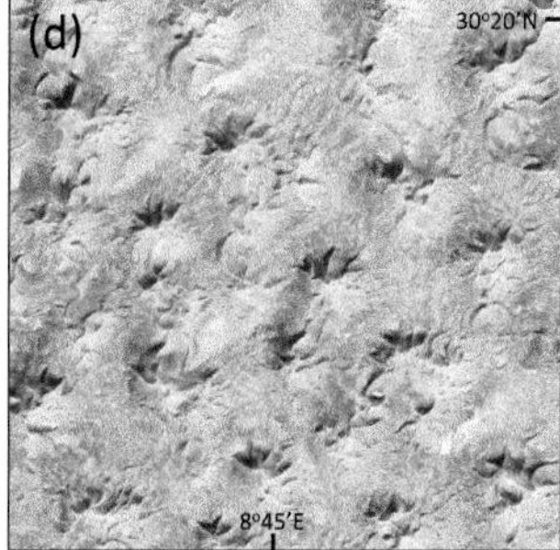

(d) Star dunes
- Width 1090-1670 m
- Peak to peak spacing 1450-2800 m

Fig. 9.3 Examples of dune types from the Grand Oriental Sand Sea (locations shown in Fig. 9.2d). Each image is 10 × 10 km. The dune types shown are discussed in detail in the text

9.4.2 Fachi-Bilma Sand Sea

The Fachi-Bilma Sand Sea is found within the Ténéré Desert. The desert itself covers an area of ~ 495,000 km^2 and is located between the Aïr and Tibesti mountains (to the east and west) and the Lake Chad basin (to the south) (Fig. 9.4). The deflation basin of the Bodélé Depression

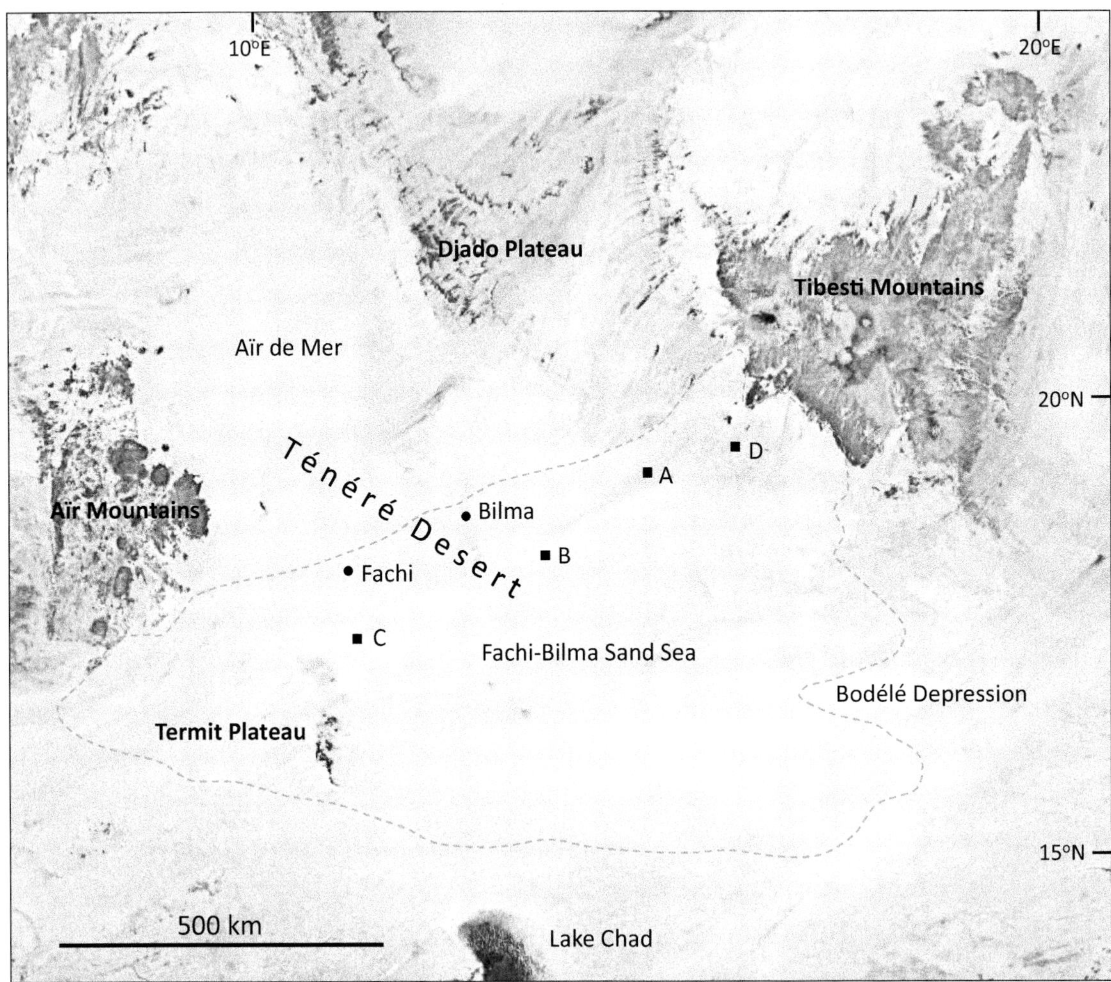

Fig. 9.4 Geography of the Ténéré desert region (base image from Google Earth, image date 31 December 2016) and indicating the generalised extent of the Fachi-Bilma Sand Sea (dotted brown line). Locations A–D are the locations of insets in Fig. 9.5

with its well-known migrating barchan dunes (Vermeesch and Drake 2008) demarcates the eastern edge of the sand sea. The climate, topography and dune systems of the Fachi-Bilma Sand Sea have been described in greatest detail by Mainguet and Callot (1978).

The Ténéré Desert is very variable in its geomorphology, dominated by thin sand sheets (Warren 1971), but outcrops of bedrock and lacustrine diatomites and gravel spreads are also present, including as the substrate between dune ridges. These surface types dominate in terms of areal coverage. The Fachi-Bilma Sand Sea itself has been mapped in slightly different locations (and thus of different sizes) in different studies (see discussion in Mainguet and Callot 1978), but the best approach to identify it is on a geomorphic basis, where sand is thickest and where dunes are best developed, which is located in the south of the Ténéré desert region (see the dashed boundary in Fig. 9.4). Also of note here is that this region lies within the watershed of the Lake Chad basin (see Knight, this volume) and thus may have been subject to changes in humidity as well as the extent of palaeo-Lake Chad itself during the Quaternary. This means that this region, currently falling into the Fachi-Bilma Sand Sea, may not have always been dry and composed of active dunes. The area of the sand sea currently lies around the 200 to 300 mm year^{-1} isohyet, thus within the arid zone, and has Sahelian grassland located to the south. The Fachi-Bilma Sand Sea is therefore located in a different climatic context to the hyperarid Grand Oriental Sand Sea, and it offers a different perspective on sand sea properties and dune dynamics. Generally, dunes at and south of the Termit Plateau are fixed (immobile under present conditions), whereas those to the north of it are free (Warren 1972). This may reflect an inheritance from the more humid Quaternary in this region. Within the Ténéré Desert as a whole, and in particular around the margins of the Fachi-Bilma Sand Sea, diatomite deposits record lacustrine phases around the early Holocene and mid-Holocene periods (Grunert et al. 1991; Gasse 2002) and are shown in particular in the Bodélé Depression. The extent of lake and aeolian deposits therefore inversely co-varies with each other and also varies with climate. Dune deposits also show condensed horizons and intraformational palaeosols, indicating more humid periods. These soils are separated by dune sand layers and have been dated in this area by the luminescence method to ~ 10 ka (Felix-Henningsen 2019), predating lake expansion, although dune sand deposition started before 15 ka (Sponholz et al. 1993).

Dunes in the Ténéré region generally were described by Warren (1971, 1972) but most attention has been paid to the Fachi-Bilma Sand Sea where dunes are best developed. This sand sea has an area of ~ 155,000 km^2 and is composed of active dunes that are present within a semi-enclosed sedimentary basin (Wilson 1973). Grunert et al. (1991) mapped the transition between fixed and active dunes in this area to be at around 18°N (which is around the position of the northern margin of the sand sea, as marked on Fig. 9.4). The areas of fixed dunes in particular are separated by gravel covered surfaces (reg and serir) and also contain wind-abraded linear bedrock ridges (yardangs). Barchan dunes are located in the southeast of this area and within the fixed dune zone (Mainguet and Callot 1978). North of the town of Bilma (Fig. 9.4), the sand layer thins and bedrock outcrops are more common.

Limited field studies show that seif and zibar dunes are found in this sand sea (Warren 1971). Seif dunes are long, straight ridges that are aligned parallel to each other (Grunert et al. 1991). Zibar dunes are low, transverse ridges that separate the seif dunes. Comparison with calculated DP based on grain size and wind regimes from these different dune types shows that resultant sand movement corresponds precisely to the orientation of the zibar dunes, whereas the seif dunes are aligned at right angles to this orientation (Warren 1972). This has also been confirmed with sediment flux modelling from this area (Lucas et al. 2015) and based on dune mapping with respect to wind direction (Mainguet and Callot 1978). These dune morphologies appear to strongly reflect their sedimentary properties, with seif dunes being much coarser than both zibar dunes and in the areas in between the seif dune ridges (Warren 1971). It is argued that the seif dunes have been subject to deflation in which the finer sediments are preferentially removed and deposited on the dune ridge leeside (Besler 1989; Callot and Crepy 2018).

Dune distributions in different sectors of the sand sea are presented by Mainguet and Callot (1978), although this is based on old (1970s) and low-quality satellite imagery and quite basic visual mapping, and no specific details about different dune types are given. Examples of these dune morphologies as shown on recent Google Earth images are given in Fig. 9.5. Linear dune types (seif dunes) dominate, and these are found mainly in the east of the sand

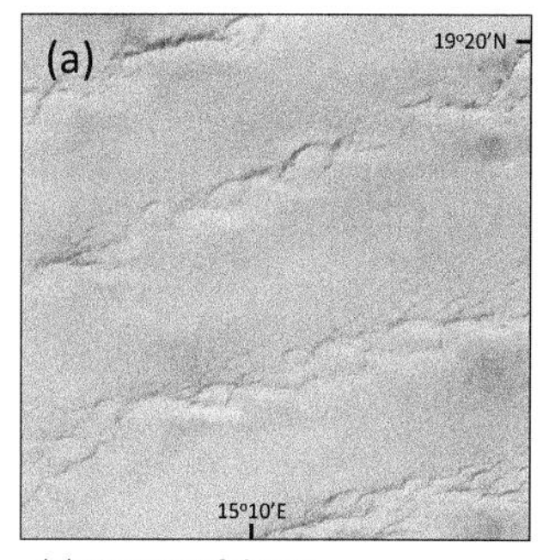

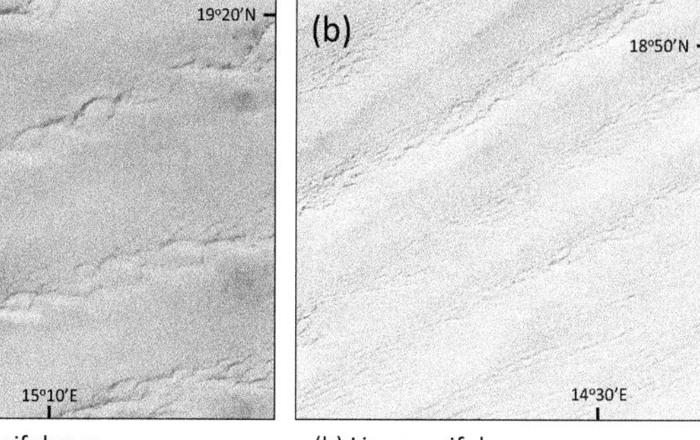

(a) Linear seif dunes
- Dune width 630-1300 m
- Crest to crest distance 1200-2900 m
- Spacing between dunes 1300-2400 m

(b) Linear seif dunes
- Dune width 1000-1300 m
- Crest to crest distance 1660-2350 m
- Spacing between dunes 250-850 m

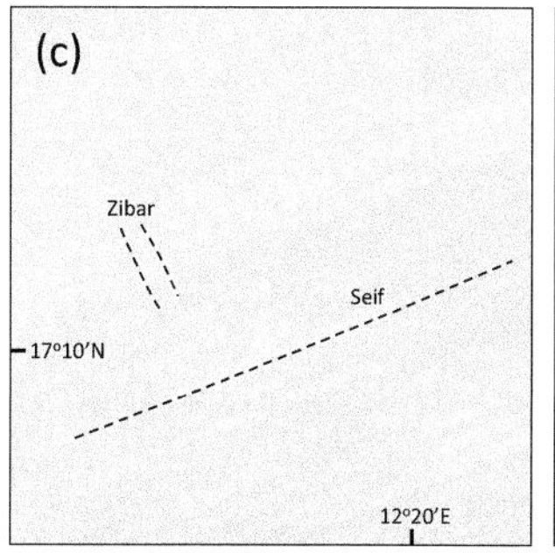

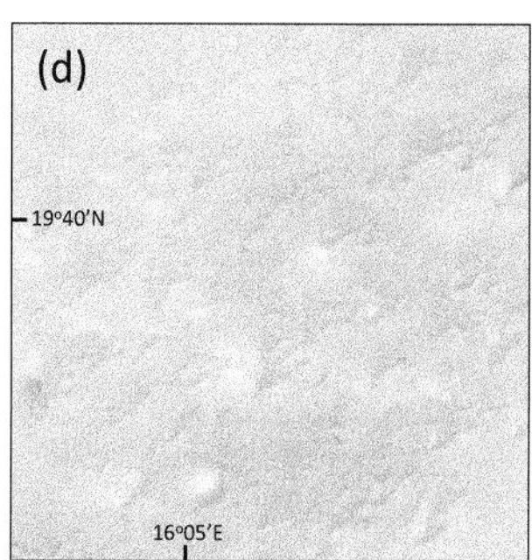

(c) Zibar dunes
- Ridge spacing 1400-1700 m

(d) Star dunes
- Dune width 440-690 m

Fig. 9.5 Examples of dune types from the Fachi-Bilma Sand Sea (locations shown in Fig. 9.4). Panels **a**, **b**, **d** are 10 × 10 km; panel **c** is 40 × 40 km. The dune types shown are discussed in detail in the text. In **c**, examples of seif and zibar dune orientations are shown by the dotted black lines

sea (Fig. 9.5a, b). The dunes in this area were examined by Lucas et al. (2015) from 50 years of aerial and satellite imagery that show these bedforms have extended in a downwind direction but show very limited lateral migration. Callot and Crepy (2018) argue that this characteristic is because of the coarse grain sizes of the dunes. Vermeesch and Leprince (2012) also suggest that very consistent dune migration rates over time in this region reflect constant wind conditions. The linear seif dunes in this area show a strong parallelism at the large scale (Fig. 9.5a, b), but in detail these dune ridges comprise smaller ridge elements that are more variable in their morphology and direction, and this reflects their ongoing downwind migration. Slip faces exist on both flanks of the seif dunes (Wilson 1971). Zibar dunes are found mainly in the southwest part of the sand sea (Fig. 9.5c), but their relationship to the seif dunes is often unclear in this region although it is most likely that the transverse zibar dunes (4–7 m high; Warren 1972) being of finer sediments are more mobile and therefore younger than the seif dunes (*contra* Warren 1972, p. 40). In the eastern Fachi-Bilma Sand Sea barchan dunes have been identified (Mainguet and Callot 1978), but in detail these appear to result from the reworking of older and larger compound ghourd dune forms (Fig. 9.5d). Mainguet and Callot (1978) identified different types of star dunes in the Fachi-Bilma Sand Sea: isolated ones, those in linear chains and those that form a checker board pattern that are separated by depressions. They proposed that the latter occur on the southwest side of the Tibesti massif where harmattan winds from the south and north meet. They called this wind zone as the 'Aire de Recontre'.

9.5 Discussion

Sand seas are a significant geomorphic element of the Sahara, but, despite a general absence of detail field studies on different dune types and locations, they appear to vary substantially in their properties and dynamics (Capot-Rey 1970). Examples of the Grand Oriental Sand Sea and the Fachi-Bilma Sand Sea show that there are different dune types and morphometric properties even within the sand sea, and some areas of the sand seas appear to be relatively active and mobile, whereas other areas and landforms appear relict (Mainguet and Chemin 1983). This may suggest that these sand seas experienced several phases of development, or that the existing dunes in different parts of the sand seas may be moving at different rates or directions, which is possible given their large size, potential for different regional wind fields and the role of topographic effects caused by bedrock massifs (Wilson 1973; Mainguet and Callot 1978; Swezey 2001).

Hesse (2019) proposed a model of sand sea development that focuses on the reworking of pre-existing fluvial sediments. This model proposes that, as conditions become more arid, pre-existing floodplain and alluvial sediments dry out and are reworked into dunes. However, this model was based on examples from Australia, Asia and North America, not from the Sahara region. It is likely in the Sahara that both (limited) fluvial and aeolian processes worked in combination in response to Quaternary climate variations (Swezey 2001). The relative interplay between these different processes in the development of Saharan sand seas is not known, but luminescence and radiocarbon dates from different sand sea dunes indicate different phases of dune activity, at least through the Holocene (Bubenzer et al. 2007; Bristow and Armitage 2016) and likely earlier in the Quaternary. However, surface geomorphological processes by wind and water observed today are only able to activate the topmost few metres of sand, or a few tens of metres of sand in the case of wholesale dune migration, leaving the sediments below this active layer untouched. This is seen clearly in the examples shown from the Grand Oriental and Fachi-Bilma Sand Seas (Figs. 9.3 and 9.5). This may mean that much of the Saharan sand seas are relict features, including not just subsurface sediments but also the variable composition of the land surface (dunes, bedrock outcrops and desert pavements) within the sand seas themselves (Mainguet and Chemin 1983). Relationships between regional wind regimes and DP (discussed by Els and Knight, this volume) for sediment transport and thus dune mobility, based on the Fryberger and Dean (1979) method, are difficult to calculate because of the lack of localised and reliable long-term wind datasets in the Sahara. This means that previously reported DP values may not be very reliable and may show high variability. For example, Hereher (2014) reported resultant DP fluxes of 2.50–24.6 $m^3\ m^{-1}\ year^{-1}$ for different dune types in Egypt. Shi et al. (2022) identified that highest values of resultant DP are found in northern and western parts of the Sahara and during winter and spring seasons. This may mean there is both spatial and interannual variations in dune migration rates (Louassa et al. 2018; Hu et al. 2021), and these may not be captured in remote sensing studies or dune migration models. Likewise, the spatial extent of sand seas may also be responsive to contemporary climate change, as it was in the past (Swezey 2001).

9.6 Summary and Future Research Directions

The Grand Oriental Sand Sea and the Fachi-Bilma Sand Sea show a variety of different dunes types, and these and their distributions reflect the multiphase evolution of the sand seas during the Quaternary (and likely earlier). Linear (seif) dunes are the most common type within the Saharan sand seas, but other types such as star, barchan, dome and zibar dunes are also found. It is likely that these reflect regional wind patterns as well as sediment grain size and availability.

Despite the availability and development of high resolution satellite data, sand dunes across the vast majority of the central Sahara have not been examined in any detail. This offers significant research opportunities in all areas, including dune morphometric patterns, migration rates, mineralogy, dating and relationships to other sand sea landforms such as hamada and pavement surfaces, palaeoriver systems and alluvial fans. There is also the potential to calculate DP for different dune fields. This may be significant where dune migration in both sand seas and in climatically marginal locations on desert fringes is a problem for sand inundation in urban, agricultural and heritage areas (Hereher 2010; Boulghobra 2016).

Acknowledgements We thank Jennifer Fitchett and Frank Eckardt for their reviewer comments on this chapter.

References

Abdelhak M, Ahmed K, Abdelkader B, Brahim Z, Rachid K (2014) Algerian Sahara sand dunes characterization. Silicon 6:149–154

Abdelkareem M, Gaber A, Abdalla F, El-Din GK (2020) Use of optical and radar remote sensing satellites for identifying and monitoring active/inactive landforms in the driest desert in Saudi Arabia. Geomorphology 362:107197. https://doi.org/10.1016/j.geomorph.2020.107197

Al-Masrahy MA, Mountney NP (2013) Remote sensing of spatial variability in aeolian dune and interdune morphology in the Rub' Al-Khali, Saudi Arabia. Aeolian Res 11:155–170

Aufrère L (1934) Les dunes du Sahara algérien (notes de morphologie dynamique). Bull Assoc Géogr Français 11:130–142

Ballantine J-AC, Okin GS, Prentiss DE, Roberts DA (2005) Mapping North African landforms using continental scale unmixing of MODIS imagery. Remote Sens Environ 97:470–483

Beddiaf S, Chihi S, Bouguettaia H, Laid Mechri M, Mahdadi N (2017) Qualitative and quantitative evaluation of quartz in different granular types of Ouargla region sand dunes: Algeria. Silicon 9:603–611

Benaafi M, Abdullatif O (2015) Sedimentological, mineralogical, and geochemical characterization of sand dunes in Saudi Arabia. Arab J Geosci 8:11073–11092

Besler H (1989) Examination of aeolian dynamics in the Ténéré, Niger Republic. Z Geomorphol Supp 74:1–12

Besler H (2008) The great sand sea in Egypt, formation, dynamics and environmental change: a sediment-analytical approach. Developments in Sedimentology, vol 59. Elsevier, Amsterdam, p 266

Bishop MA (2010) Nearest neighbor analysis of mega-barchanoid dunes, Ar Rub' al Khali, sand sea: the application of geographical indices to the understanding of dune field self-organization, maturity and environmental change. Geomorphology 120:186–194

Bland PA, Sexton AS, Jull AJT, Bevan AWR, Berry FJ, Thornley DM, Astin TR, Britt DT, Pillinger CT (1998) Climate and rock weathering: a study of terrestrial age dated ordinary chondritic meteorites from hot desert regions. Geochim Cosmochim Acta 62:3169–3184

Boulghobra N (2016) Climatic data and satellite imagery for assessing the aeolian sand deposit and barchans migration, as a major risk sources in the region of In-Salah (Central Algerian Sahara). Arab J Geosci 9:450. https://doi.org/10.1007/s12517-016-2491-x

Boulghobra N, Dridi H (2016) Fine resolution imagery and GIS for investigating the morphological characteristics, and migration rate of barchan dunes in the Erg Sidi Moussa dunefield near In-Salah (Algeria). Geogr Techn 11:14–21

Breed CS, Grow T (1979) Morphology and distribution of dunes in sand seas observed by remote sensing. In: McKee ED (ed) A study of global sand seas. USGS Professional Paper 1052. US Government Printing Office, Washington, DC, pp 253–302

Breed CS, Fryberger SC, Andrews S, McCauley C, Lennartz F, Gebel D, Horstman K (1979) Regional studies of sand seas using Landsat (ERTS) imagery. In: McKee ED (ed) A study of global sand seas. USGS Professional Paper 1052. US Government Printing Office, Washington, DC, pp 305–397

Bristow CS, Armitage SJ (2016) Dune ages in the sand deserts of the southern Sahara and Sahel. Quat Int 410:46–57

Bubenzer O, Bolten A (2008) The use of new elevation data (SRTM/ASTER) for the detection and morphometric quantification of Pleistocene megadunes (draa) in the eastern Sahara and the southern Namib. Geomorphology 102:221–231

Bubenzer O, Besler H, Hilgers A (2007) Filling the gap: OSL data expanding ^{14}C chronologies of Late Quaternary environmental change in the Libyan Desert. Quat Int 175:41–52

Bubenzer O, Embabi NS, Ashour MM (2020) Sand seas and dune fields of Egypt. Geosciences 10:101. https://doi.org/10.3390/geosciences10030101

Bullard J, Baddock M (2019) Dust: sources, entrainment, transport. In: Livingstone I, Warren A (eds) Aeolian geomorphology: a new introduction. Wiley, Chichester, pp 81–106

Busche D (1998) Die Zentrale Sahara: Oberflächenformen im Wandel. Justus Perthes Verlag, Gotha, p 284

Busche D (2001) Early Quaternary landslides of the Sahara and their significance for geomorphic and climatic history. J Arid Env 49:429–448

Callot Y, Crepy M (2018) The role of sand availability in the dynamics of linear dunes: a change in paradigm? Contributions of modeling and field observations (Erg de Fachi-Bilma, Niger, Chad). Geomorphol Relief Proc Env 24:225–235

Callot Y, Oulehri T (1996) Géodynamique des sables éoliens dans le Nord-Ouest saharien: relations entre aérologie et géomorphologie. Geodin Acta 9:1–12

Callot Y, Mering C, Simonin A (1994) Image-analysis and cartography of sand hill massifs on high resolution images: application to the Great Western Erg (NW of Algerian Sahara). Int J Remote Sens 15:3799–3822

Capot-Rey R (1970) Remarques sur les ergs du Sahara. Ann Géogr 79:2–19

Cremaschi M (1998) Late Quaternary geological evidence for environmental changes in south-western Fezzan (Libyan Sahara). In: Cremaschi M, di Lernia S (eds) Wadi Teshuinat: Palaeoenvironment and Prehistory in south-western Fezzan (Libyan Sahara). CNR, Roma-Milano, pp 13–48

deMenocal PB (1995) Plio-Pleistocene African climate. Science 270:53–59

Dong Z, Zhang Z, Qian G, Luo W, Lu J (2013) Geomorphology of star dunes in the southern Kumtagh Desert, China: control factors and formation. Environ Earth Sci 69:267–277

Drake NA, El-Hawat AS, Turner P, Armitage SJ, Salem MJ, White KH, McLaren S (2008) Palaeohydrology of the Fazzan Basin and surrounding regions: the last 7 million years. Palaeogeogr Palaeoclimatol Palaeoecol 263:131–145

Dubief J (1952) Le vent et le déplacement du sable au Sahara. Trav Inst Rech Sahar 8:123–162

El-Baz F, Maingue M, Robinson C (2000) Fluvio-aeolian dynamics in the north-eastern Sahara: the relationship between fluvial/aeolian systems and ground-water concentration. J Arid Environ 44:173–183

Elbelrhiti H (2012) Initiation and early development of barchan dunes: A case study of the Moroccan Atlantic Sahara desert. Geomorphology 138:181–188

Embabi NS (2018) Landscapes and landforms of Egypt: landforms and evolution. Springer, Cham, p 336

Ewing RC, McDonald GD, Hayes AG (2015) Multi-spatial analysis of aeolian dune-field patterns. Geomorphology 240:44–53

Felix-Henningsen P (2019) OSL-ages and paleo-climatic evidence of ancient dunes with paleosols along a SW–NE transect from the southern Sahel to the central Sahara in Niger. Z Geomorphol Suppl 62:85–118

Fryberger SC, Dean G (1979) Dune forms and wind regime. In: McKee ED (ed) A study of global sand seas. USGS Professional Paper 1052. US Government Printing Office, Washington, DC, pp 137–169

Fryberger S, Goudie AS (1981) Arid geomorphology. Progr Phys Geogr 5:420–428

Fujioka T, Chappell J (2011) Desert landscape processes on a timescale of millions of years, probed by cosmogenic nuclides. Aeolian Res 3:157–164

Gao X, Gadal C, Rozier O, Narteau C (2018) Morphodynamics of barchan and dome dunes under variable wind regimes. Geology 46:743–746

Gasse F (2002) Diatom-inferred salinity and carbonate oxygen isotopes in Holocene waterbodies of the western Sahara and Sahel (Africa). Quat Sci Rev 21:737–767

Ghellali SM, Anketell JM (1991) The Suq al Jum'ah Palaeowadi, an Example of a Plio-Quaternary Palaeo-valley from the Jabal Nafusah, Northwest Libya. Libyan Stud 22:1–6

Glennie KW, Singhvi AK (2002) Event stratigraphy, paleoenvironment and chronology of SE Arabian deserts. Quat Sci Rev 21:853–869

Goudie AS (2013) Arid and semi-arid geomorphology. Cambridge University Press, Cambridge, p 454

Goudie AS (2020) Global barchans: a distributional analysis. Aeolian Res 44:100591. https://doi.org/10.1016/j.aeolia.2020.100591

Goudie AS, Goudie AM, Viles HA (2021a) The distribution and nature of star dunes: a global analysis. Aeolian Res 50:100685. https://doi.org/10.1016/j.aeolia.2021.100685

Goudie AS, Goudie AM, Viles HA (2021b) Dome dunes: distribution and morphology. Aeolian Res 51:100713. https://doi.org/10.1016/j.aeolia.2021.100731

Grunert J, Brumhauer R, Völkel J (1991) Lacustrine sediments and Holocene climates in the southern Sahara: the example of paleolakes in the Grand Erg of Bilma (Zoo Baba and Dibella, eastern Niger). J Afr Earth Sci 12:133–146

Hadjadj K, Chihi S (2022) Rietveld refinement based quantitative phase analysis (QPA) of Ouargla (part of Grand Erg Oriental in Algeria) Dunes Sand. Silicon 14:429–437

Halsey DP, Mitchell DJ, Dews SJ (1998) Influence of climatically induced cycles in physical weathering. Quart J Eng Geol 31:359–367

Hereher ME (2010) Sand movement patterns in the Western Desert of Egypt: an environmental concern. Environ Earth Sci 59:1119–1127

Hereher ME (2011) The Sahara: a desert of change. In: Galvin CD (ed) Sand dunes: ecology, geology and conservation. Nova Science, New York, pp 101–114

Hereher ME (2014) Assessment of sand drift potential along the Nile Valley and Delta using climatic and satellite data. Appl Geogr 55:39–47

Herzog M, Henselowsky F, Bubenzer O (2021) Geomorphology of the Tafilalt Basin, South-East Morocco: implications for fluvial–aeolian dynamics and wind regimes. J Maps 17:682–689

Hesse P (2019) Sand seas. In: Livingstone I, Warren A (eds) Aeolian geomorphology: a new introduction. Wiley, Chichester, pp 179–208

Hu Z, Gao X, Lei J, Zhou N (2021) Geomorphology of aeolian dunes in the western Sahara Desert. Geomorphology 392:107916. https://doi.org/10.1016/j.geomorph.2021.107916

Khatelli H, Gabriels D (1998) Study on the dynamics of sand dunes in Tunisia: mobile barkhans move in the direction of the Sahara. Arid Soil Res Rehabilit 12:47–54

Kröpelin S, Verschuren D, Lézine A-M, Eggermont H, Cocquyt C, Francus P, Cazet J-P, Fagot M, Rumes B, Russell JM, Darius F, Conley DJ, Schuster M, von Suchodoletz H, Engstrom DR (2008) Climate-driven ecosystem succession in the Sahara: the past 6000 years. Science 320:765–768

Laity J (2008) Deserts and desert environments. Wiley, Singapore, p 342

Lancaster N (1989) Star dunes. Progr Phys Geogr 13:67–91

Lancaster N (1996) Deserts. In: Adams WM, Goudie AS, Orme AR (eds) The Physical Geography of Africa. Oxford University Press, Oxford, pp 211–237

Lancaster N, Wolfe S, Thomas D, Bristow C, Bubenzer O, Burrough S, Duller G, Halfen A, Hesse P, Roskin J, Singhvi A, Tsoar H, Tripaldi A, Yang X, Zárate M (2016) The INQUA Dunes Atlas chronologic database. Quat Int 410:3–10

Louassa S, Merzouk M, Merzouk NK (2018) Sand drift potential in western Algerian Hautes Plaines. Aeolian Res 34:27–34

Lucas A, Narteau C, Rodriguez S, Rozier O, Callot Y, Garcia A, Courrech du Pont S (2015) Sediment flux from the morphodynamics of elongating linear dunes. Geology 43:1027–1030

Mainguet M, Callot Y (1978) L'erg de Fachi-Bilma (Tchad–Niger): Contribution à la connaissance de la dynamique des ergs et des dunes des zones arides chaudes. Mémoires er Documents 18, CNRS, Paris, p 184

Mainguet M, Chemin M-C (1983) Sand seas of the Sahara and Sahel: An explanation of their thickness and sand dune type by the sand budget principle. In: Brookfield ME, Ahlbrandt TS (eds) Eolian sediments and processes. developments in sedimentology. Elsevier, Amsterdam, pp 353–363

Mainguet M, Jacqueminet C (1984) Le grand erg occidental et le grand erg oriental. Classification des dunes, balance sédimentaire et dynamique d'ensemble. Trav de l'Inst Géogr Reims 59–60:29–48

Mauz B, Felix-Henningsen P (2005) Palaeosols in Saharan and Sahelian dunes of Chad: archives of Holocene North African climate changes. Holocene 15:453–458

Maxwell TA, Haynes CV (2001) Sand sheet dynamics and quaternary landscape evolution of the Selima Sand Sheet, southern Egypt. Quat Sci Rev 20:1623–1647

McFadden LD, Eppes MC, Gillespie AR, Hallet B (2005) Physical weathering in arid landscapes due to diurnal variation in the direction of solar heating. GSA Bull 117:161–173

McKee ED (ed) (1979) A study of global sand seas. USGS Professional Paper 1052. US Government Printing Office, Washington, DC, p 429

McKee ED, Tibbitts GC (1964) Primary structures of a seif dune and associated deposits in Libya. J Sediment Petrol 34:5–17

Meftah N, Mahboub MS (2020) Spectroscopic characterizations of sand dunes minerals of El-Oued (northeast Algerian Sahara) by FTIR, XRF and XRD analyses. Silicon 12:147–153

Molaro JL, McKay CP (2010) Processes controlling rapid temperature variations on rock surfaces. Earth Surf Proc Landf 35:501–507

Moores JE, Pelletier JD, Smith PH (2008) Crack propagation by differential insolation on desert surface clasts. Geomorphology 102:472–481

Perego A, Zerboni A, Cremaschi M (2011) Geomorphological map of the Messak Settafet and Mellet (Central Sahara, SW Libya). J Maps 2011:464–475

Petitmaire N, Burollet PF, Ballais JL, Fontugne M, Rosso JC, Lazaar A (1991) Holocene paleoclimates in northern Sahara: lacustrine deposits and alluvial terraces along the Grand Erg Oriental, south Tunisia. C R Acad Sci Ser II 312:1661–1666

Robinson DA, Williams RBG (1992) Sandstone weathering in the High Atlas, Morocco. Z Geomorphol NF 36:413–429

Schlüter T (2006) Geological Atlas of Africa. Springer-Verlag, Berlin, p 307

Schuster M, Duringer P, Ghienne J-F, Vignaud P, Mackaye HT, Likius A, Brunet M (2006) The age of the Sahara desert. Science 311:821

Senut B, Pickford M, Ségalen L (2009) Neogene desertification of Africa. C R Geosci 341:591–602

Sepulchre P, Ramstein G, Schuster M (2009) Modelling the impact of tectonics, surface conditions and sea surface temperatures on Saharan and sub-Saharan climate evolution. C R Geosci 341:612–620

Shi W, Dong Z, Chen G, Bai Z, Ma F (2022) Spatial and temporal variation of the near-surface wind environment in the Sahara Desert. North Africa. Front Earth Sci 9:789800. https://doi.org/10.3389/feart.2021.789800

Smith BJ (1978) Aspect-related variations in slope angle near Béni Abbès, western Algeria. Geogr Ann 60A:175–180

Sponholz B, Baumhauer R, Felix-Henningsen P (1993) Fulgurites in the southern Central Sahara, Republic of Niger and their palaeoenvironmental significance. Holocene 3:97–104

Swezey C (2001) Eolian sediment responses to late Quaternary climate changes: temporal and spatial patterns in the Sahara. Palaeogeog Palaeoclimatol Palaeoecol 167:119–155

Taniguchi K, Endo N, Sekiguchi H (2012) The effect of periodic changes in wind direction on the deformation and morphology od isolated sand dunes based on flume experiments and field data from the Western Sahara. Geomorphology 179:286–299

Telbisz T, Keszler O (2018) DEM-based morphometry of large-scale sand dune patterns in the Grand Erg Oriental (Northern Sahara Desert, Africa). Arab J Geosci 11:382. https://doi.org/10.1007/s12517-018-3738-5

Telfer MW, Fyfe RM, Lewin S (2015) Automated mapping of linear dunefield morphometric parameters from remotely-sensed data. Aeolian Res 19:215–224

UNEP (1992) World Atlas of Desertification. Edward Arnold, London, p 69

Vermeesch P, Drake N (2008) Remotely sensed dune celerity and sand flux measurements of the world's fastest barchans (Bodélé, Chad). Geophys Res Lett 35:L24404. https://doi.org/10.1029/2008GL035921

Vermeesch P, Leprince S (2012) A 45 year time series of dune mobility indicating constant windiness over the central Sahara. Geophys Res Lett 39:L14401. https://doi.org/10.1029/2012GL052592

Vermeesch P, Fenton CR, Kober F, Wiggs GFS, Bristow CS, Xu S (2010) Sand residence times of one million years in the Namib Sand Sea from cosmogenic nuclides. Nature Geosci 3:862–865

Walker TR (1979) Red color in dune sand. In: McKee ED (ed) A study of global sand seas. USGS Professional Paper 1052. US Government Printing Office, Washington, DC, pp 61–81

Warren A (1971) Dunes in the Ténéré Desert. Geogr J 137:458–461

Warren A (1972) Observations on dunes and bi-modal sands in the Ténéré desert. Sedimentology 19:37–44

White K (1991) Geomorphological analysis of piedmont landforms in the Tunisian Southern Atlas using ground data and satellite imagery. Geogr J 157:279–294

White K, Goudie A, Parker A, Al-Farraj A (2001) Mapping the geochemistry of the northern Rub' Al Khali using multispectral remote sensing techniques. Earth Surf Proc Landf 26:735–748

Wilson IG (1971) Desert sandflow basins and a model for the development of ergs. Geogr J 137:180–199

Wilson IG (1973) Ergs. Sediment Geol 10:77–106

Zerboni A, Perego A, Cremaschi M (2015) Geomorphological map of the Tadrart Acacus massif and the Erg Uan Kasa (Libyan central Sahara). J Maps 11:772–787

Zhang Z, Ramstein G, Schuster M, Li C, Contoux C, Yan Q (2014) Aridification of the Sahara desert caused by Tethys Sea shrinkage during the Late Miocene. Nature 513:401–404

10 Sand Dune Dynamics: An Example from the Ubārī Sand Sea, SW Libya

Anja Els and Jasper Knight

Abstract

There is high spatial variability in the distribution and characteristics of sand seas in the central Sahara, and also the properties and dynamics of the sand dunes contained within them. However, there is limited information on sand dune dynamics in this region because remote sensing studies have not been undertaken in many Saharan sand seas, and the studies that exist commonly use different sensors with different resolutions and time frames of interest. This chapter presents a case study of dune dynamics from the Ubārī Sand Sea, SW Libya, where automated classification of Landsat 7 and 8 images for the period of interest 2002–2015 was used to identify and map the different morphological components of dune ridges, including dune ridge crests. The methodology used allows for changes over time in the size and location of individual dune ridges to be quantified, and thus these individual dunes can be monitored on an ongoing basis for their dynamical changes as a result of climate forcing. Results show that dune migration took place at $8.64 \pm 4.65\ \mathrm{m\ yr^{-1}}$ in this period, consistent with several other studies on sand seas. This approach to dune mapping and spatial analysis from remote sensing imagery can be employed in similar sand seas throughout the Sahara and elsewhere.

Keywords

Sand seas · Dune morphology · Remote sensing · Draa · Drift potential · Threshold velocity

A. Els · J. Knight (✉)
School of Geography, Archaeology and Environmental Studies, University of the Witwatersrand, Johannesburg 2050, South Africa
e-mail: jasper.knight@wits.ac.za

10.1 Introduction

Aeolian sediments and landforms dominate in the central Sahara, but these landscapes are highly diverse in detail. They include relatively thin but extensive sand sheets, such as the Selima Sand Sheet, and bedrock basinal depressions which offer accommodation space for much thicker but less extensive sand seas, such as the Grand Erg Oriental Sand Sea and Ténéré Sand Sea (see Knight and Merlo, this volume). Within these areas, the thickest sand accumulations are usually found in the centre of the sand seas, and the sand layer thins towards the sand sea margins. Therefore, individual sand seas are usually bounded by areas of bedrock where the sand is patchy or absent. These bedrock areas may be subdued interfluves where bedrock is exposed near the surface, or large-scale and isolated bedrock massifs that may rise hundreds of metres above the surrounding sandy plain. These massifs, such as Hoggar, Aïr, Tibesti and Ennedi, therefore compartmentalise the distribution of sand in the central Sahara, and also influence wind patterns that are manifested in dune morphology. Thus, the central Sahara region is geomorphically diverse and not simply a landscape of sand seas and sand dunes. The general physical characteristics of the central Sahara region and its dunes have been described in most detail by Lancaster (1996), Goudie (2013, his ch.4), and in German by Busche (1998, his ch.4). Sand seas, sand dunes and previous fluvial sand transport in wadis (ephemeral rivers) in these basins reflect several environmental factors. These are: (1) high sediment supply caused by long-term weathering of underlying bedrock; (2) arid conditions, giving rise to an unstable land surface with low or absent vegetation cover; and for the development of aeolian landforms; and (3) prevailing winds of sufficient strength above the sediment threshold velocity (geomorphically effective winds) to cause sediment transport and accumulation into dunes. Large dune ridges, termed *draa* in North Africa, are common across different Saharan sand seas. Sand seas cover 28% of the area of

J. Knight et al. (eds.), *Landscapes and Landforms of the Central Sahara*, World Geomorphological Landscapes, https://doi.org/10.1007/978-3-031-47160-5_10

the Sahara, and Lancaster (1996) identified 13 major sand seas in the region (see Knight and Merlo, this volume, for a regional description of these sand seas). However, in detail, the sand seas have quite different physical properties due to their locations within the Sahara as a whole, their relationship to massifs and palaeowater courses such as rivers or lakes, and differences in their underlying bedrock (from granites to sandstones and mudstones) that result in differences in sediment availability, and sediment grain size and mineralogy.

There are very few field studies that have examined Saharan sand seas and dunes in any detail, and these studies tend to be from the 1950s–1970s before dune areas became difficult to access in the field for security reasons. In recent decades, however, dune mapping from remote sensing data has supplemented this analysis (e.g. Besler 2008; Bubenzer and Bolten 2008; Ghoneim et al. 2012; Vermeesch and Leprince 2012; Boulghobra 2016; Boulghobra and Dridi 2016; Bubenzer et al. 2020; Hu et al. 2021) and has transformed the study of sand sea dunes through its ability to map large regions consistently and accurately. The aim of this chapter is to present a case study of dune mapping within one central Saharan sand sea, the Ubārī Sand Sea, SW Libya. The chapter (1) outlines how drift potential can be used to explain migration trends of sand dunes and their relationship to wind forcing. (2) The methodology employed in this study is then described in detail, including the methods of both spatial mapping and quantifying dune migration patterns. Finally, (3) the main results of this study in the Ubārī Sand Sea are presented, and these are set in the context of similar studies elsewhere in the Sahara and the broader region.

10.2 Dynamics of Dunes Within Sand Seas

Sand seas can contain mobile or fixed sand dunes. Many thousands of individual dunes can be present within Saharan sand seas, and these tend to show both systematic geomorphic patterns and a certain degree of spatial uniformity (Al-Masrahy and Mountney 2013; Telfer et al. 2015). Several dune types may be present within a single sand sea, and individual dunes may be isolated from one another, have joined or merged margins, or may be superimposed (Al-Masrahy and Mountney 2013; du Pont 2015; Telfer et al. 2015). Sand dunes migrate in response to sand grain transport by geomorphically effective winds (du Pont 2015), and thus dune morphologies in most instances reflect winds from different directions at different times. Holocene dunes are typically superimposed on Pleistocene linear draa (Bubenzer and Bolten 2008; Mercuri 2008) that were most likely formed during the hyperarid LGM (~20 ka BP). Therefore mapping of dune morphologies over time and space can inform on the relationship between climate and dune/sand sea dynamics.

Changes in dune location and morphology over time can be accounted for using the concept of drift potential (DP), which refers to the relationship between geomorphically effective wind regime and potential sediment flux (Pye and Tsoar 2009). As modified by Fryberger and Dean (1979) (after Lettau and Lettau 1977, as cited in Fryberger and Dean 1979), DP can be calculated as:

$$Q \propto V^2(V - V_{\mathrm{t}}).t$$

where Q is the proportion of sand drift, V is average wind velocity at a standard height of 10 m, V_{t} is the threshold wind velocity, and t is time of wind-blow activity. Calculating V_{t} requires calculation of threshold shear stress u_{*t} which, following Bagnold (1941), can be calculated as:

$$u_{*t} = A\left[gd\left(\frac{\rho_{\mathrm{s}} - \rho}{\rho}\right)\right]^{0.5}$$

where A is an empirical coefficient of 0.1, g is gravity, d is median grain size (D_{50}), ρ is air density and ρ_{s} is sediment density. Medium sand commonly requires V_{t} of>4 m s^{-1}. Fryberger and Dean (1979) calculated the DP for various deserts worldwide including locations from the central Sahara, based upon their wind regimes. This analysis shows strong seasonality in wind strength and direction which results in asymmetrical DP vectors. The expression of this in the landscape is the development and elongation of sand dunes, reflecting net sand drift in the direction of the most geomorphically effective prevailing wind. Dubief (1953) used wind data from 1925–1950 from across the Sahara to show that the number of such wind events do not correspond fully to the location of ergs—this in turn suggests that some ergs may be partly relict features and/or have low activity under today's climate regime. The most active dunes based on these wind data appear to be in NE Algeria/W Libya, whereas dunes south of the Hoggar and Tibesti massifs and in the Western Sahara (W Algeria/N Mali) are least active. The highest number of wind events appears to map in the Anti-Atlas and correspond to hamada (deflated boulder) surfaces. There are very few published data, however, to confirm these patterns. Net sand transport pathways, reflecting the DP of sand grains under contemporary climate regimes, were mapped by Mainguet et al. (1984) (Fig. 10.1). In the Erg Sidi Moussa dune field, central Algeria, the net resultant DP was calculated to be 39 m^3 of sand transport per 1 m of land width per year (Boulghobra 2016). A recent satellite mapping study of different dune forms from the Western Sahara showed how DP calculated from ERA5 reanalysis data can be used to confirm the sensitivity of dune morphology to changes in wind direction, and can account for the development of superimposed dunes (Hu et al. 2021).

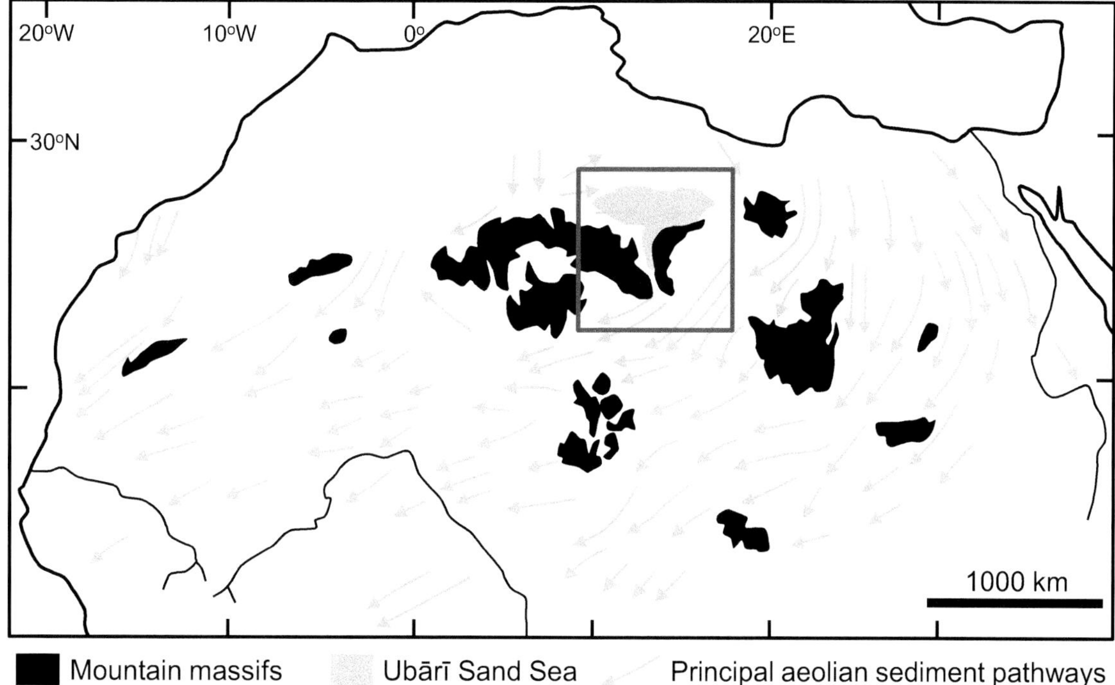

Fig. 10.1 Aeolian sand transport pathways from the central Sahara region, interpreted based on satellite imagery (redrawn from Mainguet et al. 1984). The location of the Ubārī Sand Sea is marked in yellow (Lancaster 1996). The location of the study area (Fig. 10.2) is shown in the red box

Averaged barchan dune migration rates in the Toshka region of the Western Desert of Egypt is 6 m yr^{-1} towards the SSW (Hamdan et al. 2016; Embabi 2018). This is very similar to barchan dune migration rates of 3.5 m yr^{-1} from field measurements in NW Sinai (Phillip et al. 2004) and 3–9 m yr^{-1} from SPOT imagery in the Daklha oasis, Egypt (Ghadiry et al. 2012). On the timescale of a single year, however, barchan dunes in S Tunisia migrate by 48–51 m yr^{-1} (Khatelli and Gabriels 1998). Although these are averaged values, it is unclear to what extent there is variability in these rates across all dunes within the sand seas. This is a significant unknown variable in studies of sand seas. Remote sensing-based mapping and spatial analysis has been commonly used in many dune fields globally, and although such studies have been undertaken in the central Sahara, this has not been done consistently and the morphology and dynamics of many sand seas are largely unknown.

10.3 The Ubārī Sand Sea

The Ubārī Sand Sea (~ 61,000 km^2) is located between the Wādī ash-Shātī and Wādī al-Ajāl in the broader Murzuq basin (Tawadros 2001) (Fig. 10.2). Underlying geology is mainly of continental sandstones (Cambrian and Ordovician age) overlain by Paleozoic rocks (mostly sandstones). These are in turn overlain by shallow-water and/or continental sediments of Jurassic and Lower Cretaceous age. Drake et al. (2009) used ground penetrating radar to map lake sediments beneath the Ubārī Sand Sea, and indicating a thickness of aeolian sediments of>30 m. A large part of this region is covered in recent windblown sand separated by isolated bedrock hills (Sinha and Pandey 1980). Several sand seas are also present within the Murzuq basin, which is bordered by the Tibesti (southeast) and Hoggar (southwest) massifs. Sand dunes cover 16–20% of Algeria (Goudarzi 1970).

Several different dune types are present within the Ubārī Sand Sea (Els et al. 2015; Els 2017) (Fig. 10.3). Linear dunes dominate, and these are oriented northeast to southwest, following the resultant direction of the prevailing wind (Goudarzi 1970; White et al. 2006). These dunes are 100–200 m high and 100 km long and have smaller barchan dunes on their surface (McKee 1979). In the northern and western part of the sand sea, some star dunes can also be seen (McKee 1979). Seasonal and perennial lakes are present in some interdune corridors (indicating the high seasonal water table in these areas), and several oases are found in the southern part of the sand sea (Goudarzi 1970; White et al. 2006). Duricrust deposits (consisting of calcium carbonate, silica and gypsum with abundant root

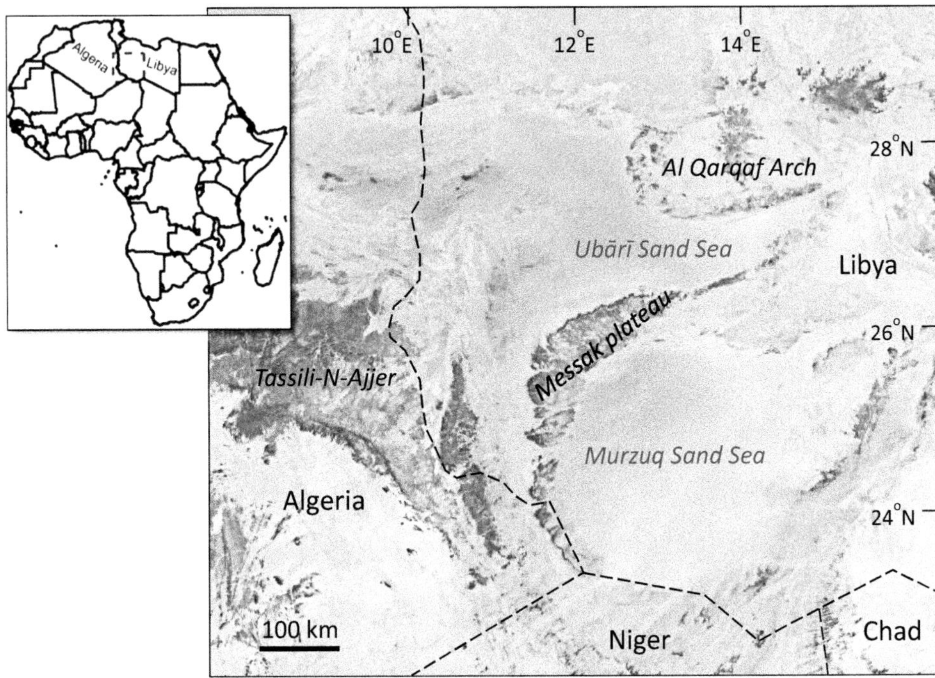

Fig. 10.2 Location map of the Ubārī Sand Sea based on a background Google Earth image. The location of the main map is shown by the red box in the inset map

casts) can also be found on the slopes and interdune areas. Lithic artefacts are often associated with duricrust outcrops, suggesting that these areas have been important at some stages of human occupation of these landscapes (White et al. 2006).

10.4 Methods

10.4.1 Image Analysis

We examined evidence for dune migration in part of the Ubārī Sand Sea using Landsat 7 ETM and Landsat 8 OLI data for September 2002 and September 2015, respectively. These sensors both have a spatial resolution of 30 m and use 8 and 11 spectral bands, respectively. The Landsat images were atmospherically corrected using FLAASH within ENVI, and then pan-sharpened, co-registered for geometric correction and mosaicked. An unsupervised classification using k-means clustering was then used to determine if dune features such as crest and trough could be identified based on spectral data only. Following this, maximum likelihood and minimum distance supervised classifications were then used in order to assess the reliability of the unsupervised classification. For all classification methods, the spectral bands chosen were the blue, green and red within the visible part of the spectrum (bands 1–3 within Landsat 7; 2–4 within Landsat 8) and the near infrared (NIR) (bands 4 and 5 for these different sensors, respectively). These bands have been used for dune mapping in previous studies (e.g. Necsoiu et al. 2009; Al-Masrahay and Mountney 2013; Telfer et al. 2015).

The unsupervised classification resulted in identification of nine different landscape feature classes. These are: water (lakes, oases), gypsum deposits, urban areas, vegetated areas, rocky outcrops, soil (non-dune), interdune areas, dune crests, and dune slopes (Fig. 10.4). In the unsupervised classification process, iteration was stopped when the pixels in a class changed by less than the specified threshold of 5%. The mapped distribution of different classes was visually inspected using red–green–blue (RGB) images in order to confirm those features that are not associated with the dunes, and those that correspond to the dunes.

Supervised classification using a maximum likelihood method was undertaken in order to evaluate the reliability of the unsupervised classification. For this purpose, the training data used were based on a high-resolution (0.85–2.07 m pixel size) pan-sharpened and mosaicked Worldview 2 image, draped over a digital elevation model (DEM) of the study area derived from Shuttle Radar Topography Mission (SRTM) data at 1 arc-second resolution. The purpose behind this was to better identify the location of dune ridge crests and basal outlines of individual dunes, using the nine different landscape feature classes as per the unsupervised classification. From this, Regions of Interest (ROI) were identified using ENVI in order to evaluate whether the nine classes have a unique spectral signature that can aid their classification. A minimum of 100 ROIs of 64×64 pixel dimensions were created per class and were tested for their spectral uniqueness using the Jeffries–Matusita

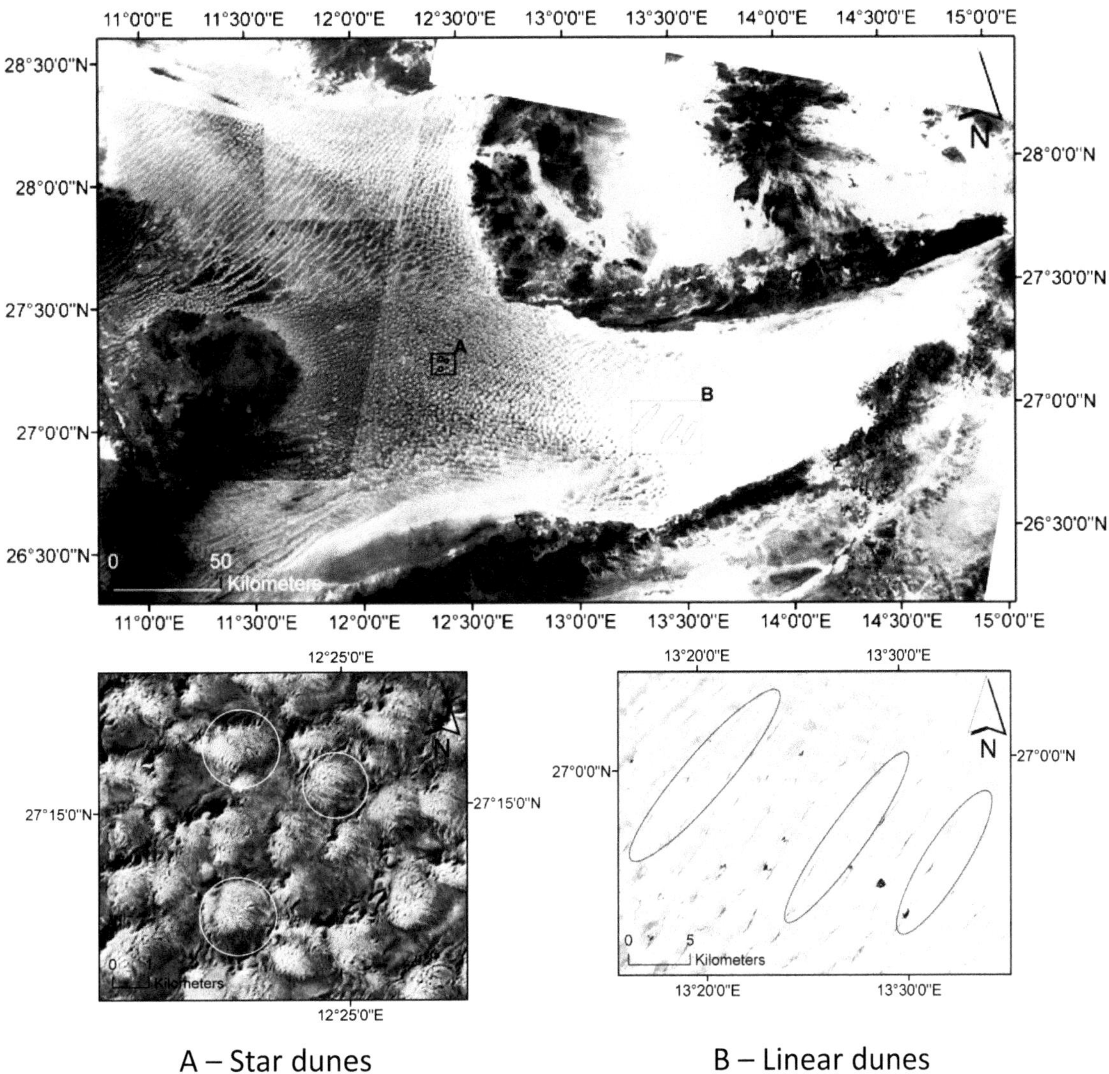

Fig. 10.3 Mosaic image of Landsat 8 tiles covering the Ubārī Sand Sea (date: October 2015), with details of star dunes (three individual star dunes are circled in yellow) and linear dunes (three individual linear dunes are circled in red) (Els 2017)

distance test (values range 0–2), in which low values indicate less difference between different combinations of classes and high values indicate greater difference. The results of this analysis are presented below.

Accuracy assessment of the supervised classification of the total dataset was undertaken using the confusion matrix module within ENVI. This can be used to evaluate the overall accuracy, producer accuracy, user accuracy, and kappa coefficient. The producer accuracy indicates how well the training samples are classified, and the user accuracy gives the probability that a pixel belongs to the class to which it was assigned (i.e. represents that class in reality). The kappa value is used to determine if there is a significant difference between the two confusion matrices.

10.4.2 Time Series Analysis of Dune Migration

The maximum likelihood classification was applied to Landsat images from October 2002 and October 2015. The accuracy of the two time series classifications was assessed with a confusion matrix. Due to its size, a subset of the Ubārī Sand Sea region (45 × 50 km, 2250 km^2) was used for time series mapping. This area is in the south-east of the sand sea and comprises linear dunes ($n = 39$) that are likely moving in single direction, hence can be usefully evaluated using the Change Detection Statistics module of ENVI. To facilitate this, dunes were digitised to form polygons within ArcGIS. The positions of dune crest and trough were determined initially by visual inspection of the Landsat images,

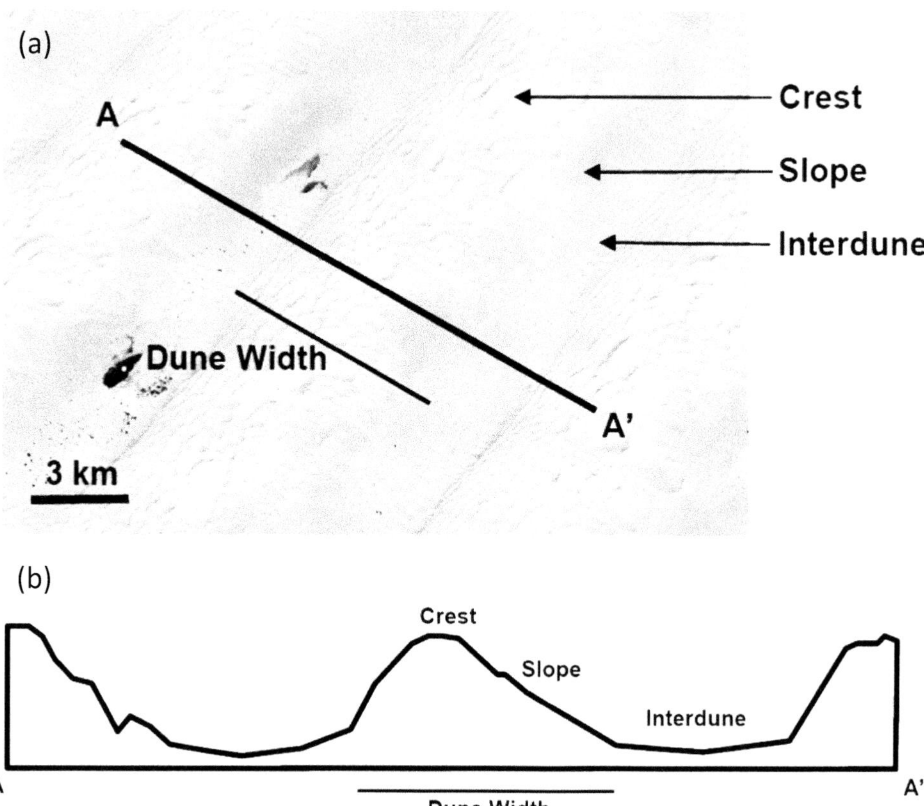

Fig. 10.4 **a** Example of a linear dune within the Ubārī Sand Sea showing the morphological components present within a single dune (base image from Google Earth), and **b** as a schematic cross section as derived from SRTM data

and were then informed by mapping of dune morphological features using the classified images. Calculation of dune migration rates is based on changes in the outline shape and crestline position of individual dunes on sequential images. This follows standard methodologies employed in dune mapping (e.g. Callot et al. 1994; Hermas et al. 2012; Telfer et al. 2015; Boulghobra and Dridi 2016; Hamdan et al. 2016; Baird et al. 2019).

10.5 Results

10.5.1 Identification of Dune Features

Supervised classification of the Landsat images shows that there is less separability between the three classes that represent dunes (interdune, slope, crest), whereas there is high separability between dunes and other identified classes

Table 10.1 Matrix of Jeffries–Matusita distance values in comparing regions of interest

	Gypsum	Interdune	Rocky outcrop	Slope	Soil	Urban	Vegetation	Water
Vegetation	–	–	–	–	–	–	–	2.0000
Urban	–	–	–	–	–	–	1.9575	1.9440
Soil	–	–	–	–	–	1.5137	1.9977	1.6941
Slope	–	–	–	–	1.9977	1.9747	2.0000	1.9989
Rocky outcrop	–	–	–	2.0000	1.9821	1.8295	1.9520	1.9995
Interdune	–	–	2.0000	0.5960	1.9961	1.9615	2.0000	1.9959
Gypsum	–	1.6119	2.0000	1.5463	1.7535	1.8453	2.0000	1.9324
Dune crest	1.6944	0.7389	2.0000	0.1395	1.9995	1.9844	2.0000	1.9998

(Table 10.1), meaning that the distribution of dunes or sand seas can be easily mapped across the region. The regions of interest were divided into training (80%) used for the supervised classification, and testing (20%) is used for accuracy assessment of the classification.

Unsupervised classification of the Landsat 8 images into the nine dune feature classes was undertaken to determine if dune features can be distinguished based on spectral signatures alone. The classes that do not represent dune features were not assigned to predefined features. Figure 10.5 shows that some dune features such as interdune areas are clearly distinguished. The crests and slopes, however, are less well defined, and some overlap occurs between these two feature classes (Table 10.1).

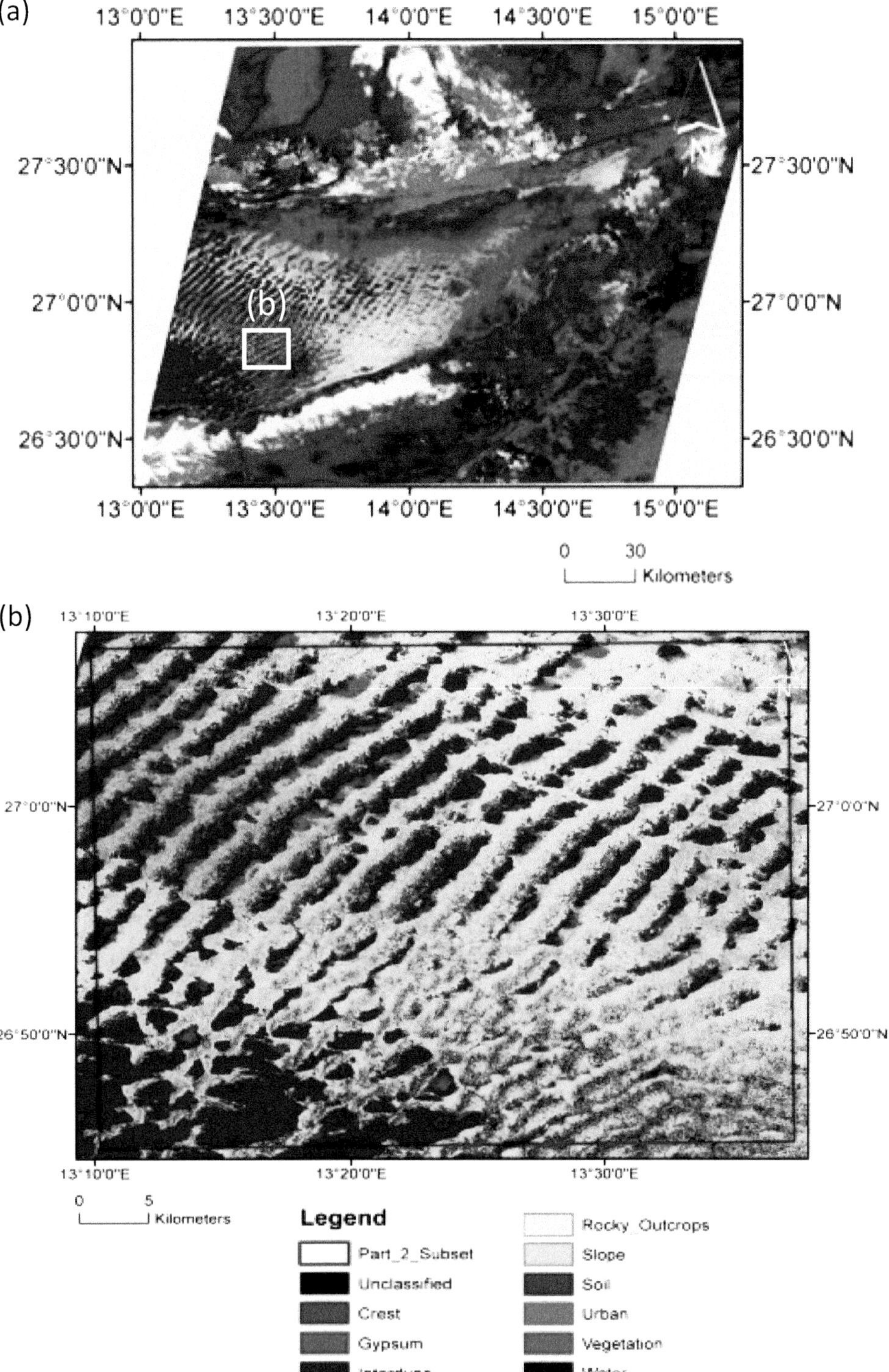

Fig. 10.5 Maximum likelihood classification for the study area in 2015 showing **a** the regional context and **b** the specific sector of the Ubārī Sand Sea analysed

Table 10.2 Error matrix for the maximum likelihood classification for the Landsat images using the RGB and NIR band combination with the highest overall accuracy (64.67%)

Class	Crest	Gypsum	Interdune	Rocky outcrop	Slope	Soil	Urban	Vegetation	Water	Total	User accuracy (%)
Crest	427	1	0	0	145	0	0	0	0	573	74.52
Gypsum	4	12	11	0	15	4	0	1	0	47	25.53
Interdune	79	0	104	0	43	0	0	0	0	226	46.02
Rocky outcrop	0	0	0	94	0	0	1	0	0	95	98.95
Slope	147	1	16	0	153	0	0	0	0	317	48.26
Soil	0	1	0	6	0	86	5	2	0	100	86.00
Urban	0	0	0	10	0	4	64	1	0	79	81.01
Vegetation	0	0	0	1	0	2	4	37	0	44	84.09
Water	0	0	0	0	0	28	2	2	4	36	11.11
Total	657	15	131	111	356	124	76	43	4	1517	
Producer accuracy (%)	64.99	80.00	79.39	84.68	42.98	69.35	84.21	86.05	100		

In terms of the accuracy assessment of the classification, the maximum likelihood classifier and the RGB and NIR band combination resulted in the highest overall accuracy (64.67%) and the highest kappa coefficient (0.5355), compared to the maximum likelihood classifier with the red (R) and NIR bands, with an overall accuracy value of 57.42% and kappa coefficient of 0.4575. A detailed illustration of the RGB and NIR band combination, resulting in this highest overall accuracy, is given in Table 10.2. These overall values are, however, very low compared to classification in vegetated environments which are usually 80% and higher (Adelabu et al. 2013). The user accuracy of the four classification combinations of the three dune feature classes for the Landsat image is as follows, in descending order of accuracy:

- Crest: maximum likelihood (R+NIR bands) (76.26%); maximum likelihood (RGB+NIR bands) (74.52%); minimum distance (RGB+NIR bands) (70.83%); and minimum distance (R+NIR bands) (69.37%);
- Slope: maximum likelihood (R+NIR) (48.91%); maximum likelihood (RGB+NIR) (48.26%); minimum distance (R+NIR) (40.79%); and minimum distance (RGB+NIR) (38.41%);
- Interdune: maximum likelihood (R+NIR) (46.48%); maximum likelihood (RGB+NIR) (46.02%); minimum distance (R+NIR) (45.24%); and minimum distance (RGB+NIR) (44.53%).

Thus, the dune crest is the most significant feature capable of being reliably identified from Landsat imagery used the methods described here. There is also the potential for some confusion between the dune features. Based on user and producer accuracies, the Landsat images show the least confusion with the maximum likelihood classification and using the RGB and NIR band combination. This results in an overall classification accuracy of 69.01% and kappa coefficient of 0.59. Greatest confusion occurs between crest, interdune and slope classes, with smaller amounts of confusion between gypsum and slopes, soil and water, urban and soil, and water and gypsum. The high confusion between crest and slope may be due to poor training samples resulting from the difficulty of identifying dune crests on the imagery.

10.5.2 Dune Migration

The final maximum likelihood classification for dunes in the region, using Landsat 8 in 2015, is shown in Fig. 10.6. Based on dune mapping from 2002 and 2015, spatial changes in dune features can be evaluated. There was an overall decrease of 41.90% (1704.42 km^2 across the entire study area) in dune slopes, an increase of 22.95% (73.57 km^2) in dune crests, and an increase of 48.12% (119.33 km^2) in interdune areas. Changes mainly took place within the dunes themselves with only small variations in peripheral areas. Percentage area changes are shown in Table 10.3.

Individual dune ridges mapped within the subsampled area ($n = 39$) yielded different properties in 2002 and 2015 (Table 10.4). Overall, dunes became longer but narrower and smaller over this time period, but there is considerable variability in these values. Likewise, calculated dune migration rates, based on the position of the dune crestline in successive images, are also variable. For migration rates of individual dunes, multiple measurements were taken at 300 m intervals along the crestline, and then averaged. Total distance moved by individual dunes varies between 29.79 and 316.65 m (average 112.32 ± 60.46 m) yielding an averaged dune migration rate of linear dunes of 8.64 ± 4.65 m yr^{-1} towards the NW, with a range value for individual dunes from 2.29 to 11.63 m yr^{-1} (Els 2017).

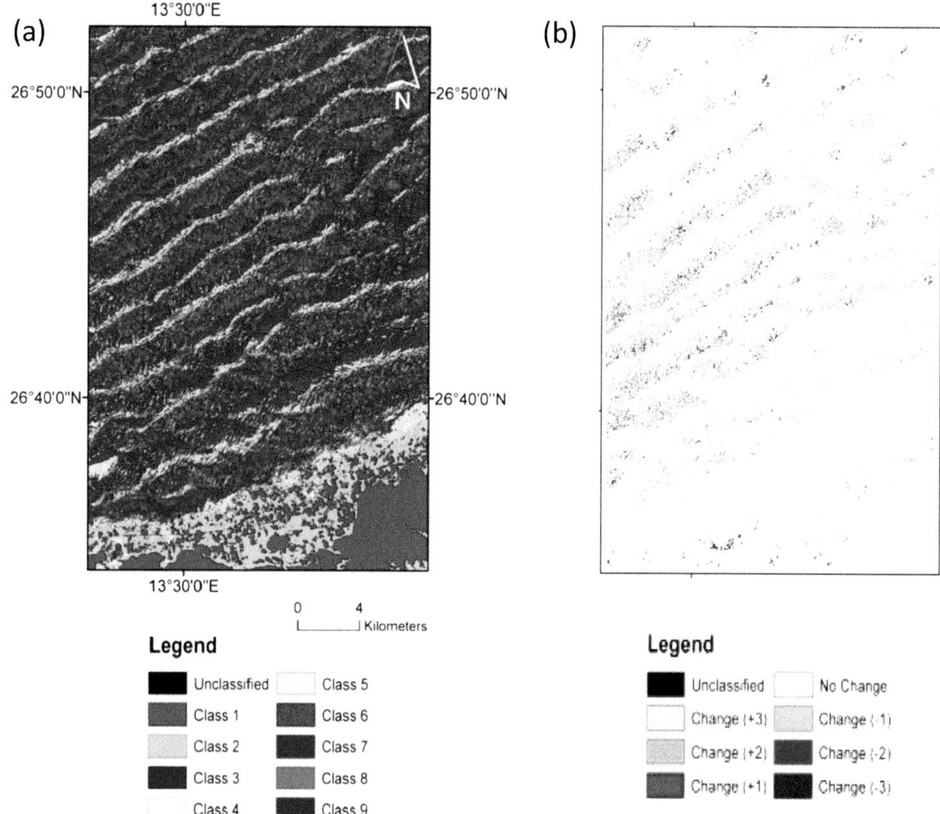

Fig. 10.6 **a** Unsupervised Landsat 8 classification of a section of the Ubārī Sand Sea using RGB and NIR bands. K-mean classes 1, 2 do not correspond to any specific dune features. Classes 3–5 are interdunes, classes 6–8 are dune slopes, class 9 is dune crest. **b** Change map of the area shown in (a) between different k-mean classes (panel a) of the comparison of the two band combinations used in the unsupervised classification module for Landsat 8

Table 10.3 Overall change in classification class between 2002 and 2015 in the study area (as % of entire mapped area)

Initial state (2002)									
Final state (2015)		Crest	Gypsum	Interdune	Slope	Soil	Urban	Vegetation	Water
	Crest	22.95	1.07	12.34	2.72	0.06	3.57	0.09	0.05
	Gypsum	4.00	72.32	8.95	41.63	33.30	0.56	50.90	72.50
	Interdune	34.47	7.02	48.12	13.04	0.16	25.86	0.23	1.15
	Slope	37.66	12.29	29.11	41.90	0.01	22.15	0.21	1.85
	Soil	0.00	1.78	0.01	0.00	30.77	0.00	13.22	2.71
	Urban	0.00	0.00	0.00	0.00	0.01	0.00	0.04	0.00
	Vegetation	0.92	4.12	1.42	0.64	34.28	0.00	33.46	13.47
	Water	0.01	1.41	0.05	0.07	0.20	0.00	1.66	8.25
Class total %	100.00	100.00	100.00	100.00	100.00	100.00	100.00	100.00	
Class changes %	77.05	27.68	51.89	58.11	69.23	100.00	66.54	91.75	
Image difference	0.57	6.96	48.12	– 22.79	2492.47	– 75.08	– 28.14	– 77.93	

10.6 Discussion

In general, the dune features in the Ubārī Sand Sea can be discriminated based on their spectral signatures alone (Fig. 10.6). Different dune features (e.g. the crest, slope and interdune) can be successfully classified from moderate-resolution Landsat imagery with overall accuracies and kappa values ranging from 52.11 to 64.67% and 0.3878 to 0.4927, respectively. This also enables dune migration rates to be calculated reasonably well with respect to the accuracy of identification of the dune crestline. The averaged linear dune migration rate of 8.64 ± 4.65 m yr^{-1} is comparable with results from previous studies. Examples include 1.3 m yr^{-1} from the Qaidam peninsula, China (Livingstone et al. 2007), 1.99 m yr^{-1} from Namibia (Besler et al. 2013), 0.7–2.0 m yr^{-1} and 2.25–13.0 m yr^{-1} from the NW Sinai,

Table 10.4 Summary of changes in dune morphology in the subsampled study area, 2002–2015

		2002	2015	Difference
Crest	Mean (km)	9.764	10.097	+0.333
	S.D	7.54	7.53	
	Range (km)	1.9 to 36.5	2.4 to 36.1	− 13.7 to + 16.4
Width	Mean (km)	1.812	1.368	− 0.132
	S.D	0.227	0.227	
	Range (km)	1.23 to 2.23	1.10 to 2.03	− 0.35 to + 0.02
Area	Mean (km^2)	20.219	18.760	− 1.459
	S.D	14.034	13.101	
	Range (km^2)	3.48 to 65.01	3.18 to 63.94	− 8.05 to + 0.15

Egypt (Phillip et al. 2004), 0.5–3.8 m yr^{-1} from the Great Kobuk Sand Dunes, Alaska (Necsoiu et al. 2009) and 9.1 ± 2.7 m yr^{-1} from NE Patagonia (del Valle et al. 2008).

These contemporary dune migration rates can be set in a geological context of longer-term changes in sand sea geomorphology and dynamics in response to climate forcing throughout the Quaternary. Kocurek et al. (1991) described the evolution of the Akchar erg, W Mauritania, and showed that stratigraphic evidence for humid and arid phases can be preserved, and that linear dunes found today are largely relict with only limited evidence for recent surface sediment activity. Variations in dune morphology reflect different windflow phases during the late Pleistocene and Holocene (Lancaster et al. 2002; Besler 2008; Hu et al. 2021). Wilson (1973) showed that dunes are more dynamic in the centre and less dynamic at the edges of sand seas. This may result from a combination of lower sediment availability as well as more humid conditions towards sand sea margins. Long-term changes in synoptic climate, affecting wind speed/direction and humidity, can cause sand seas to increase or decrease in area. Sand sea margins are thus particularly vulnerable to be affected by climate change, and most multiproxy palaeoenvironmental data tend to come from sand sea margins in areas with higher humidity (e.g. Swezey et al. 1999; El-Baz et al. 2000). Based on this idea, Haynes (1982) argued that isohyets shifted by 400 km latitudinally in the eastern Sahara between humid and arid Pleistocene climate phases, marked by development of soils (humid periods) that were then overlain by dune sand (arid periods). The effects of such climate changes are thus best expressed in marginal (Sahelian) areas, but these have not been systematically investigated. Even less is known from more central Saharan areas with lower water availability, where the influence of Pleistocene pluvial phases was presumably less significant. This is an important area of future research.

Different remote sensing methods have been used to identify and map sand seas and sand dunes in the Sahara region (Bubenzer and Bolten 2008; Ghadiry et al. 2012; Hermas et al. 2012; Boulghobra 2016; Bubenzer et al. 2020; Hu et al. 2021), but this process has been very piecemeal and the different sensors, methods and resolutions used has meant that results from different studies are not always comparable. This has limited the extent to which a regional-scale overview can be developed. Spatial mapping of sand dunes has used combinations of altimetric data (Bubenzer and Bolten 2008; Potts et al. 2008), microwave emissivity (Stephen and Long 2005; Masiello et al. 2014), and spectral analysis of sediment geochemistry (White et al. 1997, 2001; Bradley et al. 2018). Timing of dune deposition is constrained by dates obtained by luminescence methods (e.g. Bubenzer et al. 2007, 2020; Lancaster et al. 2016) and U–Pb geochronology (Garzanti et al. 2013). There is potential, however, to exploit other remote sensing products and variables and based on more advanced analytical methods. Such studies show that dunes of different ages and properties can be superimposed on one another, suggesting episodic dynamic behaviour that may provide a window into Pleistocene environments in the Sahara. This has been done successfully for palaeodrainage systems (e.g. Robinson et al. 2006; Paillou et al. 2009). Desert dune migration in response to ongoing climate and environmental change (desertification) is a threat to settlements and agricultural fields in many Sahelian areas (Mainguet and da Silva 1998). Thus, remote sensing-based approaches to dune monitoring can be used to assess future patterns of dune migration and sand encroachment risk (Boulghobra 2016).

10.7 Summary

Sand seas and the sand dunes within them are a distinctive landscape feature of the central Sahara. Sand seas are geomorphically complex and have likely evolved episodically over long timescales, although there is very little information on their development. Many recent studies have used remote sensing methods to map and monitor changes in sand dunes, including calculations of dune migration rates such as for the Ubārī Sand Sea examined in this study. Many sand seas in the central Sahara have not been well studied, however, and regional spatial patterns of, for example, the effects of changes in humidity or synoptic windflow are unknown. This is important when it comes to monitoring ongoing desertification and aridification in the Sahara under climate change.

Acknowledgements We thank Nouar Boulghobra and an anonymous reviewer for their comments on a previous version of this chapter.

References

Adelabu S, Mutanga O, Adam EE, Cho MA (2013) Exploiting machine learning algorithms for tree species classification in semiarid woodland using RapidEye image. J Appl Remote Sens 7:073480. https://doi.org/10.1117/1.JRS.7.073480

Al-Masrahy MA, Mountney NP (2013) Remote sensing of spatial variability in aeolian dune and interdune morphology in the Rub' Al-Khali, Saudi Arabia. Aeolian Res 11:155–170

Bagnold RA (1941) The physics of blown sand and desert dunes. Methuen, London, p 265

Baird T, Bristow CS, Vermeesch P (2019) Measuring sand dune migration rates with COSI-Corr and Landsat: opportunities and challenges. Remote Sens 11:2423. https://doi.org/10.3390/rs11202423

Besler H, Lancaster N, Bristow C, Henschel J, Livingstone I, Seely M, White K (2013) Helga's Dune: 40 years of dune dynamics in the Namib Desert. Geogr Ann, Ser A 95:361–368

Besler H (2008) The Great Sand Sea in Egypt, formation, dynamics and environmental change—a sediment-analytical approach. Developments in sedimentology, 59. Elsevier, Amsterdam, p 266

Boulghobra N (2016) Climatic data and satellite imagery for assessing the aeolian sand deposit and barchans migration, as a major risk sources in the region of In-Salah (Central Algerian Sahara). Arab J Geosci 9:450. https://doi.org/10.1007/s12517-016-2491-x

Boulghobra N, Dridi H (2016) Fine resolution imagery and GIS for investigating the morphological characteristics, and migration rate of barchan dunes in the Erg Sidi Moussa dunefield near In-Salah (Algeria). Geogr Techn 11:14–21

Bradley AV, McLaren SJ, Al-Dughairi A, Khalaf N (2018) A multi-scale approach interpreting sediment processes and distribution from desert sand colour in central Saudi Arabia. Earth Surf Proc Landf 43:2847–2862

Bubenzer O, Bolten A (2008) The use of new elevation data (SRTM/ASTER) for the detection and morphometric quantification of Pleistocene megadunes (draa) in the eastern Sahara and the southern Namib. Geomorphology 102:221–231

Bubenzer O, Besler H, Hilgers A (2007) Filling the gap: OSL data expanding ^{14}C chronologies of Late Quaternary environmental change in the Libyan Desert. Quat Int 175:41–52

Bubenzer O, Embabi NS, Ashour MM (2020) Sand seas and dune fields of Egypt. Geosciences 10:101. https://doi.org/10.3390/geosciences10030101

Busche D (1998) Die Zentrale Sahara: Oberflächenformen im Wandel. Justus Perthes Verlag, Gotha, p 284

Callot Y, Mering C, Simonin A (1994) Image-analysis and cartography of sand hill massifs on high resolution images: application to the Great Western Erg (NW of Algerian Sahara). Int J Remote Sens 15:3799–3822

del Valle HF, Rostagno CM, Coronato FR, Bouza PJ, Blanco PD (2008) Sand dune activity in north-eastern Patagonia. J Arid Env 72:411–422

Drake N, White K, Salem M, Armitage S, El-Hawat A, Francke J, Hounslow M, Parker A (2009) DMP VIII: Palaeohydrology and palaeoenvironment. Libyan Stud 40:171–178

du Pont SC (2015) Dune Morphodynamics. Compt Rend Phys 16:118–138

Dubief J (1953) Le ventes de sable au Sahara français. Coll Int CNRS 35:45–70

El-Baz F, Maingue M, Robinson C (2000) Fluvio-aeolian dynamics in the north-eastern Sahara: the relationship between fluvial/aeolian systems and ground-water concentration. J Arid Environ 44:173–183

Els A, Merlo S, Knight J (2015) Comparison of two satellite imaging platforms for evaluating sand dune migration in the Ubari Sand Sea (Libyan Fazzan). The International Archives of the Photogrammetry, Remote Sensing and Spatial Information Sciences, Volume XL-7/W3, 2015, 36th International Symposium on Remote Sensing of Environment, 11–15 May 2015, Berlin, Germany, pp 1375–1380

Els A (2017) Tracking sand dune movements using multitemporal remote sensing imagery: a case study of the central Sahara (Libyan Fazzān / Ubārī sand sea). MSc dissertation, University of the Witwatersrand, p 144

Embabi NS (2018) Landscapes and Landforms of Egypt – Landforms and Evolution. Springer, Cham, p 336

Fryberger SC, Dean G (1979) Dune forms and wind regime. In: McKee ED (ed) A study of global sand seas. USGS Professional Paper 1052. US Government Printing Office, Washington DC, pp 137–169

Garzanti E, Vermeesch P, Andò S, Vezzoli G, Valagussa M, Allen K, Kadi KA, Al-Juboury AIA (2013) Provenance and recycling of Arabian desert sand. Earth Sci Rev 120:1–19

Ghadiry M, Shalaby A, Koch B (2012) A new GIS-based model for automated extraction of sand dune encroachment case study: Dakhla Oases, western desert of Egypt. Egypt J Remote Sens Space Sci 15:53–65

Ghoneim E, Benedetti M, El-Baz F (2012) An integrated remote sensing and GIS analysis of the Kufrah Paleoriver, Eastern Sahara. Geomorphology 139:242–257

Goudarzi GH (1970) Geology and Mineral Resources of Libya – A Reconnaissance. Geological Survey Professional Paper 660, United States Government Printing Office, Washington, DC, p 116

Goudie AS (2013) Arid and Semi-Arid Geomorphology. Cambridge University Press, Cambridge, p 454

Hamdan MA, Refaat AA, Wahed MA (2016) Morphologic characteristics and migration rate assessment of barchan dunes in the Southeastern Western Desert of Egypt. Geomorphology 257:57–74

Haynes CV (1982) Great Sand Sea and Selima Sand Sheet, Eastern Sahara – Geochronology of Desertification. Science 217:629–633

Hermas E, Leprince S, El-Magd IA (2012) Retrieving sand dune movements using sub-pixel correlation of multi-temporal optical remote sensing imagery, northwest Sinai Peninsula. Egypt. Remote Sens Env 121:51–60

Hu Z, Gao X, Lei J, Zhou N (2021) Geomorphology of aeolian dunes in the western Sahara Desert. Geomorphology 392:107916. https://doi.org/10.1016/j.geomorph.2021.107916

Khatelli H, Gabriels D (1998) Study on the dynamics of sand dunes in Tunisia: Mobile barkhans move in the direction of the Sahara. Arid Soil Res Rehabilit 12:47–54

Kocurek G, Havholm KG, Deynoux M, Blakey RC (1991) Amalgamated accumulations resulting from climatic and eustatic changes, Akchar Erg, Mauritania. Sedimentology 38:751–772

Lancaster N (1996) Deserts. In: Adams WM, Goudie AS, Orme AR (eds) The Physical Geography of Africa. Oxford University Press, Oxford, pp 211–237

Lancaster N, Kocurek G, Singhvi A, Pandey V, Deynoux M, Ghienne J-F, Lô K (2002) Late Pleistocene and Holocene dune activity and wind regimes in the western Sahara Desert of Mauritania. Geology 30:991–994

Lancaster N, Wolfe S, Thomas D, Bristow C, Bubenzer O, Burrough S, Duller G, Halfen A, Hesse P, Roskin J, Singhvi A, Tsoar H, Tripaldi A, Yang X, Zárate M (2016) The INQUA Dunes Atlas chronologic database. Quat Int 410:3–10

Lettau K, Lettau H (1977) Experimental and micrometeorological field studies of dune migration. In: Lettau K, Lettau H (eds) Exploring the World's Driest Climate. University of Wisconsin-Madison, IES Report 101, pp 110–147

Livingstone I, Wiggs GFS, Weaver CM (2007) Geomorphology of Desert Sand Dunes: A Review of Recent Progress Earth-Sci Rev 80:239–257

Mainguet M, da Silva GG (1998) Desertification and drylands development: what can be done? Land Degrad Develop 9:375–382

Mainguet M, Borde J-M, Chemin M-C (1984) Sedimentation eolienne au Sahara et sur ses marges: Les images Meteosat et Landsat, outil pour l'analyse des temoignages géodynamique du transport eolian au sol. Travaux De L'institut Géographique De Reims 59–60:15–27

Masiello G, Serio C, Venafra S, DeFeis I, Borbas EE (2014) Diurnal variation in Sahara desert sand emissivity during the dry season from IASI observations. J Geophys Res-Atmos 119:1626–1638

McKee ED (ed) (1979) A study of global sand seas. USGS Professional Paper 1052. US Government Printing Office, Washington, DC, p 429

Mercuri AM (2008) Human influence, plant landscape evolution and climate inferences from the archaeobotanical records of the Wadi Teshuinat area (Libyan Sahara). J Arid Environ 72:1950–1967

Necsoiu M, Leprince S, Hooper DM, Dinwiddie CL, McGinnis RN, Walter GR (2009) Monitoring migration rates of an active subarctic dune field using optical imagery. Remote Sens Environ 113:2441–2447

Paillou P, Schuster M, Tooth S, Farr T, Rosenqvist A, Lopez S, Malezieux J-M (2009) Mapping of a major paleodrainage system in eastern Libya using orbital imaging radar: The Kufrah River. Earth Planet Sci Lett 277:327–333

Phillip G, Attia OEA, Draz MY, El Banna MS (2004) Dynamics of sand dunes movement and their environmental impacts on the reclamation area in NW Sinai, Egypt. Proceedings of the 7th Conference: Geology of Sinai for Development, Ismailia, pp 169–180

Potts LV, Akyilmaz O, Braun A, Shum CK (2008) Multi-resolution dune morphology using Shuttle Radar Topography Mission (SRTM) and dune mobility from fuzzy inference systems using SRTM and altimetric data. Int J Remote Sens 29:2879–2901

Pye K, Tsoar H (2009) Aeolian sand and sand dunes. Springer, Berlin, p 476

Robinson CA, El-Baz F, Al-Saud TSM, Joen SB (2006) Use of radar data to delineate palaeodrainage leading to the Kufra Oasis in the eastern Sahara. J Afr Earth Sci 44:229–240

Sinha SC, Pandey SM (1980) Hydrogeological studies in a part of the Murzuq Basin using geophysical logs. In: Salem MJ, Busrewil MT (eds) The Geology of Libya, vol 2. Academic Press, London, pp 629–633

Stephen H, Long DG (2005) Modeling microwave emissions of erg surfaces in the Sahara desert. IEEE Trans Geosci Remote Sens 43:2822–2830

Swezey C, Lancaster N, Kocurek G, Deynoux M, Blum M, Price D, Pion J-C (1999) Response of aeolian systems to Holocene climatic and hydrologic changes on the northern margin of the Sahara: a high-resolution record from the Chott Rharsa basin, Tunisia. Holocene 9:141–147

Tawadros EE (2001) Geology of Egypt and Libya. Balkema, Rotterdam, p 468

Telfer MW, Fyfe RM, Lewin S (2015) Automated mapping of linear dunefield morphometric parameters from remotely-sensed data. Aeolian Res 19:215–224

Vermeesch P, Leprince S (2012) A 45-year time series of dune mobility indicating constant windiness over the central Sahara. Geophys Res Lett 39:L14401. https://doi.org/10.1029/2012GL0552592

White K, Walde J, Drake N, Eckardt F, Settle J (1997) Mapping the iron oxide content of dune sands, Namib Sand Seas, Namibia, using Landsat Thematic Mapper data. Remote Sens Environ 62:30–39

White K, Goudie A, Parker A, Al-Farraj A (2001) Mapping the geochemistry of the northern Rub' Al Khali using multispectral remote sensing techniques. Earth Surf Proc Landf 26:735–748

White K, Charlton M, Drake N, McLaren S, Mattingly D, Brooks N (2006) Lakes of the Edeyen Awbari and the Wadi al-Hayat. In: Mattingly D, McLaren S, Savage E, al-Fasatwi Y, Gadgood K (eds) The Libyan Desert, Natural Resources and Cultural Heritage. The Society for Libyan Studies, London, pp 123–130

Wilson IG (1973) Ergs. Sediment Geol 10:77–106

11 The Hydrology and Palaeohydrology of the Central Sahara

Jasper Knight

Abstract

Present and past river systems of the Sahara reflect climate changes over 10^4–10^9 year timescales, but have considerable capacity to lead to changes in large-scale sediment and geomorphic patterns in the landscape. This chapter first reviews the key characteristics of today's fluvial activity in the central Sahara, which is limited to flash flooding within wadi systems, and the important role of groundwater in sustaining present-day agriculture, settlements, and hotspots of biodiversity in gueltas and oases. The palaeohydrology of the central Sahara is dominated by now-buried palaeochannels that represent vigorous regional-scale perennial river systems that developed during pluvial phases, mainly in the Miocene and also much later during the African Humid Period (early to mid-Holocene). The extent to which these river systems were occupied, reoccupied and abandoned during different climate phases, and the spatial and temporal patterns of river system evolution, is largely unknown, and these represent important areas for future research.

Keywords

Rainfall patterns · Hydrology · Alluvial systems · Rivers · Miocene · Palaeodrainage · TRMM · Radar · Gueltas

11.1 Introduction

A commonly cited and defining property of the Sahara as a region is its dryness (Cloudsley-Thompson 1984), and the lack of present-day rainfall is seen as both a limitation to its present ecosystems and the maintenance of human life and activity, and as a facilitator of desert processes and landforms. Both these viewpoints are mistaken. Although the central Sahara in particular is indeed dry, and can be defined as hyperarid (<100 mm precipitation yr^{-1}; UNESCO 1979), ecosystems and human activity have successfully adapted to these conditions over the late Pleistocene and in particular the Holocene, and desert system morphodynamics are strongly controlled by the presence—not the absence—of water through gueltas, oases, groundwater processes and springlines, and episodically-active wadis and ephemeral lakes (Gischler 1976; Griffin 2002; Drake et al. 2011). This highlights the general importance of water for a range of geomorphic, ecological, chemical and microclimatic processes and properties across the Sahara (Cloudsley-Thompson 1984). Several studies have examined the development of river systems and associated river geomorphic and sediment patterns in the Sahara over timescales of tens of millions of years (Le Houérou 1997; Goudie 2005) down to the event-scale of a few hours (Moawad et al. 2016). Particular focus has been paid in the literature to evidence from wetter (pluvial) periods in which new river systems were developed or older systems were reactivated (Blanchet et al. 2021). The advantage of such studies in the Sahara is that the arid environment tends to promote longevity of relict landforms, in the surface or immediate subsurface (e.g. Abotalib and Mohamed 2013; Issawi and Sallam 2017), that would not be preserved in wetter environments with more active soil development, vegetation growth, and strong chemical processes within the vadose zone. The presence of weathered and loose sediments across the central Sahara land surface also provides ideal materials that can be reworked by rivers. As a result,

J. Knight (✉)
School of Geography, Archaeology & Environmental Studies, University of the Witwatersrand, Johannesburg 2050, South Africa
e-mail: jasper.knight@wits.ac.za

J. Knight et al. (eds.), *Landscapes and Landforms of the Central Sahara*, World Geomorphological Landscapes, https://doi.org/10.1007/978-3-031-47160-5_11

there is exceptionally detailed evidence (compared to other regions) from wetter time periods in which the Sahara supported significant river systems, including in the Miocene (7–5 Ma), and the African Humid Period (15–6 kyr BP) (deMenocal 1995). Using such evidence, both present and past hydrological processes and environments in the Sahara can be identified. This evidence is also significant because of the relationships in this region between climate, water availability, geomorphic change, and development of ecosystems (Cloudsley-Thompson 1984). This sets Saharan hydrology into a wider context of synoptic circulation, monsoon and harmattan events, biodiversity status, Pleistocene hominid migration patterns including the 'Out of Africa' hypothesis, historical trade routes, and the geographical spread of its peoples and cultures (Gearon 2011).

This chapter aims to discuss the hydrology of the central Sahara focusing on (1) rainfall patterns that give rise to today's surface water systems including the episodic operation of wadis, and (2) subsurface evidence for palaeoriver systems formed during wetter periods, in particular the Miocene. The outcome of this analysis is a new viewpoint of Saharan hydrology that speaks to both macroscale controls of rivers in the geological past, such as tectonics, and present surface water systems that link to today's patterns of human settlement and biodiversity hotspots.

11.2 Today's Saharan Hydrology

The surface and near-surface (vadose zone) hydrology of the Sahara is controlled by rainfall patterns driven by climate, and by groundwater processes. These two elements give rise to very different geomorphic and sedimentary expressions, and are examined in turn.

11.2.1 Rainfall Patterns in the Sahara

The general absence of long-term and reliable weather stations has meant that rainfall patterns across much of the Sahara are not well known from field observational or instrumental evidence. This also means that smaller-scale studies of, for example, orographic forcing of rainfall around bedrock massifs or the microclimate of oases have not been widely undertaken. One exception, albeit marginal to the central Sahara, is by Slimani et al. (2007) who examined the nature of the precipitation gradient across Tunisia based on 30 years of data from 106 weather stations. This study showed increasing dryness inland from Mediterranean moisture sources but also stronger rainfall seasonality and shorter spells of rainfall in an inland direction. Thus, the character of precipitation changes considerably over space. The central Sahara region today is hyperarid (as classified based on both field and remote sensing data sources; see below for discussion of the latter), but this term does not fully capture the nature of event-scale rainfall events where dramatic landscape transformation can be effected (e.g. Megnounif et al. 2013; Abdel-Fattah et al., 2017). Walling (1996, pp. 106–107) presented a panel of maps illustrating key elements of rainfall regimes across Africa. Over the Sahara, rainfall (based on UNESCO 1979) across North Africa falls within the 100 mm yr^{-1} isohyet, with patches of the Western Desert (Egypt) and Algeria falling within 10 mm yr^{-1} (Fig. 11.1). Further, due to high potential evapotranspiration, these regions have a strongly negative hydrological balance, near zero mean annual runoff and thus a very low runoff coefficient. Gischler (1976) quantified these patterns for particular river basins that flow into the central Sahara region from wetter areas in the Sahel (see for example the Chad Basin; Knight, this volume). Several studies of flash flood events in Egypt, however, show that regional-scale and long-term rainfall and hydrological patterns do not adequately describe individual rainstorm events. For example, in hyperarid areas of the Sinai Peninsula (Egypt) with annual rainfall of 2 to 10 mm, a single storm event in January 2010 resulted in 73 mm of rainfall over 7 days, leading to significant wadi flooding (Moawad 2013). The infrequent nature of such rainstorm events also means that the geomorphic impacts of such heavy rainfall are not well described (Abdel-Fattah et al. 2017). As a consequence of these limitations, most work on rainfall patterns and climatology has been done at a regional scale using remote sensing methods.

Most such work has been done using Tropical Rainfall Measuring Mission (TRMM) data which uses satellite radar backscatter to calculate instantaneous rainfall rate at the range 1.5–2.5 km altitude at $0.1° \times 0.1°$ grid cell resolution (Kelley 2014; Liu 2015). Although there may be problems with overestimation of surface rainfall due to rain evaporation in the lower atmosphere, and through spatial averaging, TRMM data have been widely used especially in the Sahara region in the absence of reliable and high-resolution ground surface measurements. A further advantage is that the same product and methodology can be used consistently across the region, increasing confidence in spatial and temporal variations in rainfall patterns. In slightly wetter areas of the southern Sahara Sahel, there is a strong correlation between TRMM and field gauge measurements ($r = 0.98$) (Okonkwo et al. 2013). Results of TRMM data analysis show that different regions of the Sahara have different seasonal patterns, broadly reflecting its wide latitudinal range and with proximity to moisture sources in the Mediterranean and Red Sea. Winter and spring rain occur over the Mediterranean; summer rain occurs in the southern Sahara and Sahel; autumn rain takes place over the Red Sea (Kelley 2014). These seasonal patterns, however, are

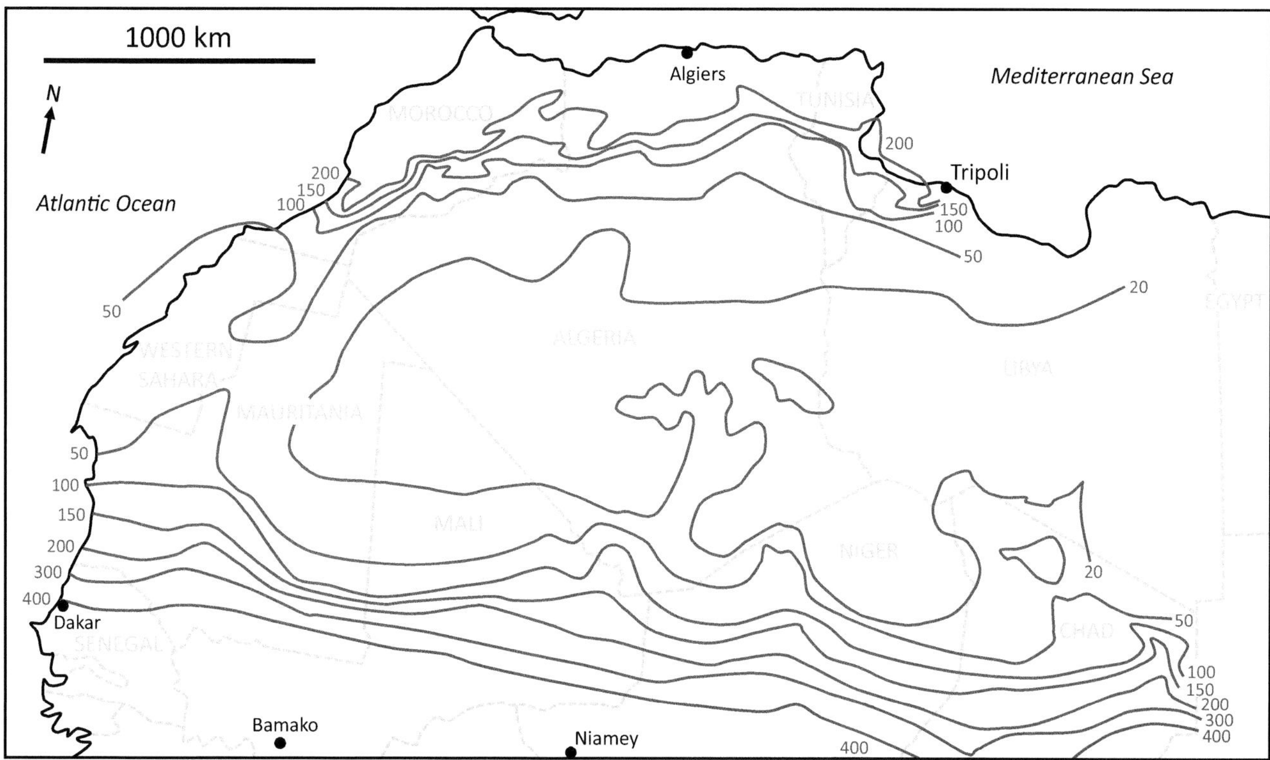

Fig. 11.1 Isohyets of mean annual precipitation (mm yr^{-1}) in the central Sahara, redrawn from Hugot (1974, p. 24)

not consistent and the low total annual rainfall means that identifying 'seasonality' is moot. Dry conditions persist in the central and Western Sahara and 'seasonal' trends are not identified. Harada et al. (2003) linked rainfall patterns to zonal winds, in which rainfall in the northern fringes of the Sahara are controlled by strength of the westerlies and Mediterranean-derived humidity, whereas the East African Jet controls rainfall patterns both seasonally and interannually in the southern Sahara and Sahel. This type of analysis based on TRMM data has mainly focused on semiarid (Sahelian) areas, however, where there are close links to vegetation and agriculture (e.g. Cho et al. 2015; Biasutti 2019), rather than the driest regions of the central Sahara. As such, it is these driest regions that remain least well known.

These remote sensing data have also been set into a wider climatological context, based mainly on Global Circulation Models (GCMs) and Medium-Range Weather Forecast models and reanalysis data. This shows that convective behaviour over the central Sahara is driven by the position and strength of the East African Jet, in particular where this interacts with mid-latitude depressions (Lavaysse et al. 2010; Schepanski and Knippertz 2011). These interactions are significant because they can give rise to enhanced dust export (Schepanski and Knippertz 2011) as well as localised rainstorm events due to air mass mixing (Vizy and Cook 2019). Several studies have used remote sensing to map areas affected by flood water inundation (e.g. Bryant and Rainey 2002; Schepanski et al. 2012; Medjani et al. 2017) but this is most applicable only to large flooded areas such as lake basins. The spatial resolution of most readily available satellite data means that smaller-scale events such as within confined wadi systems are not easily resolved.

11.2.2 Present-Day Rivers

Several studies have identified contrasting 'arid' and 'humid' fluvial phases in the Sahara region during different time periods of its history (Griffin 2002; Okonkwo et al. 2013), but this is an entirely false and unhelpful generalisation that does not account for the true hydroclimate of the region and the complex river system responses to climate forcing. For example, as a result of its negative water balance, today's central Sahara does not contain any perennial rivers (Walling 1996), but it is affected by rivers that flow into endorheic basins such as the Lake Chad basin, and by the Nile River that transverses through the eastern Sahara margins. All of these rivers are sourced from outside of the Sahara itself, mainly on highlands to the south and also the Atlas Mountains to the north-west (Gischler 1976). These exotic rivers are associated, along their channel margins, with wetlands and swamps that are areas of high biodiversity (and in places are Ramsar sites), but this also depends largely on channel belt sedimentology and regional climate. In sand-dominated areas along the Nile in particular, infiltration rates are high which, along with high evaporation rates and inward migration of active sand dunes, results in a very narrow riparian fringe. In more humid areas (Niger, Mali, Chad) with clayey substrates, moisture retention is much higher, resulting in extensive wetlands such as around Lake Chad itself or along sectors of the Niger River in Mali. There is very little information on the geomorphology and morphodynamics of these perennial extra-Saharan rivers (Gischler 1976).

By contrast, ephemeral or intermittent rivers that are active for only short periods of time, when there is intense rainfall and rapid overland flow, are the most geomorphically significant fluvial elements across much of the central Sahara (Megnounif et al. 2013; Abdel-Fattah et al. 2017) and triggered mainly by orographic rainfall over upland massifs. Overland flow is directed downslope under gravity and is thus controlled by surface topography and the presence of any pre-existing (dry) channels. As such, dry wadi systems are commonly reactivated by rainfall events. A wadi or wādī (وادي) is defined as a river channel systems that is bedrock defined, steep-sided, with a dry channel bed composed of loose sediments (Fig. 11.2a). The geometry, geomorphology and sedimentary properties of wadi systems of the Sahara have not been examined in detail, thus there is much that is not known about their dynamics, although they have been well-studied in Egypt and the Middle East (e.g. Abdel-Fattah et al. 2017; Moustafa et al. 2020). The absence of permanent water flow means that wadis do not form part of a regionally coherent drainage pattern, and when wadis are reactivated their water and sediment is rapidly dissipated, either through distributary footslope alluvial fans, or by evaporation, or by infiltration into the groundwater table (Shentsis and Rosenthal 2003; Saber et al. 2015). The higher orographic rainfall experienced over massifs means that ephemeral wadis are most likely to develop at the margins or lower slopes of mountain massifs, and may terminate in alluvial or colluvial fans or spreads in these locations. Based on field measurements in the Eastern Desert of Egypt, Foody et al. (2004) reported infiltration capacity rates of 0.07 cm h^{-1} for desert pavement substrates compared to 14.01 cm h^{-1} for the sandy bed of dry wadis. These values suggest that deflated summit and pediment surfaces, where desert pavements are found (Knight and Zerboni 2018), give rise to overland flow and rapid water transfer into wadi channels, but that water flow in wadi channels, once they accumulate water from the surrounding uplands is quickly lost through infiltration. Further, this suggests that there is rapid sediment deposition within the channel and limited sediment export from these footslope locations. The few studies on wadi sediment systems

Fig. 11.2 Photos of **a** dry wadi bed and **b** oasis fringed by vegetation in the Libyan Fazzan (*Photos* Stefania Merlo)

show that, when in flood, they experience high velocities, turbulence and suspended sediment transport, followed by rapid sediment deposition as bars or crevasse splays during waning flow phases (Balescu et al. 1998; Elfeki et al. 2020). A study of wadi flood and sediment dynamics from NW Algeria showed that sediment yield varied substantially over time and was largely dependent on antecedent conditions and thus sediment availability following the prior flood (Megnounif et al. 2013). This also suggests that flood magnitude by itself is not always indicative of its potential geomorphic impacts.

11.2.3 Groundwater Systems

Groundwater aquifers in the Sahara have been vitally important historically for the development and maintenance of ecosystems and human livelihoods because they provide assured water supplies in areas where the groundwater table naturally intersects with the land surface, forming springs and oases (Allan 1984). Sandstone strata of different ages, in particular the Nubian Sandstone found across the central and eastern Sahara including in the Kufrah, Murzuq and Chad basins, form aquifer source rocks (Gischler 1976; Lloyd 1990). Long-term development of perennial standing water bodies in the form of sand- or rock-bounded lake basins or ponds (oases and gueltas) takes place at low points in the landscape where the groundwater table is stable or groundwater flow can take place under artesian pressure (Fig. 11.2b). These surface water bodies are not well studied in terms of their physical and chemical properties, although they are critical to human occupation and are centres of biodiversity (Rzòska 1984; Alfarrah et al. 2016) and with endemic fish, amphibians, reptiles and mammals. For examples, gueltas in Mauritania cover only 43 ha in area but represent 32% and 78% of total taxa and endemic taxa, respectively (Vale et al. 2015).

Several indirect (remote sensing) methods for mapping groundwater locations and properties are also commonly used. Remote sensing using satellite gravity measurements or dielectric methods can image aquifer volume and depth (Leblanc et al. 2007; Bersi and Saibi 2020; Frappart 2020; Gonçalvès et al. 2020). These studies can help identify the spatial extent and depth of subsurface groundwater bodies, and therefore can identify the source aquifer for these bodies. Different large-scale transboundary aquifers can be identified, based mainly on their stratigraphic disposition. Frappart (2020) compared gravity measurements from the Gravity Recovery and Climate Experiment (GRACE) satellite mission for the period 2003–2016 and showed that the Tindouf Aquifer System and the North Western Sahara Aquifer System (NWSAS) are highly correlated with climate ($r=0.74$ and 0.67, respectively). The Nubian Sandstone Aquifer System experienced the greatest water loss through the period (50 km^3) and is attributed to water abstraction by population growth and increased irrigation demand in the eastern Sahara in Egypt. Scanlon et al. (2022) examined the GRACE record in the period 2002–2020 and showed that total groundwater storage decreased by 60–73 km^3 for aquifers in North Africa, mainly because of groundwater abstraction from the Nubian Sandstone and NWSAS. In addition to these large-scale aquifers, there are also smaller natural depressions, termed hollow aquifers, where shallower Quaternary fluvial, lacustrine and aeolian sediments are the host rocks (Leblanc et al. 2007). Direct water sampling from these aquifers allows for evaluation of water chemistry, flow rates/directions and contributions to surface waters (Alfarrah et al. 2016; Nadhira and Omar 2019). These analyses show that the nature and concentrations of dissolved solutes, and other water properties such as salinity, electrical conductivity and saturation index, vary significantly over space, mainly related to specific groundwater flow paths, host rock properties and local anthropogenic activities (Eddine and Belgacem 2019). Groundwater chemistry classifies the water as excellent to good quality (Azlaoui et al. 2021).

Groundwater recharge in the Sahara basin is derived mainly by precipitation. Gonçalvès et al. (2013) estimated a natural recharge rate of 1.40 ± 0.90 $km^3\ yr^{-1}$ for the NWSAS in Algeria, including the sedimentary basins occupied today by the Grand Erg Oriental and Grand Erg Occidental. This value broadly matches with contemporary rainfall in the area with a time lag of a year and corresponds to an overall groundwater renewal rate of around 40%. This suggests that, at least in this part of the Sahara, groundwater activity is more dynamic and volumetrically more variable than previously thought. Several studies have also use isotopic methods ($\delta^{18}O$ and ^{14}C values) to calculate the age of water molecules found within the groundwater system, which can inform on the timing of wetter climate periods when groundwater was being replenished. For example, in central Algeria within the NWSAS, radiocarbon dating of groundwater yields late Pleistocene ages (19–13 ^{14}C kyr BP) and cooler noble gas-derived palaeotemperatures, which together are interpreted to reflect recharge around this time period and/or the effects of water mass mixing (Darling et al. 2018). Sturchio et al. (2004) used ^{81}Kr and ^{36}Cl isotopic data from the Nubia aquifer, Egypt, to infer groundwater residence times of <1 million years.

11.3 Palaeohydrology, Palaeodrainage and Palaeorivers of the Central Sahara

The age and thus the longevity of groundwater bodies in the central Sahara suggests that past water processes may also be imprinted on Saharan landscapes in other ways, and many studies based on remote sensing data in particular show that this is indeed the case. There are two distinctive methodologies that have been used in this analysis. (1) Satellite radar data are able to identify the locations and broad-scale geometry of subsurface palaeochannels based on the ability of low frequency radar wavebands to penetrate through a dry sand cover and image the locations of coarser fluvial gravels underneath (Robinson et al. 2006; Ghoneim et al. 2012; Francke 2016; Paillou 2017). The resulting radar dataset allows for the mapping of palaeochannel patterns and hence reconstruction of past hydrological networks. Many studies have used radar to map the palaeohydrology of certain areas within the central Sahara (e.g. Robinson et al. 2006; Ghoneim and El-Baz 2007; Ghoneim et al. 2007, 2012; Paillou et al. 2009). (2) Analysis of land surface topography through Advanced Spaceborne Thermal Emission and Reflection Radiometer (ASTER) and Shuttle Radar Topography Mission (SRTM) data has been used to create digital elevation models

(DEMs), map watersheds, identify catchment characteristics, and therefore the potential for surface water systems to develop (Ghoneim and El-Baz 2007; Griffin 2011; Ghoneim et al. 2012; Paillou et al. 2012; Klokočník et al. 2017). This type of analysis therefore corresponds to the nature of the present-day land surface but it is not clear if this present topography is appropriate to reconstruct past watersheds and river patterns especially for areas affected by regional tectonic uplift over millions of year timescales. For these reasons, this approach, although technically easier, is less effective in informing on palaeohydrology.

Figure 11.3 is a composite illustration of the mapped palaeohydrological networks examined from the central Sahara region, as reported in the literature. A key element is that analysis of palaeohydrology has not been undertaken for all areas and that there are many spatial gaps, in particular in the central Sahara. In this region, there is however stratigraphic evidence for waterlain deposition and the development of both lake and river systems during the Neogene (e.g. Griffin 2006; Muftah et al. 2013; Hounslow et al. 2017; Klokočnik et al. 2017), but palaeodrainage patterns from specific outcrop locations are not clearly resolved. Mapped drainage patterns reported in the literature however are commonly directed into and through existing basins, such as the Sirte Basin, and thus linked to outflow into the Gulf of Sirte along the Mediterranean coast. This is the route taken by the palaeorivers Sahabi, Kufrah and Gilf Kebir (Griffin 2006; Ghoneim et al. 2012; Paillou et al. 2012). It is also likely that similar palaeorivers existed in other Saharan sectors, especially under a eustatic or tectonic change in base level (Goudie 2005). Blanchet et al. (2021) argued that these palaeorivers were particularly active during the interglacials of Marine Isotope Stages (MIS) 5 and 1, under precession-enhanced rainfall across the central Sahara, whereas during the glacials of MIS 4 and 3, rainfall was restricted to the Mediterranean coastal fringe. Another sector that has been examined is the contemporary drainage associated with the development of palaeo-Lake Chad which received water from the south but which may have, by exceeding the height of the Kufrah-Al-Kebir trough, flowed northwards into the Kufrah river system and thence to the Mediterranean (Ghoneim et al. 2012). This suggests that, during wetter (pluvial) climate phases, there may have been a more integrated trans-Saharan drainage network compared to present (Bussert et al. 2018). Many of the purported ancient drainage networks described in the literature are merely hypothetical, being based on DEMs of the present topography, and do not provide any evidence for rivers having flowed in topographic valleys. These studies are also at the very local scale and do not resolve well to the regional scale. For these reasons, Fig. 11.3 presents only a snapshot of elements of the palaeohydrology of the Sahara basin that have been reported in the literature, and there are many areas with no data.

Palaeolakes have also been proposed in many areas of the Sahara that are now covered by dune sand (Fig. 11.3). The evidence used for this interpretation comes from the presence of lake sediments and aquatic organisms

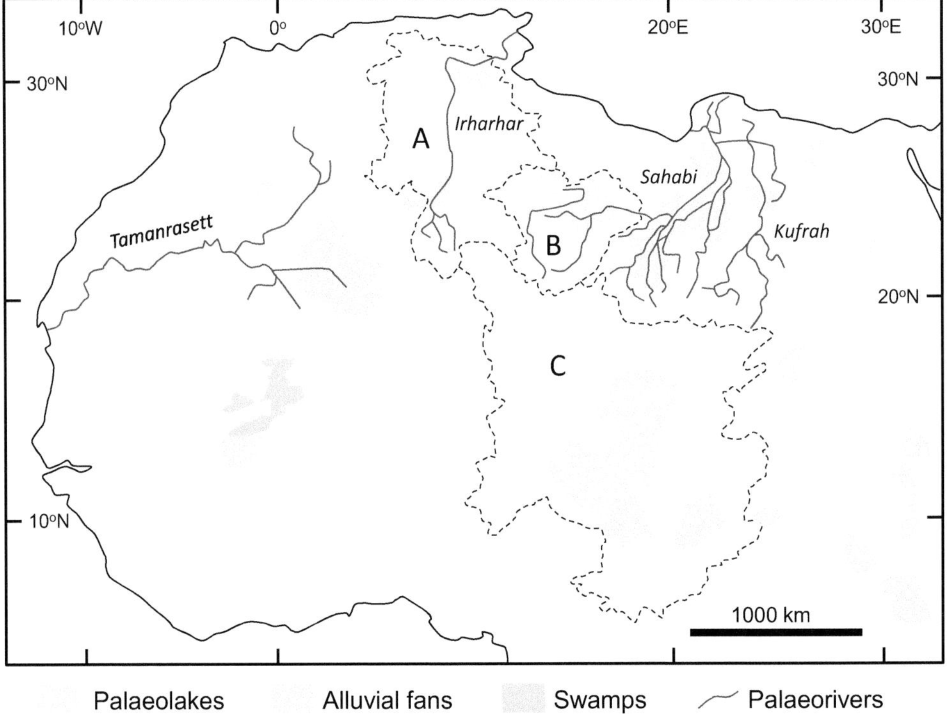

Fig. 11.3 Map of the palaeohydrology of the Sahara region, redrawn from Klokočník et al. (2017). Drainage basins for the endorheic palaeolakes are: A: The Chotts Megalake, B: Lake Megafazzan, C: Lake Megachad. The paths of the named palaeorivers are redrawn from Blanchet et al. (2021)

(Soulié-Märsche et al. 2010; Youcef and Hamdi-Aïssa 2014; Abdullah et al. 2017), radar or seismic data that indicate a depression infilled with fine sediments (Armitage et al. 2007; Drake et al. 2009; Klokočnik et al. 2017), interpretation of endorheic palaeodrainage patterns (Baumhauer 1991; Gaber et al. 2009), or by simply a 'gap' existing along a river channel (Ghoneim et al. 2012). Some of this evidence is not conclusive and the temporal phasing of different lake levels and positions is often unknown. In northeast Sahara, there is also the suggestion of drainage reversal driven by tectonics (Abdelkareem and El-Baz 2015), and this forcing factor may give rise to uncertainty in interpreting both river and lake patterns. Chorowicz and Fabre (1997) suggested that episodic tectonic uplift and faulting around the Tanezrouft and Hoggar areas, Western Sahara, led to drainage capture, although the potential timing of such event(s) is not clear.

The distribution of palaeochannels alone, from which river systems are mapped, however, does not necessarily fully inform on past hydrological patterns. This is because different channels may be of different incision ages, and existing channels may have been occupied during several different pluvial episodes throughout their history (Ghellali and Anketell 1991). In addition, the age of channel infills is not known with certainty. Only a few studies have examined fluvial sediments that correspond to palaeodrainage patterns in the Sahara. (There are more studies that have examined more general Neogene lake or basin-fill sediments; e.g. El-Shawaihdi et al. 2016; Drake et al. 2018.) Bussert et al. (2018) described Neogene channel-fill deposits from the far eastern sector of the Sahara system in Sudan. These sediments comprise fining-up successions of the Wadi Awatib Conglomerate that include imbricated to openwork conglomerate beds with minor sandstone interbeds. These sediments are interpreted as fluvial bars and sheets associated with traction or subaqueous density flows, grain flows or hyperconcentrated flows. However, the precise palaeogeography of this river stretch is not known (channel width, floodplain characteristics). Muftah et al. (2013) used mineral geochemistry to identify source areas of the fluvial As Sahabi Formation of north central Libya and related this to the course of the palaeo-Sahabi River (Fig. 11.3). This identified a sediment source area in northeast Chad, suggesting large-scale regional river system development. Even though there is limited understanding of fluvial sedimentation specifically related to palaeoriver networks, there has been no evaluation of the broader fluvial environments of these drainage systems such as the presence of meanders, terraces and floodplains or related soils or ecosystems. This is a limitation in the reconstruction of these palaeoenvironments. Griffin (2011) made some general statements of the geomorphic properties of the palaeo-Sahabi River, but this was based on SRTM data only and does not adequately resolve features buried by later blown sand. Francke (2016) showed, however, that buried channels can be well-mapped in the Murzuq Basin using ground-penetrating radar where the dry sand cover allows for radar penetration to 20–70 m depth, depending on local conditions.

11.3.1 Timing of Palaeofluvial Activity

At a regional scale and over long timescales, river responses may reflect eustatic and/or tectonic processes and events. For example, Miocene/Pliocene regression and transgression, respectively, associated with the Messinian Salinity Crisis, resulted in a change in Mediterranean base level and this may have changed incoming river profiles (Goudie 2005). Grimaud et al. (2014) examined the long profiles of West African rivers, marginal to the southern Sahara, to infer the long term history of river denudation. They suggested that denudation largely kept up with the rate of tectonic uplift of the Hoggar mantle swell, which may suggest that these presently active perennial rivers in West Africa are in broad equilibrium with respect to forcing. It may be anticipated that palaeorivers may show different patterns, although there has hitherto been no analysis of palaeoriver long profiles.

The detailed timing of initial development and then later reactivation of palaeoriver systems is not well known across the region. Some control is provided on the basis of broad, regional-scale stratigraphy, especially in a basinal context (e.g. Sirte Basin: El-Shawaihdi et al. 2016; Fazzan Basin, southern Libya: Hounslow et al. 2017) and where fluvial sediments are interbedded with K/Ar-dated lava flows (e.g. Fazzan Basin: Drake et al. 2008). Later phases of sediment deposition may be constrained by luminescence or radiocarbon dates, including where Holocene-age lake sediments, dating by luminescence, have been identified from below dune sands of the Murzuk Sand Sea in the Fazzan Basin (Drake et al. 2018). A regional stratigraphic framework for river system evolution in the Fazzan Basin was provided by Hounslow et al. (2017). Three major sediment units are identified, separated by unconformities and dated using magnetostratigraphy. The lower Shadirinah Formation including lacustrine and fluvial sediments also contains intraformational palaeosols, and this may potentially include overbank and floodplain environments, although this has not been specifically discussed as evidence for fluvial palaeogeography. Hounslow et al. (2017) suggest however that fluvial drainage into the Fazzan Basin was established by at least 23 Ma (start of the Miocene). The alternation of palustrine carbonates with fluvial and lacustrine sediments reflects variations in water availability and indicates an overall aridity trend, in particular after 11 Ma.

Other studies also suggest a similarly long timeframe of fluvial evolution in this region (e.g. Griffin 2006; Drake et al. 2008, 2018) although this evolution is not spatially or temporally well resolved. There is a general absence of radiometric age control on sediments contained within individual palaeochannels which, because of their buried nature, can only be recovered through coring or through limited field exposure. This is not a research strategy that has been widely applied and so it is not clear whether the existing age control is a useful reflection of channel activity. Thus, Fig. 11.3 shows channel spatial extent rather than the nature of channel evolution. There is much greater evidence (sedimentary, morphological, biological, dating) for pluvial activity during the African Humid Period (~11–5 kyr BP) (Tierney et al. 2017), and this period is discussed separately elsewhere (Knight, this volume).

11.4 Discussion

Water has always been a primary control on the morphodynamics and landscape evolution processes of the central Sahara, and the presence of palaeodrainage patterns attests to the preservation of this evidence in the landscape (Paillou 2017). This evidence is spatially (and likely temporally) patchy, but the existing reconstructions of regional watersheds show a correspondence to present topographic highs (e.g. Griffin 2006; Drake et al. 2008). Although this circular reasoning assumes that these high points existed at the time, palaeofluvial geochemical evidence (Muftah et al. 2013) generally support this hypothesis. This may in turn hint at the longevity of such areas in the landscape, and thus a potential low summit weathering rate in the central Sahara since the Miocene. This regional evidence (Fig. 11.3) also conceals any patterns of the invigoration or reoccupation of earlier river channels during pluvial periods and does not inform on smaller tributaries and ephemeral channels that would have also existed, or the presence of terraces, bars, meanders or levees. The development of such evidence for river system dynamics corresponds to pluvial time periods only and this would likely be separated by drier periods when river systems would be smaller, less active or absent. There is limited understanding of the spatial patterns of palaeowater availability that would impact on the presence or absence of river systems (deMenocal 1995). During the Quaternary, changes in the position of the Intertropical Convergence Zone or monsoon strength are the primary drivers of seasonal rainfall mainly in the southern Sahara Sahel (Biasutti 2019), but weather systems from the Mediterranean were also important in bringing rainfall into the northern Sahara. Areas of the central Sahara farthest from these moisture sources may have been affected by orographic rainfall, as they are today, but palaeoprecipitation patterns are largely unknown. This is a significant problem in interpretation of the climatic context of palaeodrainage patterns. This also means that there is little evidence of any sort for periods of low precipitation.

The role of palaeodrainage networks in regional sediment redistribution and geomorphic change has not been explored across the Sahara, but this will set fluvial processes into the wider context of landscape evolution. A model for long-term landscape evolution under changing palaeohydrology can include the development of topographic inversion relief (e.g. Zaki and Giegengack 2016; Williams et al. 2021). Here, clays and chemical precipitates within an active channel belt and its surrounding floodplain help cement coarse fluvial gravel deposits (Fig. 11.4a). This process is also enhanced under high evaporation rates typical of semiarid environments and creates a massive and more resistant indurated gravel body. Progressive weathering and erosion when the river system dries up under a more arid climate may produce an inverted relief, especially where deflation is active (Fig. 11.4b). Continued deflation may produce an upstanding gravel massif, either as a weathered done or as flat-topped mesas (Fig. 11.4c1), both of which have been noted in the central Sahara region (Bussert et al. 2018); or if there is accretion of either aeolian dune sand or footslope alluvial fans, the gravel body can become buried (Fig. 11.4c2). This has also been noted in the central Sahara region (e.g. Paillou 2017).

11.5 Summary and Outlook

The present and past hydrology of the central Sahara region is pivotal to a range of geomorphic processes and signatures. The preservation of palaeodrainage systems from the Miocene onwards, either on the surface or beneath a surficial sediment cover, is strong evidence for the longevity of these elements of the Saharan landscape. It is also likely that investigating the palaeohydrology of the region will yield better evidence for past climates than from more mobile and temporary landforms such as sand dunes. Palaeorivers are suggested to have been migration corridors for large animals (including hominids) (Drake et al. 2011; Coulthard et al. 2013), but the potential for fluvial sediments to contain fossils has not been investigated. Buried river gravels are also likely important as groundwater aquifers and as placer deposits, and hyperspectral remote sensing tools have potential for mapping heavy mineral anomalies, although this has also not been done. There are thus many areas of future research on Saharan palaeorivers, and involving many different types of datasets and with reference to geomorphology, sedimentology, dating and climate.

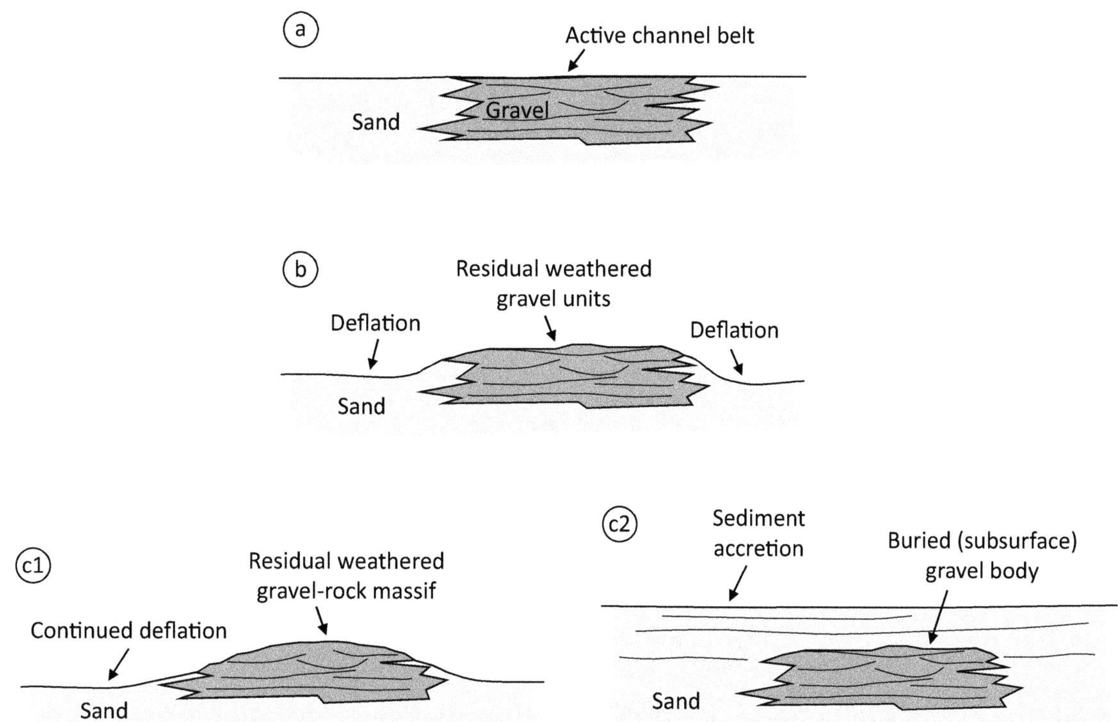

Fig. 11.4 Model for the development of inverted topographic relief in an arid fluvial environment in which there has been cementation of river gravels. Evolutionary stages (**a–c**) are explained in the text

References

Abdel-Fattah M, Saber M, Kantoush SA, Khalil MF, Sumi T, Sefelnasr AM (2017) A hydrological and geomorphometric approach to understanding the generation of wadi flash floods. Water 9:553. https://doi.org/10.3390/w9070553

Abdelkareem M, El-Baz F (2015) Evidence of drainage reversal in the NE Sahara revealed by space-borne remote sensing data. J Afr Earth Sci 110:245–257

Abdullah M, Betzler C, Frechen M, Sierralta M, Thiedig F, El Chair M (2017) Significance of calcretes for reonstruction of the Fezzan Megalake (Murzuq Basin, SW Libya). Z Dt Ges Geowiss 168:199–209

Abotalib AZ, Mohamed RSA (2013) Surface evidences supporting a probable new concept for the river systems evolution in Egypt: a remote sensing overview. Environ Earth Sci 69:1621–1635

Alfarrah N, Hweesh A, van Camp M, Walraevens K (2016) Groundwater flow and chemistry of the oases of Al Wahat. NE Libya. Environ Earth Sci 75:985. https://doi.org/10.1007/s12665-016-5796-x

Allan JA (1984) Oases. In: Cloudsley-Thompson JL (ed) Key environments: Sahara Desert. Permagon Press, Oxford, pp 325–333

Armitage SJ, Drake NA, Stokes S, El-Hawat A, Salem MJ, White K, Turner P, McLaren SJ (2007) Multiple phases of North African humidity recorded in lacustrine sediments from the Fazzan Basin, Libyan Sahara. Quat Geochronol 2:181–186

Azlaoui M, Zeddouri A, Haied N, Nezli IE, Foufou A (2021) Assessment and mapping of groundwater quality for irrigation and drinking in a semi-arid area in Algeria. J Ecol Eng 22:19–32

Balescu S, Breton J-F, Coque-Delhuille B, Lamothe M (1998) La datation par luminescence des limons de crue: une nouvelle approche de l'étude chronologique des périmètres d'irrigation antiques di Sud-Yémen. C R Acad Sci Paris, Sci Terre Planets 327:31–37

Baumhauer R (1991) Palaeolakes of the south central Sahara – problems of palaeoclimatic interpretation. Hydrobiologia 214:347–357

Bersi M, Saibi H (2020) Groundwater potential zones identification using geoelectrical sounding and remote sensing in Wadi Touil plain, Northwestern Algeria. J Afr Earth Sci 172:104014. https://doi.org/10.1016/j.jafrearsci.2020.104014

Biasutti M (2019) Rainfall trends in the African Sahel: characteristics, processes, and causes. WiRES Clim Ch 10:e591. https://doi.org/10.1002/wcc.591

Blanchet CL, Osborne AH, Tjallingii R, Ehrmann W, Friedrich T, Timmermann A, Brückmann W, Frank M (2021) Drivers of river reactivation in North Africa during the last glacial cycle. Nat Geosci 14:97–103

Bryant RG, Rainey MP (2002) Investigation of flood inundation on playas within the Zone of Chotts, using a time-series of AVHRR. Remote Sens Env 82:360–375

Bussert R, Eisawi AAM, Hamed B, Babikir IAA (2018) Neogene palaeochannel deposits in Sudan–Remnants of a trans-Saharan river system? J Afr Earth Sci 141:9–21

Cho J, Lee Y-W, Lee H-S (2015) The effect of precipitation and air temperature on land-cover change in the Sahel. Water Env J 29:439–445

Chorowicz J, Fabre J (1997) Organization of drainage networks from space imagery in the Tanezrouft plateau (Western Sahara): implications for recent intracratonic deformations. Geomorphology 21:139–151

Cloudsley-Thompson JL (ed) (1984) Key environments: Sahara Desert. Oxford, Permagon Press, p 348

Coulthard TJ, Ramirez JA, Barton N, Rogerson M, Brücher T (2013) Were rivers flowing across the sahara during the last interglacial?

Implications for human migration through Africa. PLoS ONE 8:e74834. https://doi.org/10.1371/journal.pone.0074834
Darling WG, Sorensen JPR, Newell AJ, Midgley J, Benhamza M (2018) The age and origin of groundwater in the Great Western Erg sub-basin of the North-Western Sahara aquifer system: Insights from Krechba, central Algeria. Appl Geochem 96:277–296
deMenocal PB (1995) Pio-Pleistocene African climate. Science 270:53–59
Drake NA, El-Hawat AS, Turner P, Armitage SJ, Salem MJ, White KH, McLaren S (2008) Palaeohydrology of the Fazzan Basin and surrounding regions: the last 7 million years. Palaeogeogr Palaeoclimatol Palaeoecol 263:131–145
Drake N, White K, Salem M, Armitage S, El-Hawat A, Francke J, Hounslow M, Parker A (2009) DMP VIII: Palaeohydrology and palaeoenvironment. Libyan Stud 40:171–178
Drake NA, Blench RM, Armitage SJ, Bristow CS, White KH (2011) Ancient watercourses and biogeography of the Sahara explain the peopling of the desert. PNAS 108:458–462
Drake NA, Lem RE, Armitage SJ, Breeze P, Francke J, El-Hawat AS, Salem MJ, Hounslow MW, White K (2018) Reconstructing palaeoclimate and hydrological fluctuations in the Fezzan Basin (southern Libya) since 130 ka: a catchment-based approach. Quat Sci Rev 200:376–394
Eddine SS, Belgacem H (2019) Groundwaters hydro-chemical characterization of the Tagharist basin (northern region of Khenchela, south-eastern Algeria). Desaliniz Water Treat 140:356–364
Elfeki A, Masoud M, Basahi J, Zaidi S (2020) A unified approach for hydrological modeling of arid catchments for flood hazards assessment: case study of wadi Itwad, southwest of Saudi Arabia. Arab J Geosci 13:490. https://doi.org/10.1007/s12517-020-05430-7
El-Shawaihdi MH, Mozley PS, Boaz NT, Salloum F, Pavlakis P, Muftah A, Triantaphyllou M (2016) Stratigraphy of the Neogene Sahabi units in the Sirt Basin, northeast Libya. J Afr Earth Sci 118:87–106
Foody GM, Ghoneim EM, Arnell NW (2004) Predicting locations sensitive to flash flooding in an arid environment. J Hydrol 292:48–58
Francke J (2016) Mapping paleochannels in the Libyan Sahara with ground penetrating radar. In: 16th International conference on ground penetrating radar, pp 1–5. https://doi.org/10.1109/ICGPR.2016.7572655
Frappart F (2020) Groundwater storage changes in the major North African transboundary aquifer systems during the GRACE era (2003–2016). Water 12:2669. https://doi.org/10.3390/w12102669
Gaber A, Ghoneim E, Khalaf F, El-Baz F (2009) Delineation of paleolakes in the Sinai Peninsula, Egypt, using remote sensing and GIS. J Arid Env 73:127–134
Gearon E (2011) The Sahara. A cultural history. Oxford University Press, Oxford, p 264
Ghellali SM, Anketell JM (1991) The Suq al Jum'ah palaeowadi, an example of a Plio-Quaternary palaeo-valley from the Jabal Nafisah, northwest Libya. Libyan Stud 22:1–6
Ghoneim E, El-Baz F (2007) The application of radar topographic data to mapping of a mega-paleodrainage in the Eastern Sahara. J Arid Env 69:658–675
Ghoneim E, Robinson C, El-Baz F (2007) Radar topography data reveal drainage relics in the eastern Sahara. Int J Remote Sens 28:1759–1772
Ghoneim E, Benedetti M, El-Baz F (2012) An integrated remote sensing and GIS analysis of the Kufrah Paleoriver, Eastern Sahara. Geomorphology 139–140:242–257
Gischler CE (1976) Hydrology of the Sahara. Ecol Bull (Stockholm) 24:83–101
Gonçalvès J, Petersen J, Deschamps P, Hamelin B, Baba-Sy O (2013) Quantifying the modern recharge of the "fossil" Sahara aquifers. Geophys Res Lett 40:263–2678
Gonçalvès J, Deschamps P, Hamelin B, Vallet-Coulomb C, Petersen J, Chekireb A (2020) Revisiting recharge and sustainability of the North-Western Sahara aquifers. Reg Environ Ch 20:47. https://doi.org/10.1007/s10113-020-01627-4
Goudie AS (2005) The drainage of Africa since the Cretaceous. Geomorphology 67:437–456
Griffin DL (2002) Aridity and humidity: two aspects of the late Miocene climate of North Africa and the Mediterranean. Palaeogeogr Palaeoclimatol Palaeoecol 182:65–91
Griffin DL (2006) The late Neogene Sahabi rivers of the Sahara and their climatic and environmental implications for the Chad Basin. J Geol Soc Lond 163:905–921
Griffin DL (2011) The late Neogene Sahabi rivers of the Sahara and the hamadas of the eastern Libya-Chad border area. Palaeogeogr Palaeoclimatol Palaeoecol 309:176–185
Grimaud J-L, Chardon D, Beauvais A (2014) Very long-term incision dynamics of big rivers. Earth Planet Sci Lett 405:74–84
Harada C, Sumi A, Ohmori H (2003) Seasonal and year-to-year variations of rainfall in the Sahara Desert region based on TRMM PR data. Geophys Res Lett 30:1288. https://doi.org/10.1029/2002GL016695
Hounslow MW, White HE, Drake NA, Salem MJ, El-Hawat A, McLaren SJ, Karloukovski V, Noble SR, Hlal O (2017) Miocene humid intervals and establishment of drainage networks by 23 Ma in the central Sahara, southern Libya. Gondwana Res 45:118–137
Hugot H-J (1974) Le Sahara avant le désert. Éditions des Hespérides, Toulouse, p 343
Issawi B, Sallam ES (2017) Rejuvenation of dry paleochannels in arid regions: a geological and geomorphological study. Arab J Geosci 10:14. https://doi.org/10.1007/s12517-016-2793-z
Kelley OA (2014) Where the least rainfall occurs in the Sahara Desert, the TRMM radar reveals a different pattern of rainfall each season. J Clim 27:6919–6939
Klokočník J, Kostelecký J, Cílek V, Bezděk A, Pešek I (2017) A support for the existence of paleolakes and paleorivers buried under Saharan sand by means of "gravitational signal" from EIGEN 6C4. Arab J Geosci 10:199. https://doi.org/10.1007/s12517-017-2962-8
Knight J, Zerboni A (2018) Formation of desert pavements and the interpretation of lithic-strewn landscapes of the central Sahara. J Arid Env 153:39–51
Lavaysse C, Flamant C, Janicot S, Knippertz P (2010) Links between African easterly waves, midlatitude circulation and intraseasonal pulsation of the West African heat low. Quart J Roy Met Soc 136:141–158
Leblanc M, Favreau G, Tweed S, Leduc C, Razack M, Mofor L (2007) Remote sensing for groundwater modelling in large semiarid areas: Lake Chad Basin, Africa. Hydrogeol J 15:97–100
Le Houérou HN (1997) Climate, flora and fauna changes in the Sahara over the past 500 million years. J Arid Environ 37:619–647
Liu Z (2015) Evaluation of precipitation climatology derived from TRMM multi-satellite precipitation analysis (TMPA) monthly product over land with two gauge-based products. Climate 3:964–982
Lloyd JW (1990) Groundwater resources development in the eastern Sahara. J Hydrol 119:71–87
Medjani F, Aissani B, Labar S, Djidel M, Ducrot D, Masse A, Hamilton CM-L (2017) Identifying saline wetlands in an arid desert climate using Landsat remote sensing imagery. Application on Ouargla Basin, southeastern Algeria. Arab J Geosci 10:176. https://doi.org/10.1007/s12517-017-2956-6
Megnounif A, Terfous A, Ouiloon S (2013) A graphical method to study suspended sediment dynamics during flood events in the Wadi Sebdou, NW Algeria (1973–2004). J Hydrol 497:24–36

Moawad MB (2013) Analysis of the flash flood occurred on 18 January 2010 in wadi El Arish, Egypt (a case study). Geomat Nat Haz Risk 4:254–274

Moawad MB, Aziz AOA, Mamtimin B (2016) Flash floods in the Sahara: a case study for the 28 January 2013 flood in Qena, Egypt. Geomat Nat Haz Risk 7:215–236

Moustafa AA, Salman AA, Elrayes AE, Bredemeire M (2020) Classification of wadi system based on floristic composition and edaphic conditions in South Sinai, Egypt. Catrina 20:9–21

Muftah AM, Pavlakis P, Godelitsas A, Gamaletsos P, Boaz N (2013) Paleogeography of the Eosahabi River in Libya: new insights into the mineralogy, geochemistry and paleontology of Member U1 of the Sahabi Formation, northeastern Libya. J Afr Earth Sci 78:86–96

Nadhira S, Omar S (2019) Hydrogeochemical characterization of the Complexe Terminal aquifer system in hyper-arid zones: the case of wadi Mya Basin, Algeria. Arab J Geosci 12:793. https://doi.org/10.1007/s12517-019-4917-8

Okonkwo C, Demoz B, Onyeukwu K (2013) Characteristics of drought indices and rainfall in Lake Chad Basin. Int J Remote Sens 34:7945–7961

Paillou P (2017) Mapping palaeohydrography in deserts: contribution from space-borne imaging radar. Water 9:194. https://doi.org/10.3390/w9030194

Paillou P, Schuster M, Tooth S, Farr T, Rosenqvist A, Lopez S, Malezieux J-M (2009) Mapping of a major paleodrainage system in eastern Libya using orbital imaging radar: The Kufrah River. Earth Planet Sci Lett 277:327–333

Paillou P, Tooth S, Lopez S (2012) The Kufrah paleodrainage system in Libya: a past connection to the Mediterranean Sea? Comptes Rendus Geosci 344:406–414

Robinson CA, El-Baz F, Al-Saud TSM, Jeon SB (2006) Use of radar data to delineate palaeodrainage leading to the Kufra Oasis in the eastern Sahara. J Afr Earth Sci 44:229–240

Rzòska J (1984) Temporary and other waters. In: Cloudsley-Thompson JL (ed) Key environments: Sahara Desert. Permagon Press, Oxford, pp 105–114

Saber M, Hamaguchi T, Kojiri T, Tanaka K, Sumi T (2015) A physically based distributed hydrological model of wadi system to simulate flash floods in arid regions. Arab J Geosci 8:143–160

Scanlon BR, Rateb A, Anyamba A, Kebede S, MacDonald AM, Shamsudduha M, Small J, Sun A, Taylor RG, Xie H (2022) Linkages between GRACE water storage, hydrologic extremes, and climate teleconnections in major African aquifers. Environ Res Lett 17:014046. https://doi.org/10.1088/1748-9326/ac3bfc

Schepanski K, Knippertz P (2011) Soudano-Saharan depressions and their importance for precipitation and dust: a new perspective on a classical synoptic concept. Quart J Roy Met Soc 137:1431–1445

Schepanski K, Wright TJ, Knippertz P (2012) Evidence for flash floods over deserts from loss of coherence in InSAR imagery. J Geophys Res 117:D20101. https://doi.org/10.1029/2012JD017580

Shentsis I, Rosenthal E (2003) Recharge of aquifers by flood events in an arid region. Hydrol Procs 17:695–712

Slimani M, Cudennec C, Feki H (2007) Structure du gradient pluviométrique de la transition Méditerranée-Sahara en Tunisie: determinants géographiques et saisonnalité. Hydro Sci J/J Sci Hydro 52:1088–1102

Soulié-Märsche I, Bieda S, Lafond R, Maley J, M'Baitoudji VPM, Faure H (2010) Charophytes as bio-indicators for lake level high stand at "Trou au Natron", Tibesti, Chad, during the Late Pleistocene. Glob Planet Ch 72:334–340

Sturchio NC, Du X, Purtschert R, Lehmann BE, Sultan M, Patterson LJ, Lu Z-T, Müller P, Bigler T, Bailey K, O'Connor TP, Young L, Lorenzo R, Becker R, El Alfy Z, El Kaliouby B, Dawood Y, Ahdallah AMA (2004) One million year old groundwater in the Sahara revealed by krypton-81 and chlorine-36. Geophy Res Lett 31:L05503. https://doi.org/10.1029/2003GL019234

Tierney JE, Pausata FSR, deMenocal PB (2017) Rainfall regimes of the Green Sahara. Sci Adv 3:e1601503. https://doi.org/10.1126/sciadv.1601503

UNESCO (1979) Map of the world distribution of arid regions. UNESCO, Paris, p 54

Vale CG, Pimm SL, Brito JC (2015) Overlooked mountain rock pools in deserts are critical local hotspots of biodiversity. PLoS ONE 10:e0118367. https://doi.org/10.1371/journal.pone.0118367

Vizy EK, Cook KH (2019) Understanding the summertime diurnal cycle of precipitation over sub-Saharan West Africa: regions with daytime rainfall peaks in the absence of significant topographic features. Clim Dyn 52:2903–2922

Walling DE (1996) Hydrology and rivers. In: Adams WM, Goudie AS, Orme AR (eds) The physical geography of Africa. Oxford University Press, Oxford, pp 103–121

Williams RME, Irwin RP III, Noe Dobrea EZ, Howard AD, Dietrich WE, Cawley JC (2021) Inverted channel variations identified on a distal portion of a bajada in the central Atacama Desert, Chile. Geomorphology 393:107925. https://doi.org/10.1016/j.geomorph.2021.107925

Youcef F, Hamdi-Aïssa B (2014) Paleoenvironmental reconstruction from palaeolake sediments in the area of Ouargla (Northern Sahara of Algeria). Arid Land Res Manag 28:129–146

Zaki AS, Giegengack R (2016) Inverted topography in the southeastern part of the Western Desert of Egypt. J Afr Earth Sci 121:56–61

12 Evolution and Geomorphology of Lake Chad

Jasper Knight

Abstract

Lake Chad, at the southernmost extent of the Sahara and located in the semiarid Sahelian zone, is the largest lake in the Sahara region and as such is extremely important in terms of its hydrological and ecological context and properties. However, the Lake Chad basin more widely has a complex hydrodynamic and geomorphic history reflecting its evolution in response to changes in regional climate, from the Neogene to Holocene. This chapter describes the evidence for changes in the hydrology and geomorphology taking place within the Lake Chad basin and then links this evidence to regional climate and environmental changes.

Keywords

Chad basin · Climate change · Hydrology · Shorelines · Miocene · Lake development · African humid period

12.1 Introduction

Lake Chad, an endorheic lake basin situated at the shared border between Niger, Nigeria and Chad, is located at the extreme southern margin of the Sahara. It is one of the most important inland lakes in Africa, by virtue of its large drainage basin and its ecosystems and habitats, being recognised internationally as a Ramsar site (Lemoalle et al. 2012). It has also changed dramatically in size and regional hydrologic influence throughout its geological history, and as such the geomorphology of Lake Chad can inform on these phases of development and the sensitivity of its water systems to changes in regional climate throughout the Neogene (23–2.6 million years ago) and Quaternary (Ghienne et al. 2002; Leblanc et al. 2006a).

The Lake Chad drainage basin, as defined by its surrounding topography, has an area of ~2.5 million km^2 in total and is comprised of two sub-basins (south and north) that correspond to the area feeding into the present-day Lake Chad, and the area that feeds into the topographic low existing in the northern part of the basin that was historically included within the maximal extent of the former mega-Lake Chad, respectively (Grove and Warren 1968; Lemoalle et al. 2012; Okpara et al. 2016) (Fig. 12.1b). The northern, now dry, basin is presently overdeepened by deflation and may only experience episodic water input by flash floods from the adjacent highlands of Ennedi and Tibesti (Gao et al. 2011). The modern Lake Chad basin, despite being topographically higher than the deflated basin to the north, is maintained by the perennial rivers of the Chari–Logone system which contribute ~40 million m^3 of water to Lake Chad annually from source areas in northern Nigeria and southern Chad (Gischler 1976). These rivers show peak discharge in the boreal summer associated with the West African monsoon (Armitage et al. 2015) (Fig. 12.2). By contrast, there are no perennial rivers that flow into the northern part of the basin, and any dry channels present may be only activated episodically by individual rainstorms, such as the Komodugu–Yobe system (Adeyeri et al. 2019). It is estimated that from the lake and these river subcatchments in total, around 122 million m^3 of water is lost by evapotranspiration annually (Gischler 1976). Thus, the differing climatic contexts of the northern and southern parts of the Lake Chad catchment as a whole are strongly reflected in the nature and dynamics of their contributing river systems (Bouchez et al. 2016). The lowest elevation connection between these two sub-basins is through the Bahr el Ghazal valley, which is at 287 m asl. (The lowest part of the basin today is at around 150 m asl, within the Bodélé depression.) Therefore, if Lake Chad

J. Knight (✉)
School of Geography, Archaeology & Environmental Studies, University of the Witwatersrand, Johannesburg 2050, South Africa
e-mail: jasper.knight@wits.ac.za

J. Knight et al. (eds.), *Landscapes and Landforms of the Central Sahara*, World Geomorphological Landscapes, https://doi.org/10.1007/978-3-031-47160-5_12

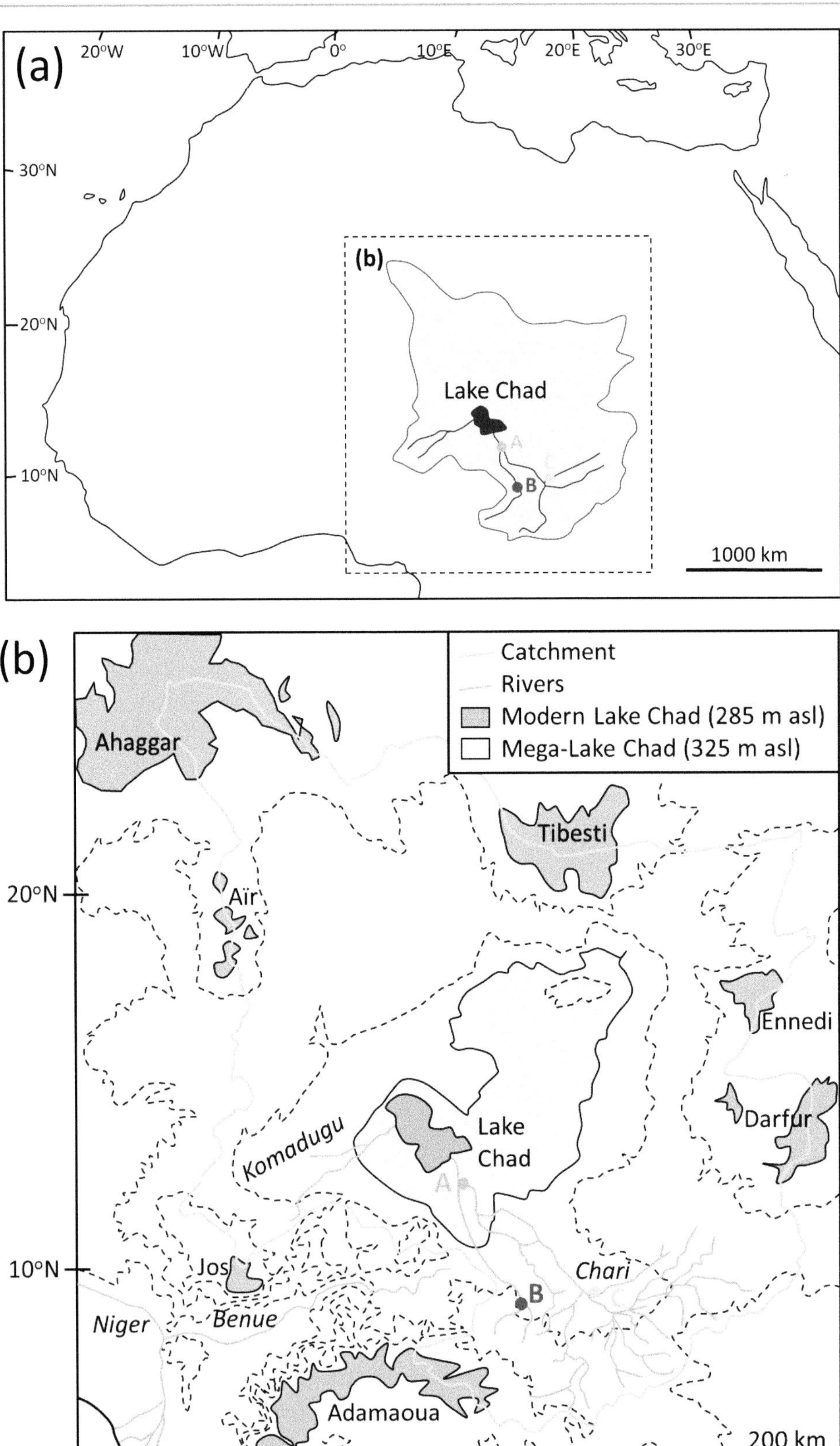

Fig. 12.1 **a** Location of Lake Chad in the south central Sahara and its catchment (shaded blue), based on Gischler (1976) and Schuster et al. (2009), including the perennial Chari–Logone river system; **b** detailed map of the Lake Chad catchment showing the position of the present and palaeo mega-Lake Chad. Contours at 200 m intervals are shown by the dotted lines. Massifs over 1000 m asl are named and shaded grey. River gauging stations A–C (Fig. 12.2) are labelled

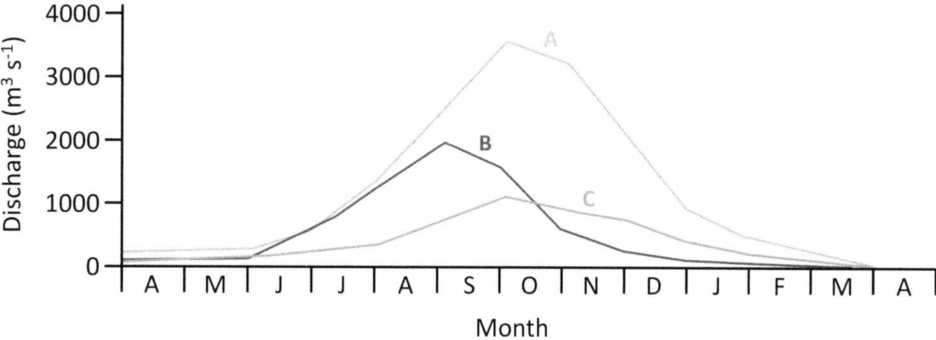

Fig. 12.2 Average monthly river discharge along the Chari–Logone river system, Lake Chad basin, over the hydrological years 1947–1965 (redrawn from Gischler 1976). Gauging stations A–C are marked on Fig. 12.1

water levels exceed this critical threshold level, water will spill over into the northern sub-basin (Schuster et al. 2009).

Leading from this introduction to the area and its hydrology, this chapter now describes the evolution of the Lake Chad basin, and the significant changes in lake area as a result of changes in climate that have affected its geomorphic evolution. This variability highlights the fragile nature of the Lake Chad environment—where there are close links between water, geomorphology, ecosystems and human activity—and the need for careful and sustainable environmental management.

12.2 Geological Development of the Lake Chad Basin

The Lake Chad basin is an intracratonic sag basin developed on a basement of igneous and metamorphic rocks dating to the Pan-African orogeny at 750–550 Ma (Schuster et al. 2009; Moussa et al. 2016) and corresponding to part of several NW–SE-trending rifts through Chad and Niger (Schlüter 2008). The basin is topographically constrained by upland massifs to all sides; thus, this is a topographic as well as a tectonic basin. These massifs define the watershed of the Lake Chad hydrological basin (Fig. 12.1b). Within this basin is > 500 m thickness of Neogene and Quaternary terrestrial sandy sediments, with lake sediments present in the area after the late Miocene. The long-term environmental and geological history of the Chad Basin is contextualised by Sylvestre et al. (2018), based on a 673 m-long core of sediments from the deepest part of the basin that sit on top of the metamorphic basement. This record comprises coarse sandstone at base, fining upwards, and then interbedded with claystone and diatomite. The clay mineral content suggests erosion of palaeosols around palaeolake margins (Moussa et al. 2016). Diatomite dominates in the period 6–2 Ma with aeolian sands thereafter. The main period of development of the Chad Basin relevant to the development of Lake Chad as is known today is reflected in the Neogene Chad Formation. This comprises a range of fluvial, deltaic, lacustrine and shoreface sediments (Shettima et al. 2018) indicative of migrating river channels (Boboye and Akaegbobi 2012). Faunal evidence in the Lake Chad basin in this period was presented by Schuster et al. (2009). This period was associated with wetter climates with soil development and containing a range of fossil faunal evidence, including fish, reptiles and birds indicative of active and varied aquatic and palustrine ecosystems, and reflects a lake extent larger than today.

This evidence shows that lake and associated lakeshore riverine and wetland environments existed in concert with the subsidence of the Chad Basin and over long time periods and before the main phase of mega-Lake Chad during the late Quaternary. Part of the deflation history of mega-Lake Chad and development of the Bodélé depression in its centre is discussed in more detail by Bristow (this volume).

12.3 Climate and Environmental Changes in the Lake Chad Basin

The key time period for development of mega-Lake Chad that is reflected in the geomorphic and sedimentary record seen in the landscape today corresponds to the African Humid Period (AHP) that existed between 14,800 and 5500 BP (deMenocal et al. 2000). This period was characterised by increased rainfall across the central Sahara region associated with intensification of the African monsoon (Shanahan et al. 2015). This in turn was triggered by insolation and ice-sheet forcing from the northern hemisphere (deMenocal et al. 2000; Timm et al. 2010). Speleothem records suggest the African monsoon extended as far north as southwest Morocco (31°N) (Sha et al. 2019). The nature of climatic variability during the AHP is not known for certain, largely due to a lack of evidence for most of this period (see Knight, this volume). Ferricrete, dune, alluvial and colluvial stratigraphy in SW Niger reflects variations in moisture availability and overbank deposition from the Niger River (Rendell et al. 2003). Luminescence dating of dune activity at the southern Saharan edge shows increased activity in different sectors at different times during the AHP (Bristow and Armitage 2016). It is notable that such evidence is from locations marginal to the driest region of the Sahara. Most preserved geomorphic evidence relates to the later part of the AHP, in which preservation was enhanced by the transition to more arid conditions during the early Holocene. It is also notable that lacustrine conditions existed prior to the AHP in the Wādi ash Shāṭi', southern Libya, in the period 26.5–22.5 kyr BP (Petit-Maire et al. 1980), which may suggest a southward migration of wetter conditions during the AHP. The Sahara was largely covered by vegetation during the AHP, giving rise to the concept of the 'Green Sahara' (Tierney et al. 2017) in which vegetation migrated across the wetter landscape (Watrin et al. 2009). There are several sites across the central Sahara where AHP vegetation archives are preserved, and this can show zonal patterns of biomes associated with rainfall shifts (Hély et al. 2014).

The end of the AHP at around 5500 BP was thought to have been relatively abrupt as a result of changes in insolation forcing giving rise to retreat of rainfall across the central Sahara (deMenocal et al. 2000; Collins et al. 2017), but other data suggest that this period was more transitional with centennial-scale wet and dry phases (Renssen et al. 2003, 2006). These mixed results can be interpreted spatially, with changes in rainfall taking place first in areas marginal to the zone influenced by monsoon-driven rainfall, but with the greatest shifts towards aridity taking place in central Saharan regions compared to areas in the wetter Sahel (Shanahan et al. 2015). This timing of the end of the AHP corresponds to the mid-Holocene thermal maximum (Berke et al. 2012) that may in turn have had implications for surface water availability. Such relationships are not explored in the present literature but these temporal patterns can only be explained in a spatial context. Different sand dune forms are present within the mega-Lake Chad footprint (Grove and Warren 1968, their Fig. 4), indicative of the dry lake bed being reworked by wind from the mid Holocene onwards.

Given that fluvial and lacustrine deposition took place before and after the AHP, it is likely that the Chad Basin and therefore development of Lake Chad itself persisted in different forms throughout this period. Development of the lake in this period is given a range of terms in the literature, including mega-lake or palaeolake, but these varied terms refer to the same thing. Quade et al. (2018) list the typical features of the geomorphology of large Saharan lake systems, and these features can be used as diagnostic criteria to identify the presence or absence of lakes at different elevations, corresponding to wetter or drier climatic periods. Key criteria include shoreline benches or erosional notches, sorted beach gravels, lake-deposited silts and deltas. Other landforms may include dune systems or spits formed by wind-driven longshore drift. Such features have been used in many studies as indicators of Lake Chad extent (e.g. Schuster et al. 2005; Drake and Bristow 2006; Leblanc et al. 2006a).

12.4 Phases of Development of Mega-Lake Chad

Different sequential phases of mega-Lake Chad are not well established through the lifetime of the lake. Several studies, however, have examined some specific phases based largely on analysis of digital elevation models (DEMs) of the region (Fig. 12.3). Here, remote sensing data, in particular high-resolution satellite radar topographic mission (SRTM) data, have been used to identify and map phases of lake development marked by shorelines that are recorded as laterally-extensive and contour-parallel benches. These studies also draw topographic profiles along which multiple benches can be identified (Ghienne et al. 2002; Leblanc et al. 2006b). The lateral continuity of such benches can confirm them to be palaeoshorelines. The highest such shoreline identified at 325 m asl yields a maximal area of mega-Lake Chad of 344,724 km^2, making it the largest lake in existence in the Sahara region during the Holocene (Quade et al. 2018). Lake elevation was controlled by an overflow channel in the Benue Trough (south-west part of the lake) at 320–325 m asl (Schuster et al. 2009; Armitage et al. 2015). This resulted in a maximum lake depth of 160 m. However, to be meaningful, dating is needed in order to position lake shorelines in time, by radiocarbon dating of aquatic organism shells or bones, or luminescence dating of shoreline sand (Leblanc et al. 2006b). Such dating evidence is sparse and therefore the detailed evolutionary history of mega-Lake Chad is not known with certainty. Armitage et al. (2015) presented some evidence for the growth and decay of the lake during the AHP, but with some phases of variability within the highstand part of the record (Fig. 12.4a). Lake stratigraphy and ostracod evidence from near the Bahr el Ghazal sill also supports a complex pattern of lake level fluctuations towards the end of the AHP (Bristow et al. 2018).

A range of geomorphic evidence is associated with lake shoreline positions, including a change in slope angle and continuity of benches (Ghienne et al. 2002), incoming incised river valleys, and pre-existing dunefields flattened by wave action (Leblanc et al. 2006a, b). Detailed mapping from SRTM data identifies sand spits and beach ridges located within cuspate forelands (Schuster et al. 2005; Drake and Bristow 2006). Longshore drift takes place as a result of seasonal winds, with fetch length increasing as the lake gets bigger. Wave refraction also takes place around these developing forelands. Thus, regional-scale coastal environments existed in the central Sahara during the Quaternary.

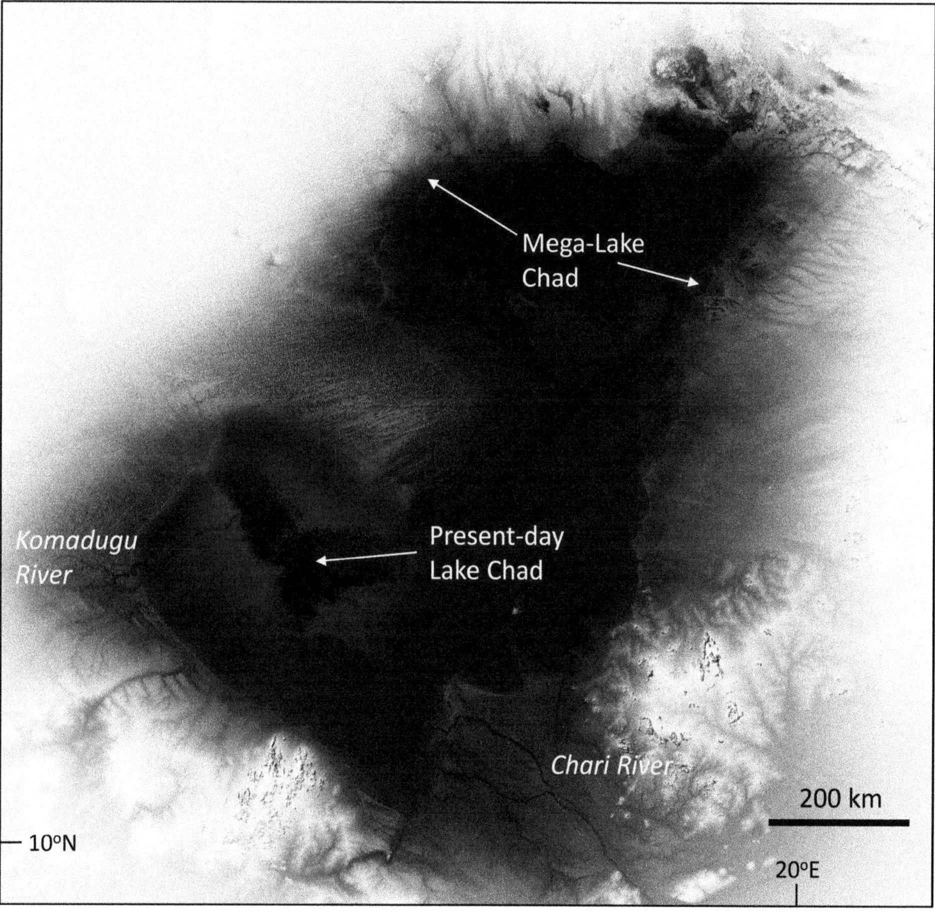

Fig. 12.3 Composite Landsat 8 image of Lake Chad (2000–2020) highlighting the position of today's Lake Chad and the zone falling within mega-Lake Chad (shaded dark). Image from NASA Earth Observatory (https://earthobservatory.nasa.gov/images/146304/remnants-of-an-ancient-lake)

A shift towards arid conditions in the central Sahara during the mid-Holocene led to progressive drying out of mega-Lake Chad and then exposure of the lake floor. Today, there are significant regional-scale patterns of dunes that indicate wind reworking of fine lacustrine sediments (Fig. 12.5), and this process has contributed to extensive deflation mainly in the northern part of the mega-Lake Chad basin and development of the Bodélé depression (Bristow et al. 2009). Spatial variations in dune forms may reflect different wind patterns, but more likely reflect spatial patterns of sediment grain size and availability. The regional mapping of dune forms in the Chad Basin undertaken by Grove and Warren (1968) using air photos (Fig. 12.5) shows denser and more complex dune patterns in the north-west of the basin (Bilma Sand Sea), where longitudinal and transverse dunes are superimposed. In this region, Callot and Crépy (2018) propose an evolutionary model in which pre-existing dunes cause a decrease in wind velocity that lead to superimposed secondary dunes being formed. It is also notable that crescentic dunes are found in the driest and finest-grained areas of the northern mega-Lake Chad basin, whereas transverse dunes are found within the coarser southern basin. Dunes are also preserved on the present Lake Chad lake floor (Fig. 12.5).

12.5 Wider Climatic and Geomorphic Evidence in the Chad Basin Region

Climate-driven changes in the hydrology of the Lake Chad basin during the late Pleistocene and Holocene are largely reflected within the wider region. For example, in the Bilma desert region near the Aïr massif and outside of mega-Lake Chad, a palaeolake existed between 10.7–6.0 ka BP with 7.2 m of lake sediments containing diatoms (Lécuyer et al. 2016). Endorheic depressions within the Bilma Sand Sea of eastern Niger developed into lake basins in particular during the early Holocene, fed by groundwater, and comprised of open water areas fringed by saltmarsh vegetation (Baumhauer 1991). Saltmarsh and associated evaporitic sabkha conditions existed because of the high potential evaporation rate, increasing salt content of the lake water. These lakes were several tens of km^2 in area and in some instances accumulated < 12 m of lake sediments. Sediments are mostly sands, but also some silts and clays were deposited within the lake, with saltmarsh/swamp organics, diatomite, and in marginal locations the presence of evaporitic crusts (calcrete, gypcrete) and ferricrete which indicate periodic changes in lake level (Grove and Warren 1968).

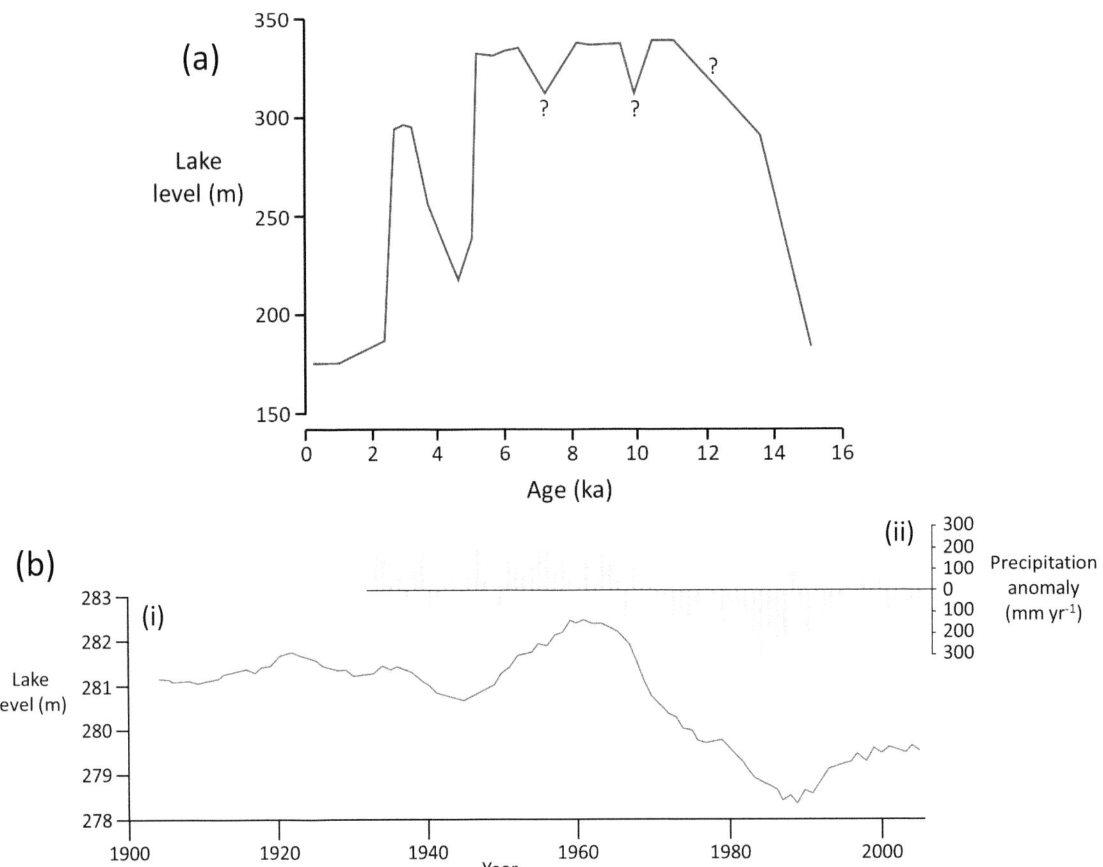

Fig. 12.4 Variations in Lake Chad water surface level, based on **a** dated geological evidence including the African Humid Period (Armitage et al. 2015), and **b (i)** instrumental, field and satellite data (1905–2005), redrawn from NASA Earth Observatory (2017), and **(ii)** precipitation anomaly in the Chari River basin (1933–2005), redrawn from Lemoalle et al. (2012)

Diatom assemblages reflect perennial deep lakes in the early Holocene that become shallower through to the mid-Holocene. In one core, Gramineae (Poaceae) pollen is present, possibly indicating human activity, around 7–6 ka. Within the wetter southern part of the Chad basin, overbank, channel-fill, floodplain and wetland deposits from the Komadugu River show luminescence ages that reflect greater river activity in the early and mid-Holocene, consistent with backing-up of water into the catchment during a highstand phase of Lake Chad (Gumnior and Preusser 2007). Likewise, palaeosols within dunes in the drier Niger part of the Chad basin also attest to wetter periods during the early to mid-Holocene (Felix-Henningsen 2000; Mauz and Felix-Henningsen 2005).

12.6 Historical Changes of Lake Chad

The water balance of Lake Chad reflects not only regional precipitation regime but also evapotranspiration loss and influence of vegetation and agriculture within the catchment. Furthermore, the surrounding sandy basin sediments promote water infiltration and thus lake level decline. Such evaporative water loss from Lake Chad can lead to significant changes in water chemistry, conductivity and water body stratification (Baumhauer 1991; Bouchez et al. 2016). Nicholson (1996) described climatic interpretations of the Bornu oral histories from the Lake Chad region from the period ~1500 AD onwards, comprising relatively wet conditions until droughts in the 1680s, then wetter conditions until ~1790s after which time the area became progressively more arid with famine conditions in the early/mid-nineteenth century. This climatic interpretation is supported by changes in lake levels and ecosystems that reflect changes in water availability within the basin.

The present Lake Chad has a surface area of ~18,000 km^2, maximum water depth of 12 m and volume of 7 km^3 (Adams 1996) (Fig. 12.3). Its shallowness means that it has an active benthic fauna and thus high benthic biomass (37 kg ha^{-1}), which is much higher than other African lakes (Adams 1996). The flat topography of this region means that only small changes in lake surface elevation can result in significant changes in lake extent: Schuster et al. (2009) used the example of a lake height decline from 284 to

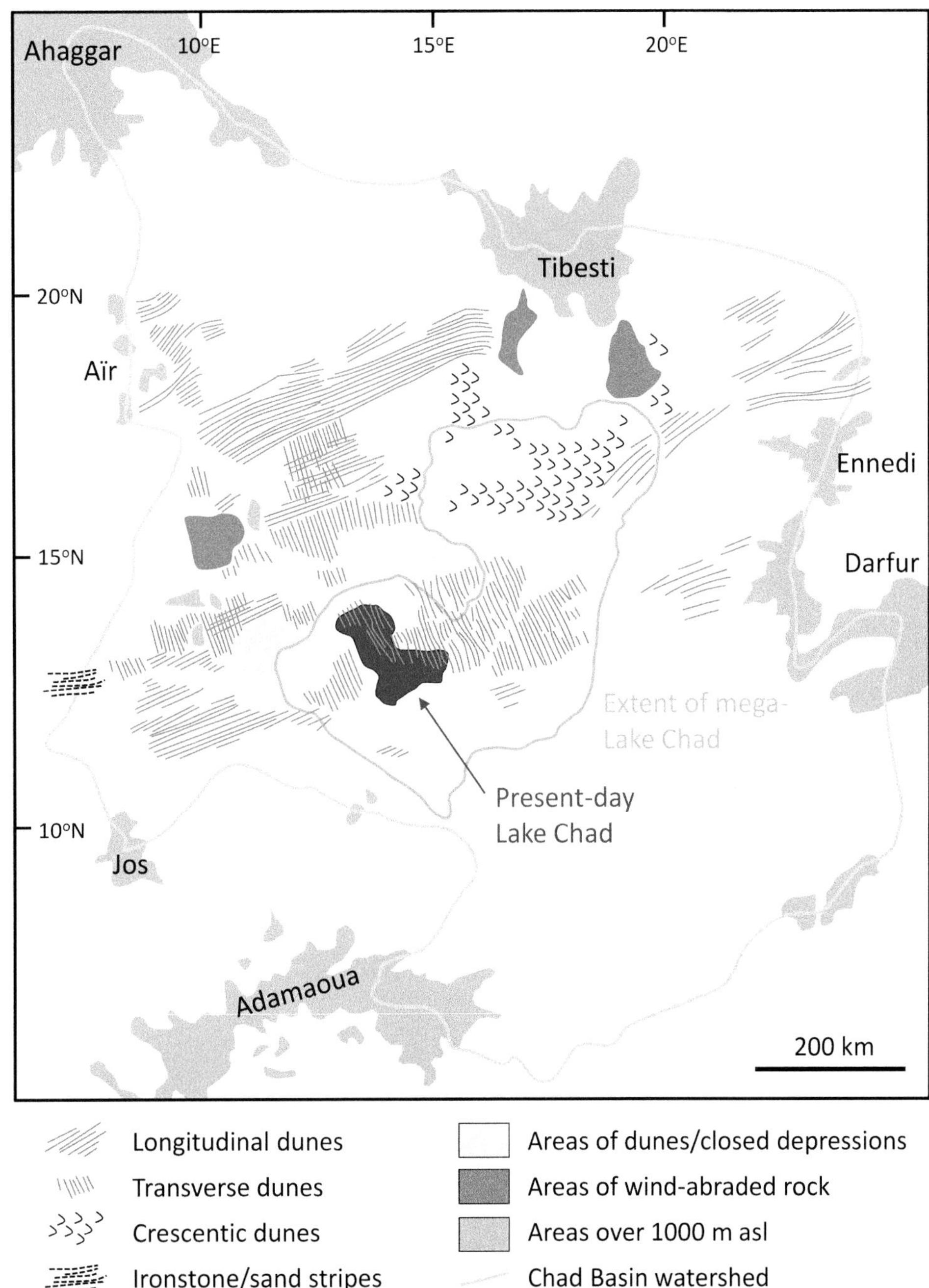

Fig. 12.5 Post-mega-Lake Chad aeolian geomorphology within the Lake Chad basin, modified and redrawn from Grove and Warren (1968, their Fig. 4)

280 m asl that results in a decreased lake area from 25,000 to 5000 km^2. The present-day lake level is ~279 m asl, likely linked to aquifer recharge, but has been undergoing a drying trend for the last decades (Fig. 12.4b) which is strongly driven by variations in the Atlantic Multidecadal Oscillation (AMO) in addition to the El Niño/Southern Oscillation (ENSO) that in turn influence monsoonal rainfall patterns. Recent changes (last few decades) in Lake Chad area reflect variations in precipitation within the catchment (Lemoalle et al. 2012) and potentially other factors such as irrigation (Gao et al. 2011). The water balance and controls on lake extent have been recently examined by Pham-Duc et al. (2020) who considered that Lake Chad today is relatively stable thanks to groundwater recharge within the catchment. Collectively, these factors are important today because ecological and hydrological conditions within the Lake Chad basin influence the viability of farming and impacts on migration and security—the region is a centre of the Islamist terror group Boko Haram (Okpara et al. 2016; Rudincová 2017). Climate models suggest decreases in rainfall and thus water input into Lake Chad over the next decades under a low emissions scenario, but

higher rainfall under higher emissions scenarios (Mahmood et al. 2020). This is likely accompanied by increasingly erratic rainfall patterns (Adeyeri et al. 2019).

12.7 Summary and Outlook

Lake Chad is an important element of today's African landscape, and is of wider regional hydrological and ecological significance. However, Lake Chad also has a complex and very long evolutionary history over at least the last 7 million years, with variations in lake area resulting from climate changes, in particular the spatial extent and strength of the West African monsoon. It was this driving factor that gave rise to the African Humid Period during which mega-Lake Chad was formed, with distinctive geomorphic signatures. The size and properties of the mega-Lake Chad basin can be explored using remote sensing imagery, from which different lake shorelines can be systematically identified. Changes in lake level also resulted in development of beaches, spits, deltas and other shoreline features that are relatively well preserved. This allows insight into past landscape dynamics at this southern edge of the central Sahara.

References

Adams WM (1996) Lakes. In: Adams WM, Goudie AS, Orme AR (eds) The physical geography of Africa. Oxford University Press, Oxford, pp 123–133

Adeyeri OE, Lawin AE, Laux P, Ishola KA, Ige SO (2019) Analysis of climate extreme indices over the Komadagu-Yobe basin, Lake Chad region: past and future occurrences. Weather Clim Extr 23:100194. https://doi.org/10.1016/j.wace.2019.100194

Armitage SJ, Bristow CS, Drake NA (2015) West African monsoon dynamics inferred from abrupt fluctuations of Lake Mega-Chad. PNAS 112:8543–8548

Baumhauer R (1991) Palaeolakes of the south central Sahara—Problems of palaeoclimatological interpretation. Hydrobiologia 214:347–357

Berke MA, Johnson TC, Werne JP, Schouten S, Sinninghe Damste JS (2012) A mid-Holocene thermal maximum at the end of the African Humid Period. Earth Planet Sci Lett 351–352:95–104

Boboye OA, Akaegbobi IM (2012) Sedimentological and palyno-environmental appraisal of the late Quaternary sediments, north-eastern Bornu Basin. Quat Int 262:14–19

Bouchez C, Goncalves J, Deschamps P, Vallet-Coulomb C, Hamelin B, Doumnang J-C, Sylvestre F (2016) Hydrological, chemical, and isotopic budgets of Lake Chad: a quantitative assessment of evaporation, transpiration and infiltration fluxes. Hydrol Earth Syst Sci 20:1599–1619

Bristow CS, Drake N, Armitage S (2009) Deflation in the dustiest place on earth: The Bodélé Depression, Chad. Geomorphology 105:50–58

Bristow CS, Armitage SJ (2016) Dune ages in the sand deserts of the southern Sahara and Sahel. Quat Int 410:46–57

Bristow CS, Holmes JA, Mattey D, Salzmann U, Sloane HJ (2018) A late Holocene palaeoenvironmental 'snapshot' of the Angamma Delta, Lake Megachad at the end of the African Humid Period. Quat Sci Rev 202:182–196

Callot Y, Crépy M (2018) Le rôle de l'alimentation en sable dans la dynamique des dunes linéaires: un changement de paradigme? Approche croisée des apports de la modélisation et des observations de terrain (Erg de Fachi-Bilma, Niger, Tchad). Géomorphol: Relief. Proc Environ 24:225–235

Collins JA, Prange M, Caley T, Gimeno L, Beckmann B, Mulitza S, Skonieczny C, Roche D, Schefuß E (2017) Rapid termination of the African Humid Period triggered by northern high-latitude cooling. Nat Comms 8:1372. https://doi.org/10.1038/s41467-017-01454-y

deMenocal P, Ortiz J, Guilderson T, Adkins J, Sarnthein M, Baker L, Yarusinsky M (2000) Abrupt onset and termination of the African Humid Period: rapid climate responses to gradual insolation forcing. Quat Sci Rev 19:347–361

Drake N, Bristow C (2006) Shorelines in the Sahara: geomorphological evidence for an enhanced monsoon from palaeolake Megachad. Holocene 16:901–911

Felix-Henningsen P (2000) Paleosols on Pleistocene dunes as indicators of paleo-monsoon events in the Sahara of East Niger. CATENA 41:43–60

Gao H, Bohn TJ, Podest E, McDonald KC, Lettenmaier DP (2011) On the causes of the shrinking of Lake Chad. Environ Res Lett 6:034021. https://doi.org/10.1088/1748-9326/6/3/034021

Ghienne J-F, Schuster M, Bernard A, Duringer P, Brunet M (2002) The Holocene giant Lake Chad revealed by digital elevation models. Quat Int 87:81–85

Gischler CE (1976) Hydrology of the Sahara. Ecol Bull (Stockholm) 24:83–101

Grove AT, Warren A (1968) Quaternary Landforms and climate on the south side of the Sahara. Geogr J 134:194–208

Gumnior M, Preusser F (2007) Late Quaternary river development in the southwest Chad Basin: OSL dating of sediment from the Komanbugu palaeofloodplain (northeast Nigeria). J Quat Sci 22:709–719

Hély C, Lézine A-M, APD contributors (2014) Holocene changes in African vegetation: tradeoff between climate and water availability. Clim Past 10:681–686

Lemoalle J, Bader J-C, Leblanc M, Sedick A (2012) Recent changes in Lake Chad: observations, simulations and management options (1973–2011). Global Planet Ch 80–81:247–254

Leblanc MJ, Leduc C, Stagnitti F, van Oevelen PJ, Jones C, Mofor LA, Razack M, Favreau G (2006a) Evidence for Megalake Chad, north-central Africa, during the late Quaternary from satellite data. Palaeogeogr Palaeoclimatol Palaeoecol 230:230–242

Leblanc M, Favreau G, Maley J, Nazoumou Y, Leduc C, Stagnitti F, van Oevelen PJ, Delclaux F, Lemoalle J (2006b) Reconstruction of Megalake Chad using Shuttle Radar Topographic Mission data. Palaeogeogr Palaeoclimatol Palaeoecol 239:16–27

Lécuyer C, Lézine A-M, Fourel F, Gasse F, Sylvestre F, Pailles C, Grenier C, Travi Y, Barral A (2016) I-n-Atei palaeolake documents past environmental changes in central Sahara at the time of the "Green Sahara": charcoal, carbon isotope and diatom records. Palaeogeogr Palaeoclimatol Palaeoecol 441:834–844

Mahmood R, Jia S, Mahmood T, Mehmood A (2020) Predicted and projected water resources changes in the Chari catchment, the Lake Chad Basin, Africa. J Hydrometeorol 21:73–91

Mauz B, Felix-Henningsen P (2005) Palaeosols in Saharan and Sahelian dunes of Chad: archives of Holocene North African climate changes. Holocene 15:453–458

Moussa A, Novello A, Lebatard A-E, Decarreau A, Fontaine C, Barboni D, Sylvestre F, Bourlès DL, Paillès C, Buchet G, Duringer P, Ghienne J-F, Maley J, Mazur J-C, Roquin C, Schuster M, Vignaud P, Brunet M (2016) Lake Chad sedimentation and environments during the late Miocene and Pliocene: new evidence from mineralogy and chemistry of the Bol core sediments. J Afr Earth Sci 118:192–204

NASA Earth Observatory (2017) The rise and fall of Africa's great lake. Available from https://earthobservatory.nasa.gov/features/LakeChad . Accessed 6 Apr 2018

Nicholson SE (1996) Environmental change. In: Adams WM, Goudie AS, Orme AR (eds) The physical geography of Africa. Oxford University Press, Oxford, pp 60–87

Okpara UT, Stringer LC, Dougill AJ (2016) Lake drying and livelihood dynamics in Lake Chad: unravelling the mechanisms, contexts and responses. Ambio 45:781–795

Petit-Maire N, Casta L, Delibrias G, Gaven C, Testud A-M (1980) Preliminary data on Quaternary palaeolacustrine deposits in the Wādi ash Shāṭi' area, Libya. In: Salem MJ, Busrewil MT (eds) The geology of Libya, vol III. Academic Press, London, pp 797–807

Pham-Duc B, Sylvestre F, Papa F, Frappart F, Bouchez C, Crétaux J-F (2020) The Lake Chad hydrology under current climate change. Sci Rept 10:5498. https://doi.org/10.1038/s41598-020-62417-w

Quade J, Dente E, Armon M, Ben Dor Y, Morin E, Adam O, Enzel Y (2018) Megalakes in the Sahara? A review. Quat Res 90:253–275

Rendell HM, Clarke ML, Warren A, Chappell A (2003) The timing of climbing dune formation in southwestern Niger: fluvio-aeolian interactions and the rôle of sand supply. Quat Sci Rev 22:1059–1065

Renssen H, Brovkin V, Fichefet T, Goosse H (2003) Holocene climate instability during the termination of the African Humid Period. Geophys Res Lett 30:1184. https://doi.org/10.1029/2002GL016636

Renssen H, Brovkin V, Fichefet T, Goosse H (2006) Simulation of the Holocene climate evolution in Northern Africa: The termination of the African Humid Period. Quat Int 150:95–102

Rudincová K (2017) Desiccation of Lake Chad as a cause of security instability in the Sahel region. GeoScape 11:112–120

Sha L, Ait Brahim Y, Wassenburg JA, Yin J, Peros M, Cruz FW, Cai Y, Li H, Du W, Zhang H, Edwards RL, Cheng H (2019) How far north did the African Monsoon fringe expand during the African Humid Period? Insights from Southwest Moroccan speleothems. Geophys Res Lett 46:14093–14102

Shanahan TM, McKay NP, Hughen KA, Overpeck JT, Otto-Bliesner B, Heil CW, King J, Scholz CA, Peck J (2015) The time-transgressive termination of the African Humid Period. Nat Geosci 8:140–144

Schlüter T (2008) Geological Atlas of Africa. Springer, Berlin, p 307

Schuster M, Roquin C, Duringer P, Brunet M, Caugy M, Fontugne M, Mackaye HT, Vignaud P, Ghienne J-F (2005) Holocene Lake Mega-Chad palaeoshorelines from space. Quat Sci Rev 24:1821–1827

Schuster M, Duringer P, Ghienne J-F, Roquin C, Sepulchre P, Moussa A, Lebatard A-E, Mackaye HT, Likius A, Vignaud P, Brunet M (2009) Chad Basin: Palaeoenvironments of the Sahara since the Late Miocene. C R Geosci 341:603–611

Shettima B, Kyari AM, Aji MM, Adams FD (2018) Storm and tide influenced depositional architecture of the Pliocene–Pleistocene Chad Formation, Chad Basin (Bornu Sub-basin) NE Nigeria: a mixed fluvial, deltaic, shoreface and lacustrine complex. J Afr Earth Sci 143:309–320

Sylvestre F, Schuster M, Vogel H, Abdheramane M, Ariztegui D, Salzmann U, Schwalb A, Waldmann N, ICDP CHADRILL Consortium (2018) The Lake CHAd Deep DRILLing project (CHADRILL)—targeting ~10 million years of environmental and climate change in Africa Sci Dril 24:71–78

Tierney JE, Pausata FSR, deMenocal PB (2017) Rainfall regimes of the Green Sahara. Sci Adv 3:e1601503. https://doi.org/10.1126/sciadv.1601503

Timm O, Köhler P, Timmermann A, Menviel L (2010) Mechanisms for the onset of the African Humid Period and Sahara greening 14.5–11 ka BP. J Clim 23:2612–2633

Watrin J, Lézine A-M, Hély C (2009) Plant migration and plant communities at the time of the "Green Sahara." C R Geosci 341:656–670

Geomorphology of the Bodélé Depression: The Dustiest Place on Earth

13

Charlie S. Bristow

Abstract

The Bodélé Depression is the site of a former lake, palaeolake Megachad, which filled the depression to a depth of 160 m in the early to middle Holocene but is now dry. The basin was carved out by wind, and strong winds continue to blow through the depression, eroding lake-bed sediments and making the Bodélé the dustiest place on Earth. The wind has sculpted the landscape creating yardangs and fast-moving sand dunes. Wind erosion has also exposed older, Miocene sediments that include important fossils of ancient hominins that used to live along the shores of an earlier lake.

Keywords

Palaeolake · Dust source · Wind erosion · Diatomite · Lake Chad · Harmattan · Monsoon

13.1 Introduction

In his account of his journey to Lake Chad, Nachtigal (1876, p. 407) described the Bodélé as "a system of shallow valleys, rich in little water holes and isolated moving dunes". He also describes it as "a fertile and pastoral plain" (Nachtigal 1876, p. 407), which makes it sound rather attractive. However, he also notes that, "this whole basin is scattered with fish bones, and seems to have been under water in the most recent times" and correctly infers that it must have been connected with (Lake) Chad (Nachtigal 1876, p. 408). While seasonal rains can turn parts of the area surrounding the Bahr el Ghazal into sparse grazing for camels and gazelle, there is little or no grazing to be had in the Bodélé today, and the few waterholes present are uninhabited and used only by nomadic camel herders. Giles (2005) described it as a place where no one else wants to be.

13.2 Location

The Bodélé Depression lies within northern Chad (Fig. 13.1). The depression, which is around 460 km long and 300 km wide and 160 m deep with an area of around 150,000 km^2, appears to have been carved out by wind erosion (Washington et al. 2006a), forming what is possibly the largest deflation basin on Earth. The base of the depression lies at an elevation of just under 170 m asl. The maximum elevation is less easy to define, but in the past, the depression was filled by a lake to a depth of around 160 m and the shorelines of that lake can be found at elevations between 220 and 230 m asl (Schuster et al. 2005; Drake and Bristow 2006). The lake shorelines provide a convenient way of defining the limits to the Bodélé in the north and in the east, but in the south and west the margins are less well defined. The western limits of the lake are partly covered by later sand dunes that have obscured the shoreline. To the south, the Bodélé Depression is linked to Lake Chad by the Bahr el Ghazal, the valley of a river which used to flow north from Lake Chad into the Bodélé when the lake level exceeded a sill at around 295 m elevation.

13.3 Geology

Erosion of the Bodélé Depression has exposed older Miocene sediments in outcrops along its southern side. These sediments appear to have been deposited in environments that were similar to those found in the Chad Basin during the Holocene and include aeolian and lacustrine deposits (Vignaud et al. 2002). The fossils of a wide range

C. S. Bristow (✉)
School of Natural Sciences, Birkbeck University of London, Malet Street, London WC1E 7HX, UK
e-mail: c.bristow@ucl.ac.uk

J. Knight et al. (eds.), *Landscapes and Landforms of the Central Sahara*, World Geomorphological Landscapes, https://doi.org/10.1007/978-3-031-47160-5_13

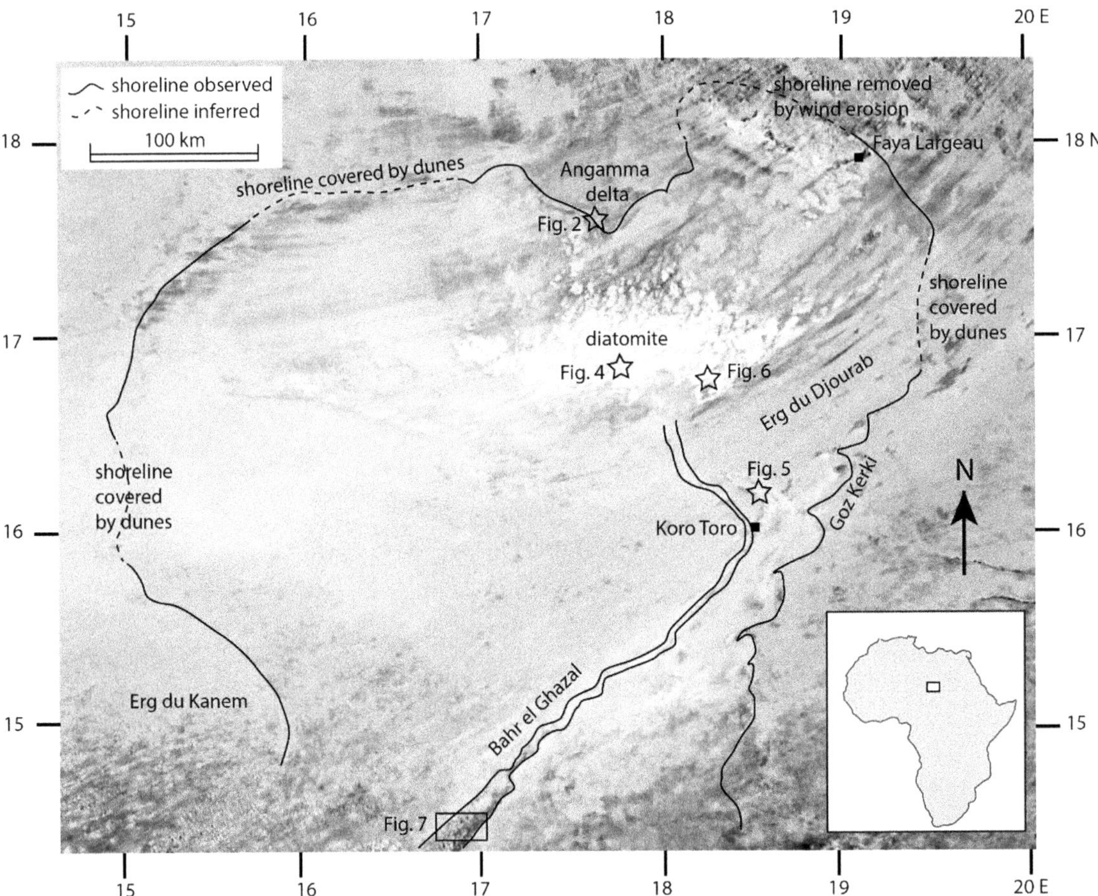

Fig. 13.1 Annotated satellite image showing the highest Holocene shorelines from the northern half of palaeolake Megachad which provides a boundary for the north, east and western limits of the Bodélé Depression. The southern boundary is less well defined but could be taken at the sill where water flows north from Lake Chad into the Bodélé Depression through the Bahr el Ghazal. The diatomite-rich sediments that cover the base of the depression show up as an area of high albedo (pale coloured area) in the middle of the image. Locations of field photographs in this article are shown by stars with the relevant figure number (*Source* Google Earth TM image Landsat/Copernicus)

of terrestrial and aquatic vertebrates have been discovered (Vignaud et al. 2002), as well as fossils of two ancient hominins, *Sahelanthropus tchadensis* (Brunet et al. 2002) known as Toumï, and the jaw bone of an australopithecine *Australopithecus bahrelghazali* known as Able (Brunet et al. 1995). Neither of the fossils was found in situ (Beauvilain 2008), and although Toumï appears to have been moved and probably reburied, it is believed to be around 7 million years old, while the australopithecine Able is around 3.5 million years old (Lebatard et al. 2008). The aeolian sediments contain sets of cross-strata that indicate a palaeowind direction from the NE (Schuster et al. 2006), a trade wind direction similar to the modern northeasterly harmattan wind. In addition, there are small exposures of Holocene and Pleistocene diatomite beds and lake sediments along the valley of the Bahr el Ghazal near Koro Toro (Servant and Servant 1970). The most dramatic outcrops in the basin are 30 m-high cliffs on the western side of the Angamma delta that form giant yardangs and exposed Holocene deltaic sediments (Bristow et al. 2018) (Fig. 13.2).

13.4 Ancient Lake Shorelines

The former shorelines of palaeolake Megachad are clearly visible on satellite images which have allowed the limits of the lake to be mapped (e.g. Schuster et al. 2005; Drake and Bristow 2006). The northern shoreline is dominated by the Angamma delta, a wave-dominated delta that was fed by rivers flowing south from the Tibesti Mountains (Schuster et al. 2005; Drake and Bristow 2006). The base of the Angamma delta sediments is underlain by volcanic breccias and tuffs (Servant et al. 1969). These are overlain by diatomite deposits with shells of *Pisidium* and *Valvata* that have been radiocarbon dated to 9260 ± 140 yr BP and 10,160 ± 160 yr BP (Servant et al. 1969). The diatomite is overlain by 20–30 m of interbedded sands, silts and clays from the delta front that were deposited between 7.5 and 4 thousand years ago, during the African Humid Period (Bristow et al. 2018). The top of the delta is marked by a low ridge of pebbles that is interpreted as a beach deposit

Fig. 13.2 Giant yardangs around 10 m high produced by wind erosion of the Holocene sediments of the Angamma delta, vehicle for scale

with an OSL age of 5.7 ± 0.3 ka (Armitage et al. 2015). This beach ridge is continuous and has not been cut by the river channels of the delta top; in fact, it appears to truncate river channels. The continuity of the beach ridge and the lack of fluvial incision have been interpreted to indicate that the river flow ceased before the lake level fell, supporting drying from the north (Armitage et al. 2015). River flow from the Tibesti had ceased prior to the mid-Holocene lake highstand, and the lake level was maintained by incoming rivers in the south, mainly the Chari River. The west side of the Angamma delta is deeply incised, with cliff sections and giant yardangs (Fig. 13.2). This erosion is attributed to a combination of groundwater sapping and wind erosion that have carved out steep sided canyons within the fine grained deltaic sediments after the lake waters receded.

At Goz Kerki, there are multiple shorelines that show spit development at different elevations along what was the eastern shore of the lake (Fig. 13.3). These spit systems were exposed to southwesterly monsoon winds as well as northeasterly harmattan winds (Schuster et al. 2005; Drake and Bristow 2006). Waves driven by the harmattan winds created longshore drift from northeast towards the southwest which formed spits, while waves driven by southwesterly monsoon winds produced a series of beach ridges (Drake and Bristow 2006). The formation of spits at different elevations is attributed to different water levels within palaeolake Megachad. OSL dating of beach deposits within the higher, 330 m shorelines (dark orange in Fig. 13.3), indicate that the lake level was at 330 m asl at 11.5 ± 0.9 ka and at 320 m asl at 6.6 ± 0.4 ka (Armitage et al. 2015). The lower shoreline at 295–310 m asl (pale orange on Fig. 13.3) might correlate

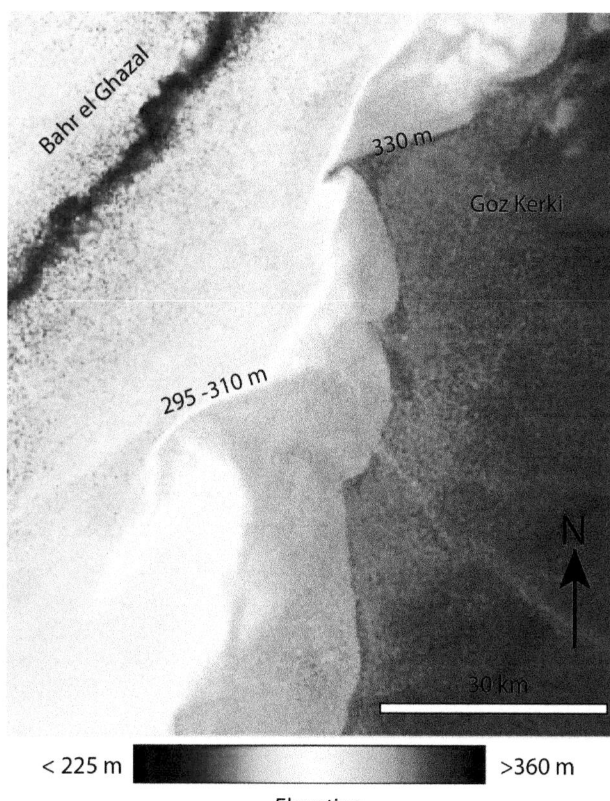

Fig. 13.3 Digital elevation model (m asl) produced by Nick Drake for part of the east side of the palaeolake at Goz Kerki showing a series of spits and embayments along the coastline. The coastal morphology indicates two sets of waves, from the northeast and the southwest, driven by harmattan and monsoon winds, respectively (Drake and Bristow 2006). The occurrence of shorelines at different elevations indicates past lake levels for palaeolake Megachad

with the 3000-year-old Ngelewa Ridge (elevation 295 m asl) on the southern side of Lake Chad. However, stratigraphic relationships suggest that this ridge might be older.

The western shore is partly covered by dunes migrating out of the depression, but can be traced into the Erg du Kanem where dunes were reworked by waves (Schuster et al. 2005; Drake and Bristow 2006). Sands excavated from the depression have probably contributed to the construction of the Erg du Kanem which lies downwind from the depression.

13.5 Lakebed Sediments

The former lakebed is covered by sediments, some of which were deposited within the lake, and others that have been blown into the depression by wind, or formed by reworking of lake sediments by wind. The lakebed sediments within the Bodélé show up clearly on satellite images due to the high albedo of very pale, almost white diatomite deposits (Fig. 13.1). Diatoms, which are unicellular algae with a siliceous cell wall called a frustule, flourished in the freshwaters of the palaeolake. When they died, the frustules accumulated on the lake bed creating the diatomite sediments, which also include fish remains (Fig. 13.4). At the outcrop scale, the diatomite is white to pale grey and locally covered by fields of granule ripples composed of eroded and windblown particles of diatomite sometimes mixed with quartz sand (Fig. 13.5). The pale coloured diatomite sediments have been mapped from satellite imagery and cover an area of around 7000 km^2 or less than 5% of the depression, while diatomite-rich sediments that have been reworked by the wind cover an additional 16,776 km^2 or 11% of the depression (Bristow et al. 2009). Most of the depression (65%) is covered by aeolian sands that have been blown into the depression from upwind (Bristow et al. 2009). These sand locally cover some areas of diatomite and, although the main areas of outcrop lie within the deeper parts of the depression (typically less than 190 m elevation), there are outcrops at higher elevations that show that the diatomite sediments were more extensive but were not eroded by the wind or covered by dune sands after the lake level fell.

The total thickness of the diatomite is not known, but outcrops in wind-sculpted yardangs show that was around 4 m thick in places (Fig. 13.6), presumably thinning towards the lake margins (Bristow et al. 2009). Erosion of the diatomite locally reveals underlying aeolian dune sands, some preserved as barchan-shaped moulds that have been unroofed by deflation of the diatomite. The contact between the diatomite and the dune sand is very sharp (Fig. 13.7), suggesting that flooding of the lakebed, which occurred around 15 ± 1.8 ka, was rapid (Armitage et al. 2015). Rapid flooding of the dunes is supported by the preservation of the barchan dune morphology, as well as a lack of pedogenesis at the contact. Beneath the lakebed sediments lies a series of gravel ridges that have been interpreted as exhumed river channels (Bristow et al. 2009), that are probably part of the Tertiary, Continental Terminal.

13.6 The Bahr el Ghazal

The Bahr el Ghazal is a river system that used to flow north from Lake Chad into the Bodélé Depression. The topographic divide between the two sub-basins lies at an elevation of around 295 m asl, approximately 15 m above the

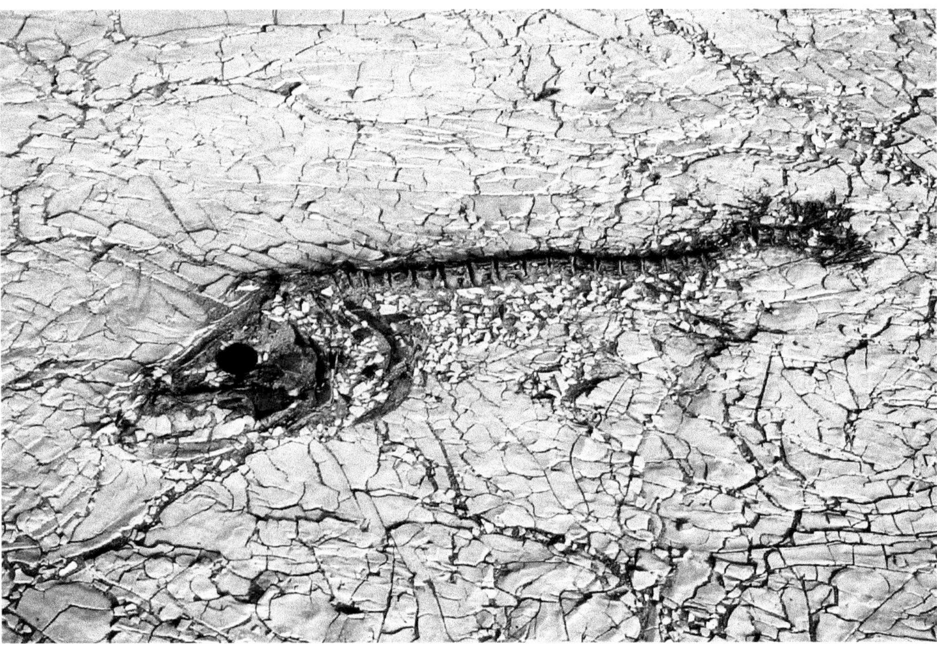

Fig. 13.4 Skeleton of a Nile Perch (*Lates niloticus*) preserved within the diatom-rich sediments of the lake bed. Lens cap for scale

Fig. 13.5 Typical scene of the desiccated lake sediments on the floor of the Bodélé Depression, pale grey to white diatomite is locally covered by fields of granule ripples composed of particles of diatomite eroded and transported by the wind

level of the modern Lake Chad. Nachtigal (1876, p. 407) quotes local people saying that "their grandfathers sailed to the north-east with boats, in expeditions of rapine or war". While it is questionable how much credence can be given to such oral histories, the satellite image (Fig. 13.8) shows that the geomorphology of the river deposits is very well preserved with the meandering river channel, point bars and scroll bar topography all visible where the river flowed between NE–SW-trending transverse dunes (Fig. 13.8). Servant and Servant (1970), who report radiocarbon ages of

Fig. 13.6 Four metres thickness of diatomite sediment exposed in a yarding near Koro Toro

Fig. 13.7 Sharp contact between yellow-coloured cross-stratified dune sands and overlying laminated diatomite indicates that the lake level rise was very rapid with little or no reworking of the dune and no pedogenic alteration of the dune before it was flooded by the lake 15,000 years ago, and covered by diatomite

1750 ± 105 yr BP for organic sediments and 140 ± 90 yr BP for shells from the Bahr el Ghazal, also support relatively recent flow through the Bar el Ghazal. While flow through the Bahr el Ghazal is usually attributed to overflow from Lake Chad, preserved channel patterns appear to show a connection to Lake Fitri. Farther north, near Koro Toro the river valley is incised exposing sections through Pleistocene sediments (Servant and Servant 1970).

13.7 Sand Dunes

There are two types of sand dunes within the Bodélé Depression, yellow-coloured and grey-coloured dunes (Fig. 13.9). The yellow dunes are dominated by quartz grains, appear to migrate into the basin from upwind and are common on the southern side of the depression where they pass into the Erg du Jourab. The grey-coloured dunes are largely composed of particles of diatomite with smaller amounts of quartz. Remote sensing studies of the dunes indicate that they are some of the fastest-moving dunes on Earth with typical migration rates of 20 to 30 m yr^{-1} and possibly over 50 m yr$-^{1}$ (Vermeesch and Leprince 2012).

13.8 Dust

The Bodélé Depression is the single biggest source of mineral dust on Earth (Prospero et al. 2002), due to the co-location of frequent, strong, near-surface winds, and a source of erodible fine-grained sediment (Warren et al. 2007). The origins of the dust were investigated in the Bodélé Dust Experiment (BoDEx) (e.g. Washington et al. 2006b; Todd et al. 2007; Warren et al. 2007; Chappell et al. 2008) (Fig. 13.10). Their major findings were that the Bodélé is a transport-limited system, where the dust output is primarily controlled by a powerful low-level jet that is topographically focussed into the Bodélé Depression (Washington and Todd 2005; Washington et al. 2006a). The dust is not derived from an accumulation of pre-existing, fine-grained (dust-sized) particles, but has to be created on site by the wind. Warren et al. (2007) hypothesised that the dust-sized particles are created by sand blasting, as well as by interactions between diatomite particles during transport by a process termed auto-abrasion, which has recently been confirmed by laboratory experiments (Bristow and Moller 2018). The barchan dunes on the lake bed act as dust mills, with flow convergence at the horns, and turbulence in the lee of the dunes adding to sediment transport and the production of dust. The northeasterly harmattan wind is focused into the Bodélé by a gap between the elevated topography of the Tibesti Mountains and the Ennedi Plateau (Washington et al. 2006a).

Dust from the Bodélé has been tracked across the Atlantic Ocean to South America (Ben-Ami et al. 2010; Yu et al. 2015) and potentially adds to the nutrients of the Amazon (Koren et al. 2006) and tropical Atlantic (Bristow et al. 2010), in part due to phosphate from fish that lived in the lake (Hudson-Edwards et al. 2014).

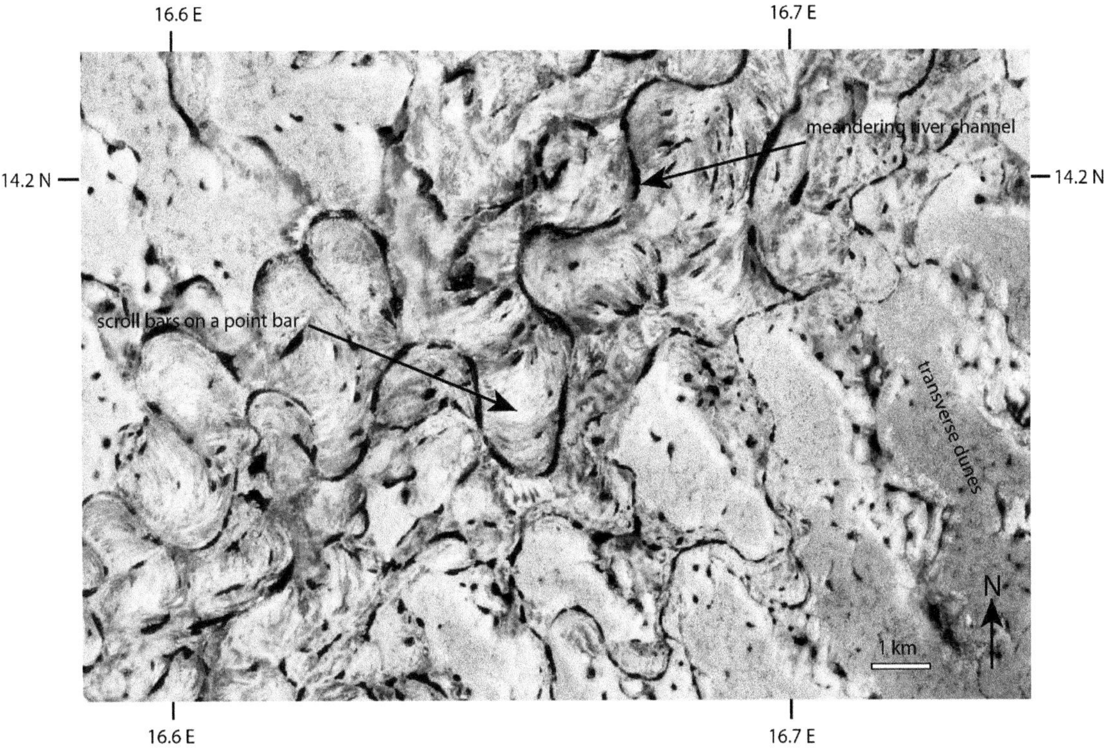

Fig. 13.8 Satellite image of the Bahr el Ghazal area showing scroll bar topography from river channels that meandered through a field of transverse dunes (*Source* Google Earth ™ image Landsat/Copernicus)

Fig. 13.9 Large compound barchan dune in the Bodélé Depression with small nebkha dunes in the foreground. The sediments are a mix of yellow-coloured quartz sand and grey diatomite particles. Small outcrops of white diatomite can be seen through gaps in the aeolian sediments

Fig. 13.10 Martin Todd and Sebastian Engelstaedter venture out into a dust storm at Chicha, a campsite with a single tree, in the Bodélé Depression during BoDEx, justifiably known as the dustiest place on Earth

References

Armitage SJ, Bristow CS, Drake NA (2015) West African monsoon dynamics inferred from abrupt fluctuations of Lake Mega-Chad. PNAS 112:8543–8548

Beauvilain A (2008) The contexts of discovery of *Australopithecus bahrelghazali* (Abel) and of *Sahelanthropus tchadensis* (Toumaï): unearthed, embedded in sandstone, or surface collected? S Afr J Sci 104:165–168

Ben-Ami Y, Koren I, Rudich Y, Artaxo P, Martin ST, Andreae MO (2010) Transport of Saharan dust from the Bodélé Depression to the Amazon Basin: a case study. Atmos Chem Phys Discuss 10:4345–4372

Bristow CS, Moller TH (2018) Testing the auto-abrasion hypothesis for dust production using diatomite dune sediments from the Bodélé Depression in Chad. Sedimentology 65:1322–1330

Bristow CS, Drake N, Armitage S (2009) Deflation in the dustiest place on earth: the Bodélé Depression, Chad. Geomorphology 105:50–58

Bristow CS, Hudson-Edwards KA, Chappell A (2010) Fertilizing the Amazon and equatorial Atlantic with West African dust. Geophys Res Lett 37:L14807. https://doi.org/10.1029/2010GL043486

Bristow CS, Holmes JA, Mattey D, Salzmann U, Sloane HJ (2018) A late Holocene palaeoenvironmental 'snapshot' of the Angamma Delta, Lake Megachad at the end of the African Humid Period. Quat Sci Rev 202:182–196

Brunet M, Beavilain A, Coppens Y, Heintz E, Moutaye AHE, Pilbeam D (1995) The first australopithecine 2,500 kilometres west of the Rift Valley (Chad). Nature 378:273–275

Brunet M, Guy F, Pilbeam D, Mackaye HT, Likius A, Ahounta D, Beauvilain A, Blondel C, Bocherens H, Boisserie JR, de Bonis L, Coppens Y, Dejax J, Denys C, Duringer P, Eisenmann V, Gongdibé F, Fronty P, Geraads D, Lehmann T, Lihoreau F, Louchart A, Mahamat A, Merceron G, Mouchelin G, Otero O, Pelaez Campomanes P, Ponce de León M, Rage J-C, Sapanet M, Schuster M, Sudre J, Tassy P, Valentin X, Vignaud P, Viriot L, Zazzo A, Zollikofer C (2002) A new hominid from the Upper Miocene of Chad, Central Africa. Nature 418:145–151

Chappell A, Warren A, O'Donoghue A, Robinson A, Thomas A, Bristow C (2008) The implications for dust emission modeling of spatial and vertical variations in the horizontal dust flux and particle size in the Bodélé Depression. Northern Chad J Geophys Res: Atmos 113:D04214. https://doi.org/10.1029/2007JD009032

Drake N, Bristow C (2006) Shorelines in the Sahara: geomorphological evidence for an enhanced monsoon from palaeolake Megachad. Holocene 16:901–911

Giles J (2005) The dustiest place on earth. Nature 434:816–819

Hudson-Edwards KA, Bristow CS, Cibin G, Mason G, Peacock CL (2014) Solid-phase phosphorus speciation in Saharan Bodélé depression dusts and source sediments. Chem Geol 384:16–26

Koren I, Kaufman YJ, Washington R, Todd MC, Rudich Y, Martins JV, Rosenfeld D (2006) The Bodélé Depression: a single spot in the Sahara that provides most of the mineral dust to the Amazon forest. Environ Res Lett 1:014005. https://doi.org/10.1088/1748-9326/1/1/014005

Lebatard A-E, Bourlès DL, Duringer P, Jolivet M, Braucher R, Carcaillet J, Schuster M, Arnaud N, Monie P, Lihoreau F, Likius A, Mackaye HT, Vignaud P, Brunet M (2008) Cosmogenic nuclide dating of *Sahelanthropus tchadensis* and *Australopithecus bahrelghazali*: Mio-Pliocene hominids from Chad. PNAS 105:3226–3231

Nachtigal D (1876) Journey to Lake Chad and neighbouring regions. J Roy Geogr Soc Lond 46:369–411

Prospero JM, Ginoux P, Torres O, Nicholson SE, Gill TE (2002) Environmental characterization of global sources of atmospheric soil dust identified with the Nimbus 7 Total Ozone Mapping Spectrometer (TOMS) absorbing aerosol product. Rev Geophys 40:1002. https://doi.org/10.1029/2000RG000095

Schuster M, Roquin C, Duringer P, Brunet M, Caugy M, Fontugne M, Mackaye HT, Vignaud P, Ghienne J-F (2005) Holocene Lake Mega-Chad palaeoshorelines from space. Quat Sci Rev 24:1821–1827

Schuster M, Duringer P, Ghienne J-F, Vignaud P, Mackaye HT, Likius A, Brunet M (2006) The age of the Sahara Desert. Science 311:821

Servant M, Servant S (1970) Les formations lacustres et les diatomées du Quaternaire Récent du fond de la cuvette Tchadienne. Rev Géogr Phys Géol Dynam XII:63–76

Servant M, Ergenzinger P, Coppens Y (1969) Datations absolues sur un delta lacustre quaternaire au Sud du Tibesti (Angamma). CR Somm Séances Soc Géol France 8:313–314

Todd MC, Washington R, Martins JV, Dubovik O, Lizcano G, M'Bainayel S, Engelstaedter S (2007) Mineral dust emission from the Bodélé depression, northern Chad during BoDEx 2005. J Geophys Res: Atmos 112:D06207. https://doi.org/10.1029/2005GL023597

Vermeesch P, Leprince S (2012) A 45 year time series of dune mobility indicating constant windiness over the central Sahara. Geophys Res Lett 39:L14401. https://doi.org/10.1029/2012GL052592

Vignaud P, Duringer P, Mackaye HT, Likius A, Blondel C, Boisserie J-R, de Bonis L, Eisenmann V, Etienne M-E, Geraads D, Guy F, Lehmann T, Lihoreau F, Lopez-Martinez N, Mourer-Chauviré C, Otero O, Rage J-C, Schuster M, Viriot L, Zazzo A, Brunet M (2002) Geology and palaeontology of the Upper Miocene Toros-Menalla hominid locality, Chad. Nature 418:152–155

Warren A, Chappell A, Todd MC, Bristow C, Drake N, Engelstaedter S, Martins V, M'bainayel S, Washington R (2007) Dust-raising in the dustiest place on earth. Geomorphology 92:25–37

Washington R, Todd MC (2005) Atmospheric controls on mineral dust emission from the Bodélé Depression, Chad: the role of the low level jet. Geophys Res Lett 32:L17701. https://doi.org/10.1029/2005GL023597

Washington R, Todd MC, Lizcano G, Tegen I, Flamant C, Koren I, Ginoux P, Engelstaedter S, Bristow CS, Zender CS, Goudie AS, Warren A, Prospero JM (2006a) Links between topography, wind, deflation, lakes and dust: the case of the Bodélé Depression, Chad. Geophys Res Lett 33:L09401. https://doi.org/10.1029/2006GL025827

Washington R, Todd MC, Engelstaedter S, Mbainayel S, Mitchell F (2006b) Dust and the low level circulation over the Bodélé Depression, Chad: observations from BoDEx 2005. J Geophys Res: Atmos 111:DO3201. https://doi.org/10.1029/2005JD006502

Yu H, Chin M, Yuan T, Bian H, Remer LA, Prospero JM, Omar A, Winker D, Yang Y, Zhang Y, Zhang Z, Zhao C (2015) The fertilizing role of African dust in the Amazon rainforest: a first multiyear assessment based on data from Cloud-Aerosol Lidar and infrared pathfinder satellite observations. Geophys Res Lett 42:1984–1991

Dust from the Central Sahara: Environmental and Cultural Impacts

14

Jasper Knight

Abstract

Dust export from the central Sahara region, and in particular from source region hotspots such as the Bodélé Depression, is well known from geological records, European documentary evidence and local oral histories and traditions over long timescales and is well imaged using remote optical sensors. However, the varied environmental and cultural impacts of dust transport and deposition, both around the Sahara margins and farther afield along atmospheric transport paths, have not been integrated together. This chapter examines some of the most significant environmental and cultural impacts of Saharan dust dynamics in local and regional contexts and thus teleconnections between the central Sahara and regional- to hemispheric-scale physical and human systems.

Keywords

Deflation · Dust export · Harmattan · Optical sensors · Red rain · Seasonal wind patterns · Teleconnections · High-pressure cell

14.1 Introduction

The dry land surface throughout the Sahara region and the presence of fine sediment grain sizes and strong winds make it a key area of dust export globally (Evan et al. 2016). The radiative forcing and dust veil effects of such aerosols also mean that land surface properties and dynamics in the Sahara have potential to affect global climates. The total dust flux from North Africa in total is 1087 Tg (1087 million tonnes) yr^{-1}, compared with a fluvial flux of 168 million tonnes yr^{-1} (Bullard and Baddock 2019). Total estimated dust export from the Sahara region alone is on the order of 60–200 million tonnes yr^{-1} (Morales 1979), and of Saharan dust export in 1981–1982, 60% by volume was released south to the Gulf of Guinea, 28% west to the North Atlantic and 12% north to Europe (D'Almeida 1986). The distinctive properties of the Saharan land surface make it susceptible to be affected by wind. These include: (1) land surface aridity: the absence of surface moisture reduces inter-grain adhesion and therefore allows grains to be blown around more easily; (2) the absence of stabilising vegetation in many areas: this means that the land surface is not protected by spreading root systems or the reduced surface wind speeds afforded by vegetation cover; (3) the presence of unconsolidated and loose sediment particles, in particular of fine grain sizes: this is generated by strong physical weathering, and to a lesser extent chemical and biological weathering, over long timescales; and (4) strong and dominantly unidirectional seasonal winds: where winds are above the threshold velocity for sediment transport, this allows for regional-scale sediment export from one area to another and therefore the development of net erosional and depositional landforms in certain areas, depending on their resultant sediment budgets. Seasonal wind patterns in the Sahara are determined by synoptic circulation and the position and strength of high-pressure cells, which drives harmattan winds (discussed below). Thus, changes in environmental and climatic conditions in central Sahara source areas and synoptic circulation patterns have potential to impact on total dust flux and transport directions. *Dust* is usually defined as particles and aggregates that are up to 100 μm in diameter and which can be transported long distances from source by suspension (Bullard and Baddock 2019). Although mineral grains are the most common constituents of dust, it can also include biological materials (bacteria, diatoms, pollen). Wind transport results in sorting

J. Knight (✉)
School of Geography, Archaeology and Environmental Studies, University of the Witwatersrand, Johannesburg 2050, South Africa
e-mail: jasper.knight@wits.ac.za

J. Knight et al. (eds.), *Landscapes and Landforms of the Central Sahara*, World Geomorphological Landscapes, https://doi.org/10.1007/978-3-031-47160-5_14

Fig. 14.1 Sandstorm (haboob) showing the advancing nose of the event, Al Asad, Iraq. Image reused under Wikimedia Commons licence, image available from http://www.defenselink.mil/photos/newsphoto.aspx?newsphotoid=6469 and https://en.wikipedia.org/wiki/File:Sandstorm_in_Al_Asad,_Iraq.jpg#filelinks

effects such that particle size and total volumetric dust flux decrease with increased transport distance. Dust events themselves also have different definitions. *Blown dust* refers to atmospheric dust transport that can be observed by the naked eye but where visibility is greater than 1 km. Where visibility is reduced to less than 1 km, these events are termed *dust storms* (Laity 2008) (Fig. 14.1). However, published studies as well as different documentary sources may use different definitions or be referring to different types of events. This must be borne in mind when trying to compare events of different types or when based on different documentary accounts.

Although many remote sensing studies have identified and modelled recent dust patterns, documentary data sources of different types, including agricultural and ecclesiastical records from different locations in the European Mediterranean, have also been used to discuss spatial and temporal patterns of dust and similar climatic events over decadal to centennial timescales (e.g. Bücher and Lucas 1975; Lamb 1982; Goudie and Middleton 1992; Middleton 2019). This latter data type focuses on the impacts that dust events have on human activity and identifies the fact that blown dust and dust storm events can be considered as natural hazards (Tozer and Leys 2013; Middleton 2017; Middleton et al. 2019). This chapter draws together a diverse range of viewpoints and approaches, including such studies mentioned above, in order to consider the wider context of dust export from the central Sahara region. In detail, the chapter (1) describes some previous studies of dust export and their methodologies and the nature of harmattan winds that are commonly associated with near-field dust fluxes; (2) outlines the environmental, human and cultural impacts of dust events from different records; and (3) discusses these events in the context of late Holocene environmental and climatic change in the central Sahara region.

14.2 Previous Studies of Dust Export

Several Saharan dust export areas have been identified (Middleton and Goudie 2001; Crouvi et al. 2012; Gherboudj et al. 2017). The most well-known site is the Bodélé Depression, Chad, where the presence of fine-grained and now-dry lake beds and diatomite provides very small particles capable of being taken up into wind transport (Abouchami et al. 2013). A summary of Bodélé Depression properties and dynamics is given by Bristow (this volume). Based on dated diatomite horizons, rates of long-term deflation are on the order of 1.6 mm yr^{-1} and can be calculated (Bristow et al. 2009). However, such calculations likely hide significant short-term variations in dust fluxes.

Within the wider Sahara region, other dust source sites also exist (Crouvi et al. 2012; Gherboudj et al. 2017) (Fig. 14.2). These include the Qattara Depression (Egypt), Chott el-Jerïd (southern Tunisia), Chott Melghir (northeast Algeria), Grand Erg Occidental (Algeria), Great Sand Sea (Libya), NW Tibesti region (Libya/Chad) and Sebkhet te-n-Dghâmcha (Mauritania/Western Sahara) (Gherboudj et al. 2017). Some of these are silt, whereas others are sandier. Many of these are located within basinal geomorphic settings and (like the Bodélé Depression) could also have been

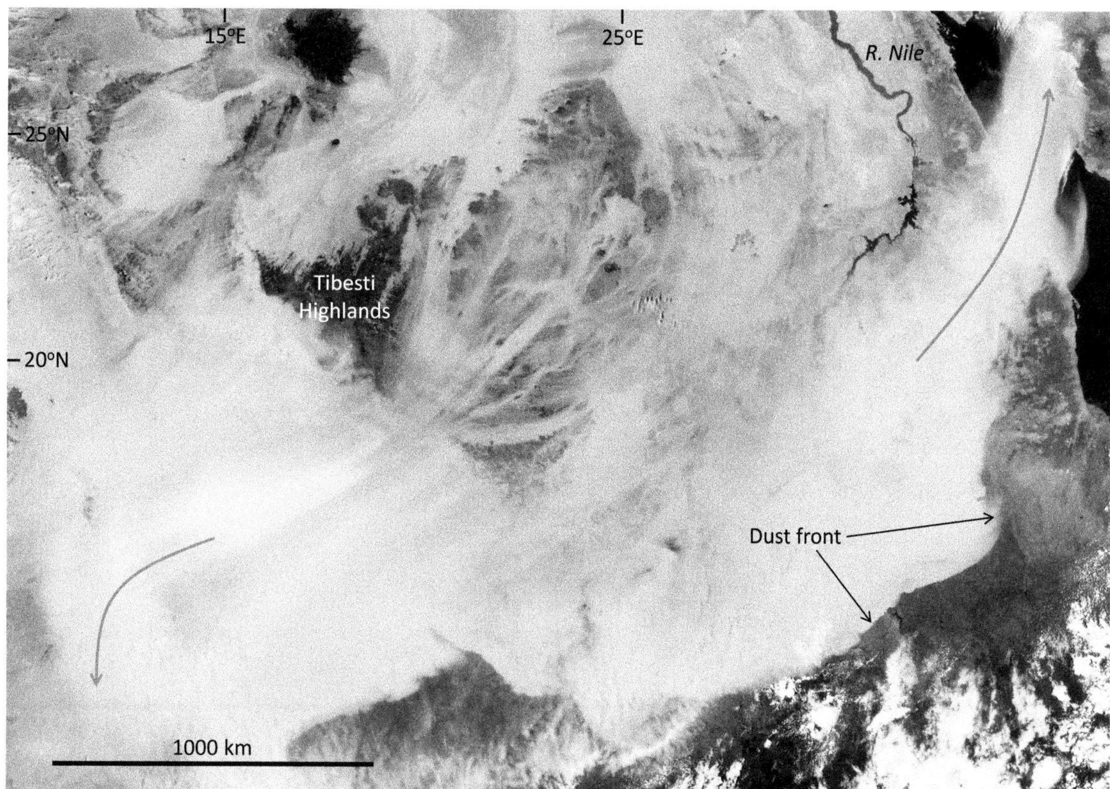

Fig. 14.2 Satellite image of a regional dust event in the southeast-central Sahara region, showing the dust front as well as export across the Red Sea (top right). VIIRS image from the Suomi NPP satellite, available from NASA Earth Observatory, date of image: 29 March 2018

locations of fine sediment deposition during lateglacial/early Holocene pluvial phases, although this is not known for certain and then activated as dust sources when these locations dried out (Crouvi et al. 2012). In addition, variations in wind strength can also impact on the locations of deflation activity and the size/volume of total exported sediment (Friese et al. 2017). Regional surface wind speeds required for Saharan dust deflation are on the order of 6.5–13.0 m s^{-1}, but this varies according to the nature of the land surface (Helgren and Prospero 1987). Increased wind strength can lift larger grains into the windstream and can reactivate sediment export from source areas that would otherwise not be available (Crouvi et al. 2012). Conversely, lighter winds can decrease the total sand flux and concentrate this flux from very fine substrates only. For example, dust export to Europe from the Sahara is favoured where there are strong mid-troposphere winds and slow-moving circulation cells (Wheeler 1986; Stuut et al. 2009). Bergametti et al. (1989) argue that northwest Africa is a key dust source during winter and summer, whereas Libya (north Africa) is most active in spring. However, variations in sediment flux as a consequence of changes in wind climate are not well understood, and much sediment that is already in transit in the atmosphere is difficult to link back to different source areas using remote sensing data alone, and sometimes geochemical signatures are inconclusive.

The most commonly used modern method to examine dust sources and fluxes from the central Sahara is remote sensing, and this has been deployed using different sensors, over different time periods and with different spatial and temporal resolutions (e.g. Bristow et al. 2009). These studies have commonly used aerosol optical depth (AOD) as a key indicator of dust concentration and also particle size (Fig. 14.3). Optical depth represents the total particle concentration through the entire depth of the atmosphere, but potentially includes other aerosols as well as dust. Thus, measures of AOD do not just refer specifically to dust in long-distance transport at the height of the tropopause (which is the approximate height to which large-scale dust export occurs). It also potentially includes near-surface dust and sand concentrations in the form of dust storms that arise as a result of different meteorological conditions, including harmattan and similar seasonal wind flow events. Atmospheric dust concentration has been commonly measured by an optical sensor on the Total Ozone Mapping Spectrometer (TOMS) satellite (Washington et al. 2003); SeaWiFS data have also been used (Koren et al. 2003). These concentrations are measured in terms of the

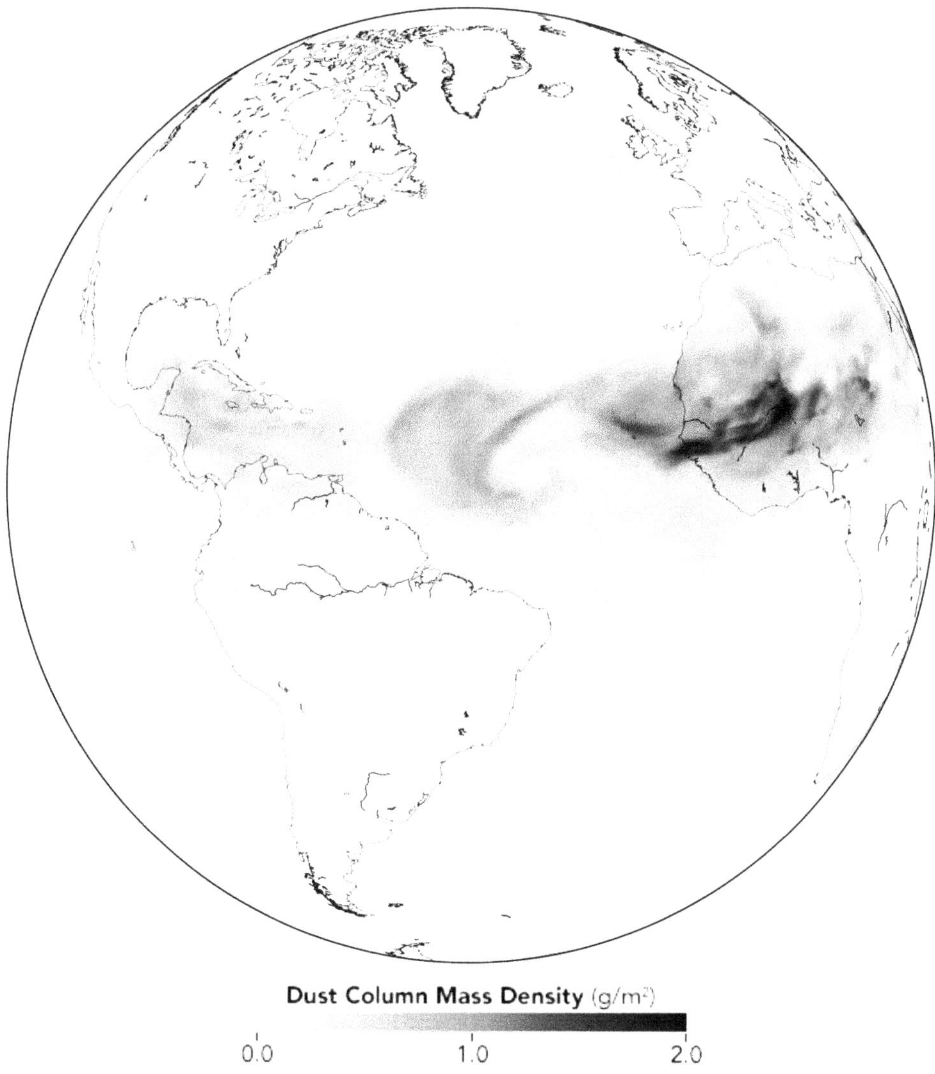

Fig. 14.3 Image of westward dust transport from the central Sahara region across the Atlantic and to central America. Data analysis from the Goddard Earth Observing System Model, Version 5 (GEOS-5), available from NASA Earth Observatory, 28 June 2018

Aerosol Index (AI) which reflects aerosol optical thickness. Very large AI values of > 30 are found in the Bodélé depression, with slightly lower values of > 24 found in the western Sahara. This compares with AI values of > 8–11 in the Etosha and Mkgadikgadi pans in southern Africa. AI values are generally inversely proportional to rainfall (Laity 2008).

14.2.1 Harmattan Wind Flow Patterns

The most well-known seasonal wind pattern leading to near-field sediment export from the Sahara region is the harmattan. This refers to a seasonal pattern of surface winds in the southwest Sahara, mainly in Mali, Niger and Nigeria, and directed from northeast to southwest towards the Gulf of Guinea (Afeti and Resch 2000) (Fig. 14.4). The climatology of harmattan events has been examined in several studies (Schwanghart and Schütt 2008; Schepanski et al. 2017). The harmattan is associated with dry, low-humidity air flow and high near-surface dust concentrations that can give rise to sandstorm events. This is because gusty conditions at wind fronts (termed *haboobs*) have a high sediment concentration as a result of land surface deflation (Adeniran et al. 2018) (Fig. 14.5). These sediments can act as cloud condensation nuclei, contributing to cloud development and rapid wet deposition at the haboob front. Variability in harmattan wind strength and thus near-surface dust concentration is linked to Atlantic sea surface temperatures (SST) and intertropical convergence zone (ITCZ) position.

In terms of larger-scale synoptic patterns, dust concentrations in Barbados (central Atlantic) peak in the northern hemisphere summer (months JJA) corresponding to maximum Sahara dust export at 700 mb (hPa) height by easterly waves (Tetzlaff and Wolter 1980). Red rain events (see below) in Europe show different seasonal patterns. In western Europe (Spain, France, UK) and eastern Europe (Turkey, Greece, Russia), these events are most common in spring (months MAM); central Europe shows peaks in

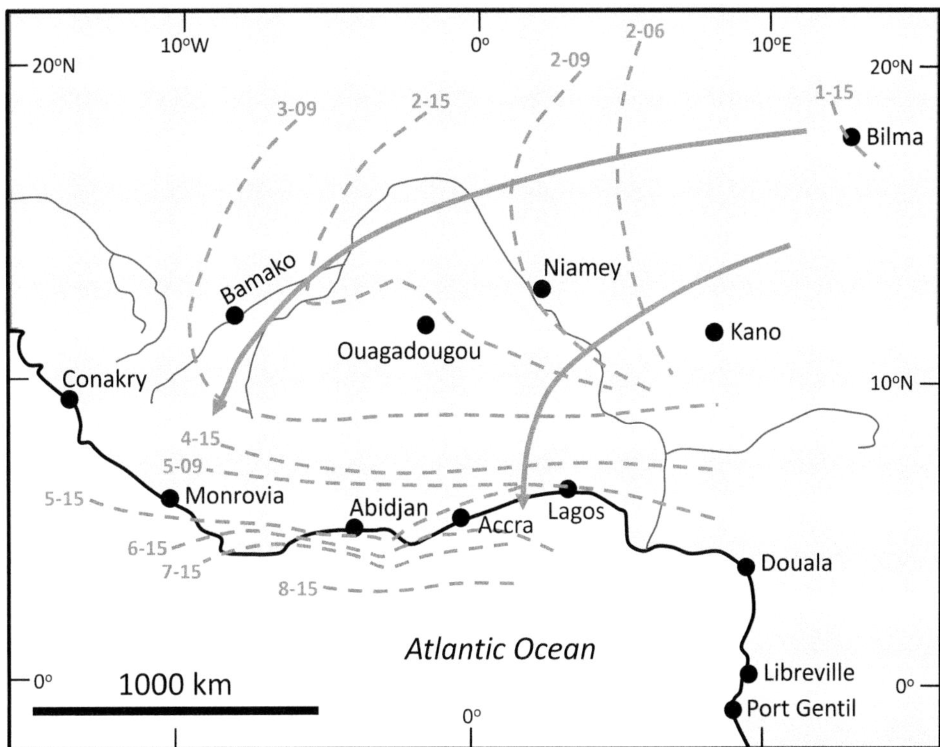

Fig. 14.4 Schematic map of the migration of the harmattan dust front (dotted brown line), in March 1973 in West Africa (redrawn from Morales 1979, p. 46). Numbers refer to the position of the dust front (day-hour)

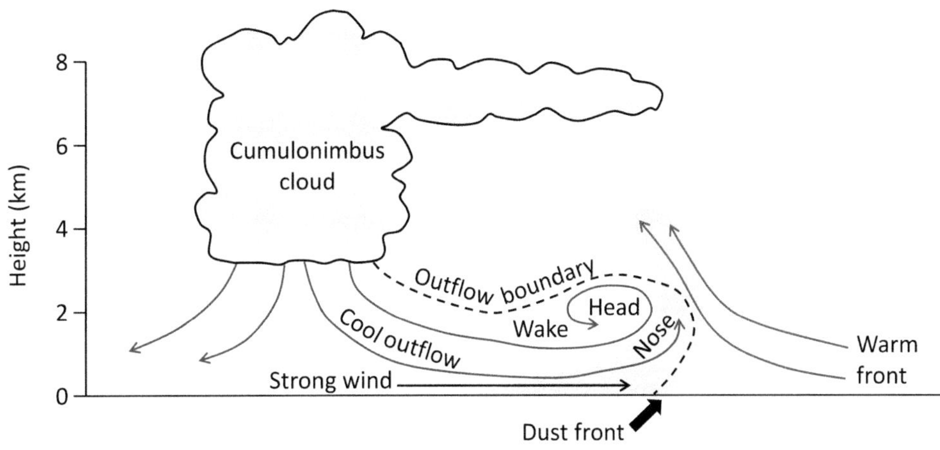

Fig. 14.5 Schematic cross-section of a haboob (redrawn after Bullard and Baddock 2019)

both spring and autumn (months SON) (White et al. 2012). Variations in timing in the different regions are due to variability in Rossby waves.

The seasonal high-pressure cell responsible for the harmattan also causes similar katabatic wind patterns elsewhere around the central Sahara, although these are much less well described in the literature. Warren (2019, p. 11) lists many of these local names. Land breezes are recorded seasonally along the coasts of Libya and Algeria, and these can contribute to dust export into the Mediterranean basin. An interesting early description in English of a haboob was made by T.E. Lawrence ('Lawrence of Arabia') in 1917 (Rooney 2017).

14.3 Environmental Impacts of Dust Transport and Deposition

There are varied environmental impacts of dust transport, both locally in areas adjacent to source and more distally where atmospheric rainout of fine particles takes place (Goudie and Middleton 2001; Middleton 2017, 2019). This section summarises some of the major impacts, noting that there are close relationships between some of these physical effects and also impacts on people.

Atmospheric effects. Atmospheric dust can act as cloud condensation nuclei, and thus, their concentration is decreased when they are scavenged by water droplets. 'Red rain', described below, is the outcome of this process. Near the land surface, dust haze can contribute to reduced AOD and dust haze events decrease in number with increased humidity and land surface vegetation density (Anuforom 2007). Mean visibility decreases by 18.9%, and AI values increase by 24% during harmattan events (Anuforom et al. 2007); however, there is significant interannual variability in harmattan strength which can influence atmospheric properties. The presence of atmospheric dust can impact on patterns of net spectral irradiance as well as be manifested differently at different wave bands. Spectral irradiance decreases by 50% upon passage of a dust event as dust concentration increases (Adeyefa and Holmgren 1996). Spectral diffusion also increases significantly as a result of atmospheric turbidity and by light scattering at certain wavelengths (Horvath et al. 2018). Electrical effects in the atmosphere have also been noted, resulting from negatively charged dust particles being lofted at haboob fronts (Williams et al. 2009).

Atmospheric rainout on land surfaces. 'Red rain' (blood rain) describes the wet deposition (rainout) of atmospheric Sahara dust, observed in particular over Europe and marked in documentary records over the last centuries (Sala et al. 1996). The red colour refers to the oxide-stained dust present within the raindrops which, upon evaporation, leaves a red residue over windows, cars, vegetation, etc. In southern Spain, a marked increase in such events has been noted since the 1980s, possibly attributed to changes in atmospheric circulation (Sala et al. 1996). Volumetrically, dust deposition from individual atmospheric events is highly variable, with up to 19 g m^2 reported for a single event in northeast Spain (average 5.2 g m^2 for the period of 1983–1994) (Avila et al. 1997). Windblown dust, by dry and wet deposition, contributes to sediment supply to the land surface and soils (Simonson 1995). Model simulations suggest deposition values of 0.133 g m^2 y^{-1} (dry) and 0.085 g m^2 y^{-1} (wet) in Hungary, central Europe, with higher values located nearer to source (Varga et al. 2016). Field measurements from Libya give values of 82–276 g m^2 yr^{-1} (O'Hara et al. 2006) although there are very few field measurements across the region for comparison. Red rain samples recorded from the Canary Islands (January 1999) had a unimodal grain size distribution, mean grain size of 17–21 μm and a Munsell colour of 7.5 YR 6/6 (reddish yellow). Sample composition comprised 50% of SiO_2, 12% of Al_2O_3 and 10% of CaO (Criado and Dorta 2003). Dust contribution to soil varies in quantity and affects soil structure, geochemistry and nutrient status (Simonson 1995). Dust contribution by seasonal harmattan winds in Côte d'Ivoire also introduces significant mineral nutrient input to soils: 0.11 kg ha yr^{-1} of phosphorus, 2.5 kg ha yr^{-1} of potassium, 3.5 kg ha yr^{-1} of calcium and 0.4 kg ha yr^{-1} of magnesium (Stoorvogel et al. 1997). However, different source areas have different compositions, controlled by underlying geology (Avila et al. 1997), and isotopic methods have been used to distinguish between soil components that are derived from local and Saharan dust sources (Erel and Torrent 2010). The presence of dissolved calcite and sea salt from winds crossing the Mediterranean makes the red rain alkaline: Sala et al. (1996) reported a pH of 7.86, Ca content of 21 ppm, K content of 6 ppm and Na content of 3 ppm.

The impacts of dust fallout on land surface ecosystems have been well studied. Annual dust export to the Amazon basin, following the trajectory of upper level easterly winds, is ~40 million tonnes, around half of which is derived from the Bodélé depression alone (Koren et al. 2006). This dust provides additional nutrients to highly leached tropical soils. Pathogens, bacteria and algae can also be transported as organic components as part of the dust load and can potentially spread diseases to ecosystems. Recent Saharan dust deposition episodes within Swiss glaciers are recorded as distinctive but thin observable layers, and these have marked geochemical anomalies, in particular of Ca^{2+}, Mg^{2+}, Na^+, Cl^- and SO_4^{2-} (Sodermann et al. 2006). Similar records exist from within the snowpack (Meola et al. 2015). An ice core from Colle Gnifetti (Swiss/Italian border, European Alps) shows increased dust content around the early mediaeval period (1100–1200 AD), corresponding to increased Saharan dust flux at this time (Bohleber et al. 2018).

Atmospheric rainout into water bodies. The neutralising effect of dust in water droplets makes acidic rainwater more alkaline but also impacts on lake and river water pH and water quality properties. There are also impacts on water turbidity and aquatic ecosystems. However, these changes are not well understood from many freshwater lake and river systems. Decreased dust flux during the 1990s resulted in reduced alkalinity (less $CaCO_3$ input) of rainwater and thus acidification of surface water bodies in southern Europe (Loÿe-Pilot and Martin 1996).

Under harmattan conditions, 146,000 t yr^{-1} and 42,000 t yr^{-1} of mineral and organic components respectively are deposited in Lake Volta dam in Ghana (Breuning-Madsen et al. 2012). However, despite this large total volume, this represents only 1% of total suspended sediment flux into the lake, and likewise, the contribution of bioavailable Ca, Mg and P is very limited. Saharan dust rainout into the Mediterranean Sea is estimated to be 10×10^9 t yr^{-1}, and rainout in the northwest Mediterranean to be 4×10^6 t yr^{-1}, equivalent to the sediment flux from the Rhône River (Loÿe-Pilot and Martin 1996). This dust flux, including from alkaline dust sources in the Sahara such as Bodélé Depression lake beds, can change seawater pH. Dust fallout effects are well known and studied, from the marine environment where this dust can contribute to 'ocean fertilisation' of planktic organisms, such as coccolithophorids (Neuer et al. 2004). The mechanism behind this is based on increased bioavailability of iron, derived from minerogenic dust particles. Ocean surfaces that have high nitrate but low chlorophyll content are particularly sensitive to variations in dust fluxes (Schulz et al. 2012). Other aquatic effects have not been well studied.

Human health effects. The most recent review of the human health effects of dust is given by Goudie (2014) which includes examples from around the central Sahara region. High dust concentrations decrease air quality and can lead to a broad range of health impacts including on respiratory health, cardiovascular disease, pneumonia and hypertension (Laity 2008). High dust concentrations during sand storms, commonly associated with gusty frontal conditions, result in reduced visibility, wind abrasion of standing structures, increased incidents of vehicle accidents and respiratory impacts on people (Middleton 2017). In more detail, dry winds and harmattan dust can reduce immunological risk and may be a contributory factor to meningitis (García-Pando et al. 2014). In urban areas, dust can also amplify existing levels of aerosols from diesel engines, biomass burning and savanna fires (Naijda et al. 2018). In addition to sediment, different micro-organisms and chemicals from the land surface can also be transported from dust source areas, and these can potentially impact on human and environmental health in the areas of dust deposition (Laity 2008). During a dust storm event in New South Wales, Australia (23 September 2009), dust concentrations were 34–48 times the values permitted by National Air Quality Standards (Leys et al. 2011).

14.4 Historical and Cultural Records of Dust Transport and Deposition

The occurrence and impacts of dust export from the Sahara (Fig. 14.2) have been well examined from central Europe and the eastern Mediterranean from a range of documentary evidence. These records include weather diaries, ships' logs and merchants' records, newspaper reports, records of agricultural production and prices, ecclesiastical and liturgical records, rent books and travellers' journals. Such records are useful because they provide first-hand accounts of dust events and can inform on their cultural significance. In addition, long temporal records can inform on changes in dust occurrence and location over time and thus can be linked to climate forcing.

Lamb (1988) described how variations in rainfall in the central Sahara during the last 1000 years influenced meteorological events in Europe, as documented in different written records. During the early Middle Ages, the Sahara was wetter than present, with seasonal rain reaching to southern Algeria and Libya and facilitating more active use of trans-Saharan trade routes through the Fazzan region. Lake Chad experienced a highstand at this time. This period was followed by more arid conditions during the thirteenth and fourteenth centuries, in which enhanced southerly and southeasterly winds from the Sahara brought more frequent events of locusts and dust to central Europe (Lamb 1988, p. 241). During the Little Ice Age (LIA, ~1550–1850 AD), the Sahara was wetter than today, reported by travellers in the region, with rain in the Sahel in the southern Sahara extending farther north (Lamb 1988, p. 197). Low lake levels existed in the region in 1450–1550 AD, but this was followed in the two centuries after by higher water levels in Lake Chad due to more restricted northward migration of the ITCZ (Lamb 1982, p. 226). A dry phase across the Sahara and Middle East also took place in the 16th–17th centuries (Lamb 1982). Decadal-scale variations in aridity, in particular during the 1530s–1550s and just prior to the LIA, are also recorded in southern Spain (Rodrigo et al. 1995). Thevenon et al. (2012) presented a time series of Saharan dust deposition from the Colle Gneifetti glacier from the 1430s to present, based on dust concentrations within individual ice layers. Significant periods of dustiness are shown during ~ 1570–1670 and ~ 1870–1970 (Fig. 14.6). It is also notable that there is a significant short dusty period around 1770, predating the Laki volcanic eruption of 1783. A table of dust fallout events across Europe, from 1669 to 1980, was presented by Mainguet (1983), including the fallout location and chemical analysis of the dust.

The phenomenon of 'red rain' has been noted for centuries in Europe. For example, during the LIA, this phenomenon was suppressed as a result of a southerly shift of Atlantic westerlies (Lamb 1982). Further, optical effects of high dust concentrations (red sunsets) were noted in many archives and diaries and by artists. Dessens and van Dinh (1990) described the local climatology of dust events in southern France. These take place mainly in summer and are associated with warm southerly winds and hazy atmospheric conditions ('brume sèche' or dry haze). Sunrises and sunsets are strongly yellow to red. Heavy rain falls a day or two after the haze events, leading to 'red rain' wet deposition.

An interesting potential connection between climate, dust and culture is shown through the example of the etymology of a word associated with weather phenomena in the Mediterranean region. The southern French word *giboul* (now giboulée) emerged in dialect during the mediaeval period (the first record is from 1548; Hatzfeld and Darmesteter 1920) and refers to periods of sudden and intense showers. Although French etymological dictionaries record this as 'origine inconnue' ('origin unknown'), it is argued here that the word derives from the Libyan Arabic word *ghibli* (قِبْلِيّ) meaning the land breeze developed from

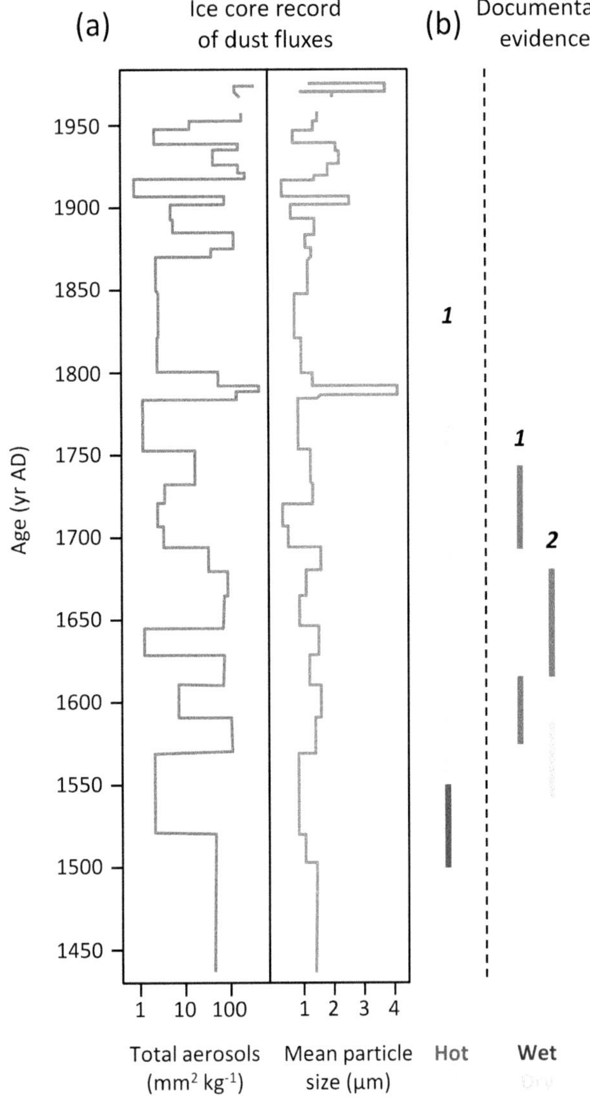

Fig. 14.6 **a** Decadal dust fluxes into the Colle Gnifetti glacier, Swiss/Italian border, European Alps, for the period 1430–1980 AD (redrawn from Thevenon et al. 2012), with **b** additional regional climate information based on documentary data. *Source 1* Lamb (1982); *Source 2* Rodrigo et al. (1995)

Libya and Algeria. This wind flows northwards towards the Mediterranean especially during the spring and summer. The ghibli winds are therefore directly comparable to the harmattan and thus associated with dust transport from the northern Sahara towards southern Europe. In the Mediterranean, this brings gusty, warm, strong wind conditions and 'red rain' events. It also causes high sea water levels on southern European coasts due to high wind stress (Conte et al. 1996).

An interesting question is why this word appears in mediaeval French at this time. An answer could be the development or invigoration of trading routes during this period (Braudel 1972), bringing new ideas, cultures and thus words into southern France. Braudel (1972) notes that there was increased nomadic migration and trade route development associated with increased regional aridity during the late 15th and early sixteenth centuries, in particular trade from north Africa to Marseilles and Lyons. Warms summer winds from the Sahara brought stable conditions for seafaring (Braudel 1972, p. 257).

14.5 Discussion

The export of Saharan dust and its transport through the global atmosphere can be identified when dust is rained out onto the land or ocean surface and preserved within various sediment archives. Unlike similar modes of transport and deposition undertaken by volcanic tephra, the fate and properties of this dust are poorly studied. In addition, the preservation potential or any post-depositional (diagenetic) changes that this dust undergoes are not well known. This is an important area of future research.

The relationship of Saharan dust fluxes to periods of land surface aridity, identified over different spatial and temporal scales, provides a link between dust export and climate forcing (Stuut et al. 2009). For example, McGee et al. (2013) showed that dust transport to Atlantic marine cores increased during arid periods of Heinrich Event 1 and the Younger Dryas and after the termination of the African Humid period (AHP) at 5000 yr BP. They show that dust fluxes over the last 2000 years were five times higher than during the AHP. An interesting question is whether increased aridity and dust activity at the end of the AHP caused changes in human activity in the region (see discussion in Knight, this volume). A recent study by Carolin et al. (2019) notes the coincidence in timing between increased dust activity and social–cultural changes in the Middle East around 4200 yr BP. This is also likely to have taken place across north Africa, associated with changes in surface water availability, agricultural patterns and trade routes (Manning and Timpson 2014) and evidenced in different archaeological contexts. Explicit relationships between people, culture and dust have not yet been fully explored although Middleton (2019) discusses some relationships between climate, human activity and increased dust storm activity and dust fluxes.

Alternatively, there is also evidence that global dust fluxes, of which the central Sahara is the biggest source area, have impacts on climate through amplification and biophysical feedbacks (Washington et al. 2009; Evan et al. 2016). A study of AOD at Cape Verde (Atlantic Ocean) showed that high dust concentrations have a radiative forcing efficiency of 90 W/m^2 at a wavelength of 380 nm (Gehlot et al. 2015), suggesting there are significant feedbacks and forcing that are not yet fully considered in climate models. Models of future regional climate change suggest that dust episodes may decrease as a result of weakening of tropical circulation patterns (Evan et al. 2016), but there may also be other land surface feedbacks to consider in these projections.

14.6 Summary and Outlook

Dust exported from the central Sahara, a globally important source area, can impact on the environment and people outside of the region during and after rainout (Middleton 2017, 2019). These impacts may be highly variable over time and space, depending on dust flux density, but have not been fully examined, especially within the context of the epidemiology of human health effects (mainly of respiratory diseases) or for transport of pathogens. Important unknown elements include any feedbacks that the presence of dust in the air, soil or water may have on climate and environmental changes in specific regions, both in sub-Saharan cities and beyond. Likewise, the association of dust events with human culture has not been fully examined in anthropological contexts, and this is an important area of future work. Although climate projections and recent instrumental evidence suggest continued dryness and thus potential for dust events from the Sahara region, increased rainstorm activity over recent decades in the southern Saharan Sahel (Taylor et al. 2017) may reduce the likelihood of dust export by harmattan winds. Wind advection models do not fully consider implications of dust export, and this is also an area of future research.

Acknowledgements I thank Jennifer Fitchett for comments on the previous version of this chapter.

References

Abouchami W, Näthe K, Kumar A, Galer SJG, Jochum KP, Williams E, Horbe AMC, Rosa JWC, Balsam W, Adams D, Mezger K, Andreae MO (2013) Geochemical and isotopic characterization of the Bodélé depression dust source and implications for transatlantic dust transport to the Amazon Basin. Earth Planet Sci Lett 380:112–123

Adeyefa ZD, Holmgren B (1996) Spectral solar irradiance before and during a harmattan dust spell. Sol Energy 57:195–203

Adeniran JA, Aremu AS, Saadu YO, Yusuf RO (2018) Particulate matter concentration levels during intense haze event in an urban environment. Environ Monitor Assess 190:41. https://doi.org/10.1007/s10661-017-6414-4

Anuforom AC (2007) Spatial distribution and temporal variability of Harmattan dust haze in sub-Sahel West Africa. Atmos Environ 41:9079–9090

Anuforom AC, Akeh LE, Okeke PN, Opara FE (2007) Inter-annual variability and long-term trend of UV-absorbing aerosols during harmattan season in sub-Saharan West Africa. Atmos Environ 41:1550–1559

Afeti GM, Resch FJ (2000) Physical characteristics of Saharan dust near the Gulf of Guinea. Atmos Environ 34:1273–1279

Avila A, Queralt-Mitjans I, Alarcón M (1997) Mineralogical composition of African dust delivered by red rains over northeastern Spain. J Geophys Res-Atmos 102:21977–21996

Bergametti G, Gomes L, Remoudaki E, Desbois M, Martin D, Buat-Ménard P (1989) Present transport and deposition patterns of African dusts to the north-western Mediterranean. In: Leinen M, Sarnthein M (eds) Paleoclimate and paleometeorology: modern and past patterns of global atmospheric transport. NATO ASI Series 282C, Kluwer, Dordrecht, pp 227–252

Bohleber P, Erhardt T, Spaulding N, Hoffmann H, Fischer H, Mayewski P (2018) Temperature and mineral dust variability recorded in two low-accumulation Alpine ice cores over the last millennium. Clim Past 14:21–37

Braudel F (1972) The Mediterranean and the Mediterranean World in the Age of Phillip II, 2 vol. Collins, London

Breuning-Madsen H, Lyngsie G, Awadzi TW (2012) Sediment and nutrient deposition in Lake Volta in Ghana due to harmattan dust. Catena 92:99–105

Bristow CS, Drake N, Armitage S (2009) Deflation in the dustiest place on Earth: The Bodélé depression, Chad. Geomorphology 105:50–58

Bücher A, Lucas C (1975) Poussières Africaines Sur L'europe. La Meteorol 5:53–69

Bullard J, Baddock M (2019) Dust: sources, entrainment, transport. In: Livingstone I, Warren A (eds) Aeolian Geomorphology—a new introduction. Wiley, Chichester, pp 81–106

Carolin SA, Walker RT, Day CC, Ersek V, Sloan RA, Dee MW, Talebian M, Henderson GM (2019) Precise timing of abrupt increase in dust activity in the Middle East coincident with 4.2 ka social change. PNAS 116:67–72

Conte M, Colacino M, Piervitali E (1996) Atlantic disturbances deeply penetrating the African continent: effects of Saharan regions and the Mediterranean basin. In: Guerzoni S, Chester R (eds) The impact of desert dust across the Mediterranean. Kluwer, Dordrecht, pp 93–102

Criado C, Dorta P (2003) An unusual 'blood rain' over the Canary Islands (Spain). The storm of January 1999. J Arid Environ 55:765–783

Crouvi O, Schepanski K, Amit R, Gillespie AR, Enzel Y (2012) Multiple dust sources in the Sahara Desert: the importance of sand dunes. Geophys Res Lett 39:L13401. https://doi.org/10.1029/2012GL052145

D'Almeida GA (1986) A model for Saharan dust transport. J Clim Appl Meteorol 25:903–916

Dessens J, van Dinh P (1990) Frequent Saharan dust outbreaks north of the Pyrenees: a sign of a climatic change? Weather 45:327–333

Erel Y, Torrent J (2010) Contribution of Saharan dust to Mediterranean soils assessed by sequential extraction and Pb and Sr isotopes. Chem Geol 275:19–25

Evan AT, Flamant C, Gaetani M, Guichard F (2016) The past, present and future of African dust. Nature 531:493–495

Friese CA, van Hateren JA, Vogt C, Fischer G, Stuut J-BW (2017) Seasonal provenance changes in present-day Saharan dust collected in and off Mauritania. Atmos Chem Phys 17:10163–10193

García-Pando CP, Stanton MC, Diggle PJ, Trzaska S, Miller RL, Perlwitz JP, Baldasano JM, Cuevas E, Ceccato P, Yaka P, Thomson MC (2014) Soil dust aerosols and wind as predictors of seasonal meningitis incidence in Niger. Environ Health Perspect 122:679–686

Gehlot S, Minnett PJ, Stammer D (2015) Impact of Sahara dust on solar radiation at Cape Verde Islands derived from MODIS and surface measurements. Remote Sens Environ 166:154–162

Gherboudj I, Beegum SN, Ghedira H (2017) Identifying natural dust source regions over the Middle-East and North-Africa: Estimation of dust emission potential. Earth Sci Rev 165:342–355

Goudie AS (2014) Desert dust and human health disorders. Environ Int 63:101–113

Goudie AS, Middleton NJ (1992) The changing frequency of dust storms through time. Clim Ch 20:197–225

Goudie AS, Middleton NJ (2001) Saharan dust storms: nature and consequences. Earth Sci Rev 56:179–204

Hatzfeld A, Darmesteter A (1920) Dictionnaire general de la langue française, 2 vol. Librairie Delagrave, Paris

Helgren DM, Prospero JM (1987) Wind velocities associated with dust deflation events in the western Sahara. J Clim Appl Meteorol 26:1147–1151

Horvath H, Alados Arboledas L, Olmo Reyes FJ (2018) Angular scattering of the Sahara dust aerosol. Atmos Chem Phys 18:17735–17744

Koren I, Joseph JH, Israelevich P (2003) Detection of dust plumes and their sources in northeastern Libya. Can J Remote Sens 29:792–796

Koren I, Kaufman YJ, Washington R, Todd MC, Rudich Y, Martins JV, Rosenfeld D (2006) The Bodélé depression: a single spot in the Sahara that provides most of the mineral dust to the Amazon forest. Environ Res Lett 1:014005. https://doi.org/10.1088/1748-9326/1/1/014005

Laity J (2008) Deserts and desert environments. Wiley, Chichester, p 342

Lamb HH (1982) Climate, history and the modern world. Methuen, London, p 387

Lamb HH (1988) Weather, climate & human affairs. Routledge, London, p 364

Leys JF, Heidenreich SJ, Strong CL, McTainsh GH, Quigley S (2011) PM_{10} concentrations and mass transport during "Red Dawn"—Sydney 23 September 2009. Aeolian Res 3:327–342

Loÿe-Pilot MD, Martin JM (1996) Saharan dust input to the western Mediterranean basin: an eleven years record in Corsica. In: Guerzoni S, Chester R (eds) The impact of desert dust across the Mediterranean. Kluwer, Dordrecht, pp 191–199

Mainguet M (1983) Tentative mega-geomorphological study of the Sahara. In: Gardner R, Scoging H (eds) Mega-geomorphology. Clarendon Press, Oxford, pp 113–133

Manning K, Timpson A (2014) The demographic response to Holocene climate change in the Sahara. Quat Sci Rev 101:28–35

McGee D, deMenocal PB, Winckler G, Stuut JBW, Bradtmiller LI (2013) The magnitude, timing and abruptness of changes in North African dust deposition over the last 20,000 yr. Earth Planet Sci Lett 371–372:163–176

Meola M, Lazzaro A, Zeyer J (2015) Bacterial composition and survival of Sahara dust particles transported to the European Alps. Front Microbiol 6:1454. https://doi.org/10.3389/fmicb.2015.01454

Middleton NJ (2017) Desert dust hazards: a global review. Aeolian Res 24:53–63

Middleton N (2019) Variability and trends in dust storm frequency on decadal timescales: climatic drivers and human impacts. Geosciences 9:261. https://doi.org/10.3390/geosciences9060261

Middleton NJ, Goudie AS (2001) Saharan dust: sources and trajectories. Trans Inst Br Geogr, NS 26:165–181

Middleton N, Tozer P, Tozer B (2019) Sand and dust storms: underrated natural hazards. Disasters 4:390–409

Morales C (ed) (1979) Saharan dust, mobilization, transport, deposition. Wiley and Sons, Chichester, p 297

Naijda L, Ali-Khodja H, Khardi S (2018) Sources and levels of particulate matter in North African and Sub-Saharan cities: a literature review. Environ Sci Poll Res 25:12303–12328

Neuer S, Torres-Padrón ME, Gelado-Caballero MD, Rueda MJ, Hernández-Brito J, Davenport R, Wefer G (2004) Dust deposition pulses to the eastern subtropical North Atlantic gyre: Does ocean's biogeochemistry respond? Glob Biogeochem Cycl 18:GB4020. https://doi.org/10.1029/2004GB002228

O'Hara SL, Clarke ML, Elatrash MS (2006) Field measurements of desert dust deposition in Libya. Atmos Environ 40:3881–3897

Rodrigo FS, Esteban Parra MJ, Castro Diez Y (1995) Reconstruction of total annual rainfall in Andalusia (southern Spain) during the 16th and 17th centuries from documentary sources. Theoret Appl Climatol 52:207–218

Rooney GG (2017) Haboobs, dust spouts and Lawrence of Arabia. Weather 72:107–110

Sala JQ, Cantos JO, Chiva EM (1996) Red dust rain within the Spanish Mediterranean area. Clim Ch 32:215–228

Schepanski K, Heinold B, Tegen I (2017) Harmattan, Saharan heat low, and West African Monsoon circulation: modulations on the Saharan dust outflow towards the North Atlantic. Atmos Chem Phys 17:10223–10243

Schulz M, Prospero JM, Baker AR, Dentener F, Ickes L, Liss PS, Mahowald NM, Nickovic S, García-Pando CP, Rodríguez S, Sarin M, Tegen I, Duce RA (2012) Atmospheric transport and deposition of mineral dust to the ocean: implications for research needs. Environ Sci Tech 46:10390–10404

Schwanghart W, Schütt B (2008) Meteorological causes of harmattan dust in West Africa. Geomorphology 95:412–428

Simonson RW (1995) Airborne dust and its significance to soils. Geoderma 65:1–43

Sodermann H, Palmer AS, Schwierz C, Schwikowski M, Wernli H (2006) The transport history of two Saharan dust events archived in an Alpine ice core. Atmos Chem Phys 6:667–688

Stoorvogel JJ, van Breemen N, Janssen BH (1997) The nutrient input by harmattan dust to a forest ecosystem in Côte d'Ivoire, Africa. Biogeochemistry 37:145–157

Stuut J-B, Smalley I, O'Hara-Dhand K (2009) Aeolian dust in Europe: African sources and European deposits. Quat Int 198:234–245

Taylor CM, Belušić D, Guichard F, Parker DJ, Vischel T, Bock O, Harris PP, Janicot S, Klein C, Panthou G (2017) Frequency of extreme Sahelian storms tripled since 1982 in satellite observations. Nature 544:475–478

Tetzlaff G, Wolter K (1980) Meteorological patterns and the transport of mineral dust from the north African continent. Palaeoecol Afr 12:31–42

Thevenon F, Chiaradia M, Adatte T, Hueglin C, Potél J (2012) Characterization of modern and fossil mineral dust transported to high altitude in the Western Alps: Saharan sources and transport patterns. Adv Meteorol, 674385. https://doi.org/10.1155/2012/674385

Tozer P, Leys J (2013) Dust storms—what do they really cost? Rangeland J 35:131–142

Varga C, Cserháti C, Kovács J, Szalai Z (2016) Saharan dust deposition in the Carpathian Basin and its possible effects on interglacial soil formation. Aeolian Res 22:1–12

Warren A (2019) Global frameworks for aeolian geomorphology. In: Livingstone I, Warren A (eds) Aeolian geomorphology—a new introduction. Wiley, Chichester, pp 1–26

Washington R, Todd M, Middleton NJ, Goudie AS (2003) Dust-storm source areas determined by the total ozone monitoring spectrometer and surface observations. Ann Assoc Am Geogr 93:297–313

Washington R, Bouet C, Cautenet G, Mackenzie E, Ashpole I, Engelstaedter S, Lizcarno G, Henderson GM, Schepanski K, Tegen I (2009) Dust as a tipping element: the Bodélé depression, Chad. PNAS 106:20564–20571

Wheeler DA (1986) The meteorological background to the fall of Saharan dust, November 1984. Meteorol Mag 115:1–9

White JR, Cerveny RS, Balling RC (2012) Seasonality in European red dust/"blood" rain events. Bull Am Meteorol Soc 93:471–476

Williams E, Nathou N, Hicks E, Pontikis C, Russell B, Miller M, Bartholomew MJ (2009) The electrification of dust-lofting gust fronts ('haboobs') in the Sahel. Atmos Res 91:292–298

The African Humid Period and the 'Green Sahara'

15

Jasper Knight

Abstract

The African Humid Period (AHP, ~14.8–5.5 ka BP) is a key time period of the lateglacial and early Holocene when, under a wetter climate regime, areas of the central Sahara that are now dry once experienced flowing rivers and permanent lakes. More extensive vegetation and fauna then developed and this was able to sustain a greater intensity of and likely more sedentary patterns of human activity. For this reason, this period is known informally as the 'Green Sahara'. Evidence for the AHP comes from a range of geomorphic, sedimentary, biological, and archaeological data from across North Africa in particular. In the central Sahara, this evidence provides insight into AHP climates, environments, and the ways in which prehistoric human activity exploited these environmental conditions. This chapter reviews the AHP as a time period, including its onset and termination, the evidence for climatic and environmental conditions during this period, and the richness and diversity of the archaeological record. The AHP provides a good example of the close interrelationships between climate change, land surface processes and human activity, and as such may usefully inform on the sensitivity of both environmental and human systems to climate forcing.

Keywords

Archaeological record · River reactivation · Rock shelter occupation · Pastoralism · Younger Dryas · Pollen record · Dust flux

J. Knight (✉)
School of Geography, Archaeology, Environmental Studies, University of Witwatersrand, Johannesburg 2050, South Africa
e-mail: jasper.Knight@wits.ac.za

15.1 Introduction

The African Humid Period (AHP, ~14.8–5.5 ka BP; Shanahan et al. 2015) is defined according to its major climatic and environmental property—that of greater moisture availability that existed in most areas of Africa during the lateglacial and early Holocene. However, this time period is also significant for the different ways in which climate of the period was imprinted on the geomorphological, ecological, and archaeological record, in particular in the Sahara. As such, the AHP of the Sahara has captured the imagination of archaeologists and the public alike for the richness and diversity of this record, reflecting a regional climate and environment that is so different to today's (Gatto and Zerboni 2015; Coutros 2019). The climatic and environmental context of the AHP is summarised in Fig. 15.1. Despite the palaeoenvironmental imprints of this period and the relatively large literature dealing with human occupation and cultural activities across different parts of Africa (e.g. Burrough and Thomas 2013; Gatto and Zerboni 2015; Coutros 2019), the climatological context of the AHP is not clear. This includes the climatological forcing factors leading to the AHP, any climatic phases within the AHP, the amount and seasonality of annual rainfall, the causes of the termination of the AHP (deMenocal et al. 2000; Timm et al. 2010), and any climate feedbacks between the atmosphere and land surface (Hopcroft et al. 2017; Chen et al. 2020). There is also no agreement on the time span of the AHP as this depends mainly on how the AHP's start and end points are determined and the palaeorecord under consideration. For example, a $\delta^{18}O$ record of African monsoon rainfall from speleothems in Morocco suggests the AHP terminated abruptly at 4 ka (Sha et al. 2019), which is later than that suggested by other records and in other areas (e.g. Amaral et al. 2013; Roubeix and Chalié 2019). The nature of any time-transgressive behaviour in the start/end of the AHP has not been well resolved. These points highlight that there

J. Knight et al. (eds.), *Landscapes and Landforms of the Central Sahara*, World Geomorphological Landscapes, https://doi.org/10.1007/978-3-031-47160-5_15

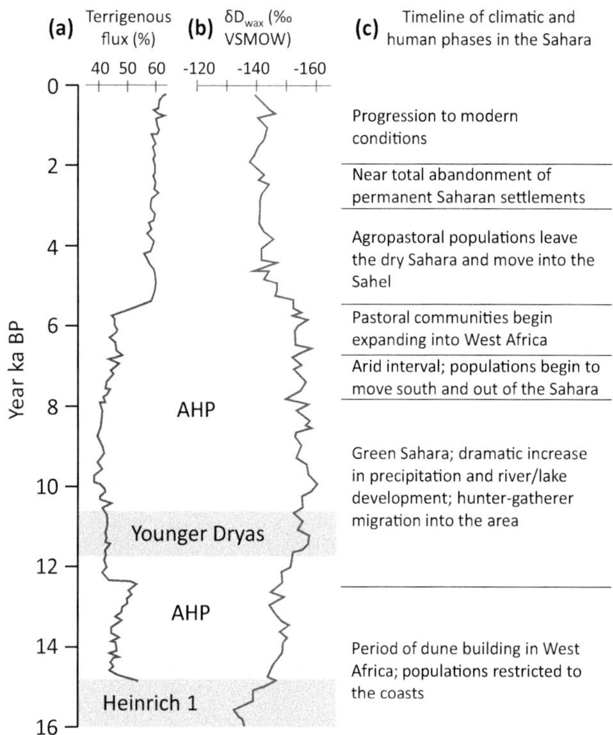

Fig. 15.1 Climatic, environmental and human context of the African Humid Period (time period defined according to Shanahan et al. (2015). **a** terrigenous sediment fluxes (%) in North Atlantic marine core 658C (deMenocal et al. 2000), **b** δD_{wax} from core GeoB4905-4 in the Gulf of Guinea (Collins et al. 2017), **c** timeline of climatic and human phases in the Sahara (Coutros 2019)

are several areas of uncertainty in current knowledge of this time period.

This chapter describes and discusses the evidence for climatic and environmental conditions during the AHP in the central Sahara region, using a range of palaeorecords. Synergies between the physical environment and its resources, and human activity and occupation in the landscape, are critically discussed. This demonstrates the complex and entangled nature of human–environmental relations, and raises interesting research questions on the reasons why and how the central Sahara became seemingly abandoned upon increasingly arid conditions in the mid-Holocene, and the nature of human adaptations to these changing conditions.

15.2 Palaeorecords of the African Humid Period

The AHP is considered to have initially developed across North Africa with strengthening of monsoon rains across Sahelian areas at ~15–14.5 ka, but wetter conditions spread across the region from this period after the Younger Dryas (~11.5–10.7 ka) (deMenocal et al. 2000). This spatial pattern of rainfall was linked to a latitudinal expansion of the intertropical convergence zone (ITCZ) and more persistent monsoon circulation developing in Sahelian areas. Tropical cyclones were also more common during the AHP (Pausata et al. 2017; Marks et al. 2019). As a result, regional-scale river runoff and thus sediment yield to the marine environment also increased, evidenced by more negative $\delta^{18}O$ seawater values and by changes in sediment mineralogy and provenance and sediment yield to marine cores (deMenocal et al. 2000; Arbuszewski et al. 2013). The period of the AHP encompasses the Younger Dryas and this can be seen as a temporary climatic deterioration (cooler, dryer) marked by changes to land surface properties (vegetation, water availability, soil moisture) and by more subdued monsoon conditions (Garcin et al. 2007). The Younger Dryas period is not well resolved in palaeorecords in North Africa.

The progressive development of environmental conditions during the AHP is seen in different palaeorecords across the region. Terrestrial records (e.g. rivers, lakes, wetlands, speleothems) can be considered as in situ and location-specific records that reflect the nature of climate and environment in their immediate area. There are also marine, fluvial deltaic, and aeolian records formed by the deposition of sediments outside of the region. These different types of records are therefore found in different environmental and depositional settings, and this frames how such records can be interpreted. The most commonly used records found in the literature are now presented and discussed thematically.

15.2.1 River and Lake Records

Higher rainfall totals experienced during the AHP led to a greater availability of surface water as well as higher groundwater table positions. This availability of water was manifested in a number of different ways throughout the central Sahara region. Higher rainfall led to increased overland flow, sediment erosion and transport, and accumulation of water and sediment within reactivated river channel systems. Thus, the AHP was characterised by development of both perennial and ephemeral river channels networks, in some cases extending regionally as integrated systems across the Sahara (e.g. Pachur and Kröpelin 1987; Matter et al. 2016; Wu et al. 2017). Palaeodrainage systems were formed in the central Sahara mainly during the Miocene but also in later periods including the AHP, and are discussed by Knight (this volume). Those potentially reactivated during the much more recent AHP are likely to be better preserved on the land surface than earlier systems. However, in most instances, evidence for recent rivers (palaeochannels, fluvial sediments) is indirect and based on the presence of riparian or wetland vegetation as interpreted from palaeoecological records (e.g. Lézine et al. 2011), or from patterns of presently-dry wadi systems that are assumed to have been active during the AHP (e.g. Zerai 2009; Matter et al. 2016). Wadi systems have not been fully mapped across the region, hence the degree to which wetter conditions during the AHP reinvigorated such systems is still largely unknown. Increased river discharge and more extreme flood events along the large perennial Nile River during the AHP have also been reconstructed using geochemical proxies of sediment provenance such as neodymium radiogenic isotopes (εNd), and commonly constrained by radiometric dating (mainly ^{14}C). Results typically show a phased onset and termination of the AHP as it progressed that are largely controlled by the latitudinal extent of ITCZ and monsoon rains (Ménot et al. 2020; Mologni et al. 2020). Thus, atmospheric circulation patterns such as the position of the Congo Air Boundary, and the interplay between Atlantic and Indian Ocean moisture sources, are important (Burrough and Thomas 2013; Castañeda et al. 2016). Dated periods of alluvial aggradation also took place in the NE Atlas Mountains of Morocco (see Depreux et al. 2021, their Fig. 10) spanning the lateglacial and Holocene, but there is not a clear signal of the AHP in the timing of these events. This may indicate the role of episodic Atlantic- or Mediterranean-sourced cyclones in this record. Modelled variations in rainfall oxygen isotopic values during the AHP, calibrated against proxy data from different locations in East and North Africa, suggest changes in the volumetric balance of Atlantic and Indian ocean moisture sources (Tierney et al. 2011). Lake records in East Africa however suggest that there was local thermal amplification during the AHP, as a result of albedo feedbacks between the land surface and vegetation that are unaccounted for by $\delta^{18}O$ and δD records (Berke et al. 2012; Tierney et al. 2017; Chen et al. 2020).

Lake records were also formed during the AHP in some locations, as rainfall increased and the groundwater table rose in the profile. Quade et al. (2018) reviewed the evidence for megalakes within the central Sahara, including megalake Chad (see Knight, this volume). Based on the hydrological requirements to sustain megalakes of the sizes calculated (from 32,000 km^2 for megalake Darfur to 345,000 km^2 for megalake Chad), Quade et al. (2018) considered that this reflects mean annual rainfall of > 1200 mm and displacement of tropical rainbelts by some 1000 km. A significant issue is that the evolution of individual lakes over time is not well constrained, and many palaeoshorelines are undated. This means that it is not clear whether different lakes experienced similar evolutionary phases, in correspondence with rainfall variations, or whether they evolved independently of climate or each other. This may be especially the case for lakes whose catchments draw on rainfall outside of the region. Tracking lake level changes in individual lakes, through a combination of luminescence and radiocarbon dating, sediment patterns and altimetric surveys, can inform on snapshots of megalake environments during the AHP. For example, a northern sector of megalake Chad shows a delta formed at 7000 BP with in situ bioclastic sediment layers containing molluscs and ostracods (Bristow et al. 2018). Isotopic analysis shows that high evaporation rates existed and that it is likely that lake level position here reflects the regional climatic-controlled water budget. The dynamics of these megalakes, however, also reflect their topographic contexts, including land surface slope (which has implications for lake hypsometry) and the presence of any low-elevation sills that allowed for lake water spillage into adjacent catchments to flow in different directions at different times. This can result in significant and dramatic changes in lake surface elevation and lake area, such as Lake Turkana which repeatedly varied in surface elevation by < 70 m during the early Holocene and within the AHP (Forman et al. 2014). In palaeolake records in the Fazzan Basin, Libya, intraformational gypsum layers attest to dry lakebed conditions whereas bioturbated soil horizons suggest a vegetated lake or wetland fringe alongside a lake or river (Drake et al. 2011). Similar evidence is found in a Holocene footslope record near the Mediterranean coast of Libya where palustrine and aeolian sediments are variously interbedded with calcareous duricrusts, buried soils, and lithic artefacts, with the stratigraphy constrained by radiocarbon ages (Giraudi et al. 2012). Taken together, this evidence indicates wetter conditions from 9.4 to 5.0 ka with the presence of footslope wetlands sustained by high water tables and downslope runoff. Short arid episodes marked by aeolian sediment pulses took place at 8.2 ka and 5.5–5.4 ka. The period 5.0–4.4 ka was transitional towards late Holocene aridity when both wet and dry environments coexisted. A palaeolake record from NE Niger, in the southern part of the central Sahara, shows deposition of organic-rich diatomite from 10.6 to 7.3 cal ka BP followed by a significant shift towards dryer conditions at 7.3 ka (Brauneck et al. 2013). This period is marked by increased clastic sediment input and transition to a full sabkha environment sometime after 6 ka. It is notable that evidence for aridity from 7.3 ka at this southernmost site is earlier than for other sites farther north (e.g. Shanahan et al. 2015).

15.2.2 Biome and Pollen Records

Different sedimentary and ecological records provide evidence for changes in land surface processes and environments during the AHP, in particular ecosystem responses to wetter climate conditions. Specific evidence comes from pollen records from individual wetland or lake sites. For example, around megalake Chad maximal humid conditions (rainfall ~800 mm per year) and woodland development took place between about 6.7 and 6.0 ka (Amaral et al. 2013). A key indicator here the presence of pollen grains from the genus *Uapaca*, a native Afrotropical tree typically found in miombo woodlands of central Africa today, and therefore indicating wetter conditions. Different North African lake records through the AHP show different timings in the increase, peak, and decrease of *Uapaca* pollen but with peaks at ~8.2 ka and 6.2 ka (Amaral et al. 2013, their Fig. 10). This may indicate time-progressive biome responses to the transit of rainbelts across North Africa.

Phelps et al. (2020) described covariability between forest (wetter conditions) and grass biomes (dryer conditions) across North Africa through the AHP. They show that changes in biome area are relatively insensitive to climate variability, and instead biomes persisted throughout the AHP and through periods such as the Younger Dryas under lower rainfall regimes. This persistence of biomes is argued to reflect autogenic processes taking place within individual biomes themselves, rather than in response to external forcing. Watrin et al. (2009) argued that, as different biomes developed, biodiversity would have also increased as a result of progressive development of more complex gallery forest and supporting secondary plant understoreys as well as animal species migration. It is also likely that biodiversity hotspots, then as now, were focused on areas of permanent water and wetlands. Based on phytolith evidence from the Fazzan Basin, Libya, Parker et al. (2008) showed that C_4 grassland existed during the AHP but that there was a transition to C_3 grassland after the end of the AHP as a response to changes in rainfall seasonality as well as annual rainfall totals (e.g. Tierney et al. 2017).

15.2.3 Dust Records

The central and southern Sahara region is today a primary site of global dust production which is exported from the region by high-level winds (see Bristow, this volume). An increase in surface wetness and vegetation during the AHP suggests decreased or more episodic and variable dust fluxes existed. Palchan and Torfstein (2019) report a flux decrease of 50% during the AHP based on dust contained within marine cores in the Red Sea and Gulf of Aden. However, data resolution in the examined cores is low. A similar dust record from the Ionian Sea shows peaks corresponding to Heinrich Event 1 (16.2–15.1 cal ka BP; Hodell et al. 2017) and the Younger Dryas, but more subdued dust export during the period ~ 10–8 ka, in the main period of the AHP and corresponding to deposition of Sapropel 1 in the Mediterranean basin (Ehrmann et al. 2017). It is notable that the dust record, reflecting dry land surface conditions, has a broadly antiphase relationship with many other proxies during the AHP that reflect humidity rather than aridity. Using a climate model, Pausata et al. (2016) showed that the West African monsoon during the AHP extended 500 km farther north than present and that this was amplified by vegetation feedbacks that also helped to suppress dust fluxes. A lake record from the Moroccan Atlas Mountains shows millennial-scale pulses in dust fluxes through the AHP and into the Holocene that are interpreted to reflect North Atlantic Oscillation mode in combination with North Atlantic Heinrich Event 1 and weakened West African monsoon phases (Zielhofer et al. 2017).

15.2.4 Archaeological Records

Wetter conditions, the presence of perennial water and plant and animal food sources meant that sedentary human activity thrived in the central Sahara during the AHP as well as provided the resources to sustain migrating populations. Much archaeological work has examined evidence for human activity and cultural patterns at this time, mainly from cave and rock shelter sites where evidence is well preserved. There is by contrast very little evidence for or from open-air sites such as at the margins of rivers and lakes (e.g. Khalidi et al. 2020) where sites may have been seasonal or where human groups may have been more vulnerable. Cremaschi and Zerboni (2009) mapped the distribution of Middle Pastoral-Neolithic (~7.4–5.0 ka) hearths on the margins of a lowland palaeodelta in SW Libya, and these exactly match the margins of a river palaeochannel, suggesting that people made temporary use of water and food sources available in this location.

Caves and rock shelters containing archaeological records are extensively found around the margins of bedrock massifs throughout the central Sahara, such as the Tadrart Acacus, SW Libya (see Balbo, this volume). Here, cave-floor stratigraphies < 3 m thick comprise aeolian sand with cave roof-collapsed blocks, unconformably overlain by colluvial sand beds with humic soils, charcoal and thin aeolian interbeds (Cremaschi 1998). The basal aeolian sands are luminescence dated to 70–90 ka (MIS 4 and 5a–c). The overlying units interpreted as human occupation layers date to the Holocene. It is notable that there is a significant depositional hiatus between these periods, possibly related to absence of humid conditions. Pollen evidence through the Holocene part of the profile shows a dominance of Poaceae-Cyperaceae with some shrubs present (Asteraceae + Chenopodiaceae) particularly in the early Holocene, suggesting dryer conditions (Cremaschi et al. 2014). This does not indicate any significant change in biomes or in human occupation through the end of the AHP. Cremaschi et al. (2014) argue that there were changes in water bodies in the surrounding environment although this is not based on any clear evidence. However, human occupation of cave and rock shelter sites throughout the central Sahara during the AHP is marked by the presence of rock art and petroglyphs, most commonly of animals (see Gallinaro, this volume). These depictions can be interpreted to reflect the animals, and thus environments, of the surrounding areas at the time of their creation (Guagnin 2014; Ben Nasr and Walsh 2020). Rock art depictions and the location of rock art sites in the landscape appear to reflect herding and animal migration routes (Barnett and Guagnin 2014), and this is likely to have been environmentally determined by topography, food, and water.

Increased evidence for human presence during the AHP, facilitated by greater water availability, has also provided a richer range of evidence for cultural and subsistence practises (Coutros 2019). There is therefore a vigorous debate in the literature regarding the extent of hunter-gatherer activities *versus* pastoralism, thus to what extent human populations were sedentary or mobile through the Saharan landscape (di Lernia 2002; Cancellieri and di Lernia 2014). This debate is informed by radiocarbon dating that indicates the longevity of occupation of individual sites, the presence of lithic toolkits, pottery styles, and the range of food materials present (e.g. cereal grains, bones of key species such as Barbary sheep *Ammotragus lervia*). Sedentary or mobile hunter-gatherer lifestyles are difficult to evaluate because of a lack of evidence for human activity in the region prior to the wetter AHP, and wetter conditions may have simply enhanced mobility (Castañeda et al. 2009), irrespective or any change in lifestyle. Stojanowski and Knudson (2014) argue that mid-Holocene aridity was a driving factor for changes in mobility strategies in the region. Manning and Timpson (2014) highlight that wetter conditions also increase the population carrying capacity, and this in itself can lead to different social and cultural responses if human populations are competing for resources and occupation of key sites. Gatto and Zerboni (2015) explore this theme in detail, arguing that incipient social and economic complexity arises as a result of intensified food production in the landscape, and as resources and territory are exploited and fought-over. Brass (2019) argued that more complex social structures and communities developed in the central Sahara during the late AHP, as a result of change resource usage and locations of settlements. First (~11.3–10.0 ka), small lowland and mountain sites were exploited, with specialised hunting and Barbary sheep kept in cave enclosures. Then (~10.0–8.3 ka), more diverse resources were exploited, primary settlements were located in mountains, and cultural practises related to pottery and sheep herding were developed. Later (~8.3–7.4 ka), settlements in both mountains and lowlands were maintained with seasonal herding patterns. Finally (~7.0–5.75 ka), a mixed hunting and pastoral economy developed with a settlement focus in lowland areas, and vertical transhumance was practised (where sheep and goats were herded into the mountains during the dry season) (Brass 2019, his Table 3) (see also Blench, this volume).

15.3 The Termination of the AHP

The onset of the AHP is much less well constrained than the termination of the AHP. Some records show a relatively abrupt termination based on the individual proxy examined (e.g. Sha et al. 2019) whereas other proxies show more gradual changes from which it is more difficult to identify any start and/or end points (e.g. Castañeda et al. 2016). The greater availability of mid-Holocene age records across North Africa, however, allows for a more detailed and higher-resolution analysis of the termination of the AHP (e.g. deMenocal et al. 2000; Renssen et al. 2003; Garcin et al. 2007; Shanahan et al. 2015; Collins et al. 2017). A progressive change to more arid conditions took place from ~7.2 ka onwards (Shanahan et al. 2015) but different records can be used to illustrate the spatial and temporal patterns of this termination period. For example, Mediterranean marine records show a gradual development towards the peak of the AHP at 9 ka, but $\delta^{18}O$ proxies indicative of ocean ventilation appear relatively insensitive to a dramatic decrease in Nile River discharge at 7 ka (Tachikawa et al. 2015). A leaf wax record from south of the Sahara in the Gulf of Guinea shows the AHP termination after 6 ka but very little variability within the AHP itself (Collins et al. 2017). Farther north, marine core 658C from offshore North Africa and adjacent to Saharan source areas shows a rapid increase in terrigenous and decreased carbonate materials marking the end of the AHP at 5.5 ka (deMenocal et al. 2000), but biogenic opal decreases progressively after the Younger Dryas and $\delta^{18}O$ values show no variability at all across this period. Shanahan et al. (2015) argue that a southward shift of the monsoon rainbelt in West Africa triggered a time-transgressive termination of the AHP, in a north to south direction, between 6 and 3 ka. However, this was based on record wiggle-matching only, without any clear support. Further, a numerical model simulating the AHP termination period of 7.5–5.5 ka shows decadal to centennial climate fluctuations throughout this time (Renssen et al. 2003), typical of a climate transition. This is likely a more realistic representation of the end of the AHP. Wright (2017) argued that widespread pastoralism and changes in land surface vegetation and water availability led to climate feedbacks, although whether human activity forced a rapid termination of the AHP, or whether agricultural adaptation prolonged the AHP, is a matter of debate (Gatto and Zerboni 2015; Brierley et al. 2018).

15.4 Discussion

The AHP is a critical time period in the development of the contemporary Sahara because it describes the co-relationships between climate and environmental change, and responses by human societies and cultures that are imprinted in the Saharan landscape. It also represents the most recent wet period of the central Sahara before the onset of mid-Holocene aridity that therefore allowed for the preservation of evidence for wet geomorphic systems, such as wadi channels that can be observed in the Sahara today. Evidence found in the central Sahara for climate and environmental conditions during the AHP are quite diverse in terms of record type, but large areas of the Sahara do not contain any significant records and there is also a general absence of long (millennial-scale) records. These limitations mean that the detailed terrestrial signatures of the AHP are not well resolved, and marine proxy records from the adjacent Atlantic and Mediterranean have therefore been focused on instead, despite these providing only a regional averaging of environmental conditions. For example, dust fluxes into marine cores are not well constrained in terms of either geochemical and isotopic constraints on sediment source areas, or in their timing, because of the sampling resolution and low accumulation rates of these marine records. Other proxy records elsewhere in Africa also resolve aspects of the AHP, its start and termination, as well as some phases within the AHP as a whole (Chase et al. 2010; Burrough and Thomas 2013; Truc et al. 2013).

Climate modelling and isotopic data for the AHP indicate zonal changes in synoptic circulation, rainfall patterns, and water balance (Tierney et al. 2011; Hopcroft et al. 2017). This highlights that the AHP was most likely forced by insolation, where the 30°N insolation curve exactly matches the duration and peak timing of the AHP (Collins et al. 2017). On top of this was climate amplification through land surface feedbacks (e.g. Chandan and Peltier 2020; Chen et al. 2020) rather than by ocean-driven perturbation. The large size of North Africa and thus its continentality effects helped ensure its relative insensitivity to direct ocean forcing. What is significant, however, is that the AHP is interrupted by shorter dryer phases of the Younger Dryas and the 8.2 ka event (Garcin et al. 2007). Although evidence for these dryer phases exists in various palaeorecords in the region, the fact that conditions of the AHP apparently became rapidly re-established suggests (1) only limited impacts of these events in North Africa when compared to the midlatitude Northern Hemisphere, or that (2) the low-latitude West African monsoon and ITCZ circulation at this time was relatively insensitive to higher latitude forcing. However, based on a $\delta^{15}N$ record including the AHP from a rock hyrax (*Procavia capensis*) midden in Namibia, Chase et al. (2013) argued against a low-latitude insolation forcing model for the AHP. Evaluating the impacts of events such as the Younger Dryas, 8.2 ka event and Heinrich Event 1 is an important area of future research. Progressive weakening of the AHP in the period ~7–5 ka reflects a weakening of insolation forcing and thus zonal circulation and rainfall patterns (Garcin et al. 2007; Shanahan et al. 2015; Collins et al. 2017). It also begs the question as to whether Saharan palaeorecords show similar evidence during other climate transitions (e.g. MIS 6 to 5, in the Middle Pleistocene).

The AHP has been argued to be critical to the understanding of human migration patterns from central and eastern Africa to the Middle East (e.g. Coulthard et al. 2013). Perennial water availability and riparian vegetation allowed for human and animal populations to exist throughout the central Sahara, and for these rivers to act as migration corridors. These elements of AHP development and the nature of feedbacks between different elements of climate, the land surface and human activity are shown in Fig. 15.2. Coutros (2019) described the human settlement and cultural responses in the central Sahara to climate during the AHP, with a focus on subsistence strategies such as the development of pastoralism. Understanding such relationships can inform on interconnections between the physical and human environments of the Sahara, and raises interesting research questions on the reasons why and how the central Sahara became seemingly abandoned upon increasingly arid conditions in the mid-Holocene.

15.5 Summary and Outlook

The AHP is a period of rapid change in the recent history of the central Sahara, and it demonstrates the close interplay between climate and environmental changes in landscape resources, and how these were exploited or adapted to by Saharan peoples (Gatto and Zerboni 2015). Although the concept of the 'Green Sahara' is an attractive proposition, the existing palaeorecords of the AHP offer only a partial snapshot of the period, and there are many spatial and temporal gaps. Many research questions thus remain, including river dynamics; slope and floodplain processes away from river channels; megalake development and ecosystems; and detailed patterns of human occupation in the region—Cremaschi and Zerboni (2009, their Fig. 6) mapped hundreds of potential occupation sites in just a small region of the SW Fazzan; many tens of thousands must still exist across the Sahara as a whole. The wider hemispheric-scale climatic setting of the AHP is also under-researched, but this offers opportunities to explore teleconnections between

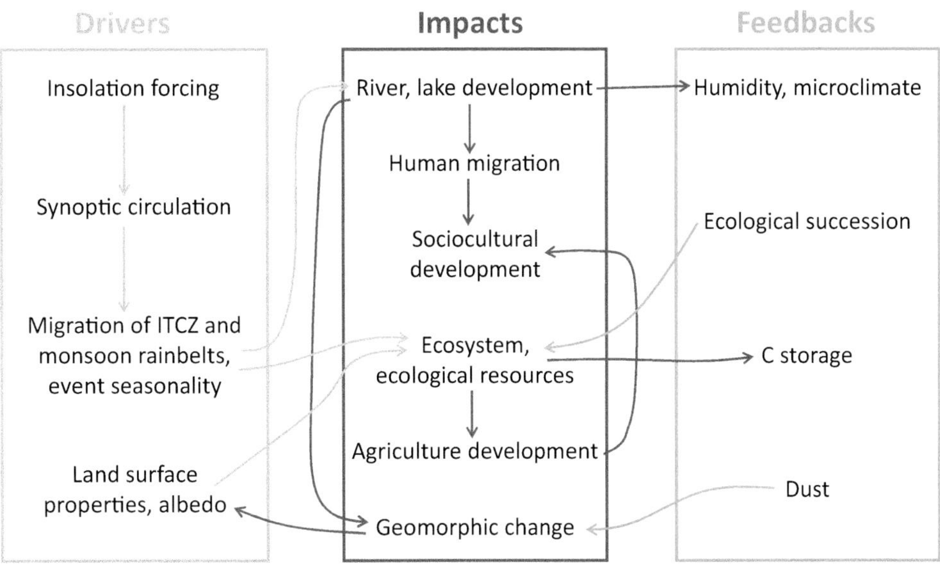

Fig. 15.2 Illustration of the drivers and land surface responses and feedbacks found within the central Sahara region during the AHP. The arrows indicate the forcings and responses between different variables

low- and midlatitudes, the dynamics of the ITCZ and the West African monsoon, and the low-latitude reach of high-latitude cool episodes such as the Younger Dryas (e.g. Chase et al. 2010; Truc et al. 2013).

Acknowledgements Stefano Biagetti and Jennifer Fitchett are thanked for their comments on an earlier version of this chapter.

References

Amaral PGC, Vincens A, Guiot J, Buchet G, Deschamps P, Doumnang J-C, Sylvestre F (2013) Palynological evidence for gradual vegetation and climate changes during the African Humid Period termination at 13° N from a Mega-Lake Chad sedimentary sequence. Clim Past 9:223–241

Arbuszewski JA, deMenocal PB, Cléroux C, Bradtmiller L, Mix A (2013) Meridional shifts of the Atlantic intertropical convergence zone since the last glacial maximum. Nat Geosci 6:959–962

Barnett T, Guagnin M (2014) Changing places: rock art and holocene landscapes in the Wadi al-Ajal, South-West Libya. J Afr Archaeol 12:165–182

Ben Nasr J, Walsh K (2020) Environment and rock art in the Jebel Ousselat, Atlas mountains, Tunisia. J Med Archaeol 33:3–28

Berke MA, Johnson TC, Werne JP, Schouten S, Sinninghe Damsté JS (2012) A mid-Holocene thermal maximum at the end of the African Humid Period. Earth Planet Sci Lett 351–352:95–104

Brass M (2019) The emergence of mobile pastoral elites during the middle to late Holocene in the Sahara. J Afr Archaeol 17:53–75

Brauneck J, Mees F, Baumhauer R (2013) A record of early to middle Holocene environmental change inferred from lake deposits beneath a Sabkha sequence in the Central Sahara (Seggedim, NE Niger). J Paleolimnol 49:605–618

Brierley C, Manning K, Maslin M (2018) Pastoralism may have delayed the end of the green Sahara. Nat Comm 9:4018. https://doi.org/10.1038/s41467-018-06321-y

Bristow CS, Holmes JA, Mattey D, Salzmann U, Sloane HJ (2018) A late Holocene palaeoenvironmental 'snapshot' of the Angamma Delta, Lake Megachad at the end of the African Humid Period. Quat Sci Rev 202:182–196

Burrough SL, Thomas DSG (2013) Central southern Africa at the time of the African Humid Period: a new analysis of Holocene palaeoenvironmental and palaeoclimate data. Quat Sci Rev 80:29–46

Cancellieri E, di Lernia S (2014) Re-entering the central Sahara at the onset of the Holocene: a territorial approach to early Acacus hunter-gatherers (SW Libya). Quat Int 320:43–62

Castañeda IS, Mulitza S, Schefuß E, Lopes dos Santos RA, Sinninghe Damsté JS, Schouten S (2009) Wet phases in the Sahara/Sahel region and human migration patterns in North Africa. PNAS 106:20159–20163

Castañeda IS, Schouten S, Pätzold J, Lucassen F, Kasemann S, Kuhlmann H, Schefuß E (2016) Hydroclimate variability in the Nile River Basin during the past 28,000 years. Earth Planet Sci Lett 438:47–56

Chandan D, Peltier WR (2020) African Humid Period precipitation sustained by robust vegetation, soil, and lake feedbacks. Geophys Res Lett 47:e2020GL088728. https://doi.org/10.1029/2020GL088728

Chase BM, Meadows ME, Carr AS, Reimer PJ (2010) Evidence for progressive Holocene aridification in southern Africa recorded in Namibian hyrax middens: Implications for African Monsoon dynamics and the "African Humid Period." Quat Res 74:36–45

Chen W, Ciais P, Zhu D, Ducharne A, Viovy N, Qiu C, Huang C (2020) Feedbacks of soil properties on vegetation during the Green Sahara period. Quat Sci Rev 240:106389. https://doi.org/10.1016/j.quascirev.2020.106389

Collins JA, Prange M, Caley T, Gimeno L, Beckmann B, Mulitza S, Skonieczny C, Roche D, Schefuß E (2017) Rapid termination of the African Humid Period triggered by northern high-latitude cooling. Nat Comm 8:1372. https://doi.org/10.1038/s41467-017-01454-y

Coulthard TJ, Ramirez JA, Barton N, Rogerson M, Brücher T (2013) Were rivers flowing across the Sahara during the last interglacial? implications for human migration through Africa. PLoS ONE 8:e74834. https://doi.org/10.1371/journal.pone.0074834

Coutros PR (2019) A fluid past: socio-hydrological systems of the West African Sahel across the long durée. WiRES Water 6:e1365. https://doi.org/10.1002/wat2.1365

Cremaschi M (1998) Late Quaternary geological evidence for environmental changes in south-western Fezzan (Libyan Sahara). In: Cremaschi M, di Lernia S (eds) Wadi Teshuinat—Palaeoenvironment and Prehistory in south-western Fezzan (Libyan Sahara). CNR, Roma-Milano, pp 13–48

Cremaschi M, Zerboni A (2009) Early to Middle Holocene landscape exploitation in a drying environment: two case studies compared from the central Sahara (SW Fezzan, Libya). C R Geosci 341:689–702

Cremaschi M, Zerboni A, Mercuri AM, Olmi L, Biagetti S, di Lernia S (2014) Takarkori rock shelter (SW Libya): an archive of Holocene climate and environmental changes in the central Sahara. Quat Sci Rev 101:36–60

deMenocal P, Ortiz J, Guilderson T, Adkins J, Sarnthein M, Baker L, Yarusinsky M (2000) Abrupt onset and termination of the African Humid Period: rapid climate responses to gradual insolation forcing. Quat Sci Rev 19:347–361

Depreux B, Lefèvre D, Berger J-F, Segaoui F, Boudad L, El Harradji A, Degeai J-P, Limondin-Lozouet N (2021) Alluvial records of the African Humid Period from the NW African highlands (Moulouya basin, NE Morocco). Quat Sci Rev 255:106807. https://doi.org/10.1016/j.quascirev.2021.106807

di Lernia S (2002) Dry climatic events and cultural trajectories: adjusting middle holocene pastoral economy of the Libyan Sahara. In: Hassan FA (ed) Droughts, food and culture: ecological change and food security in Africa's later prehistory. Kluwer Academic Publishers, New York, pp 225–250

Drake N, Salem M, Armitage S, Francke J, Hounslow M, Hlal O, White K, El-Hawat A (2011) Palaeohydrology and palaeoenvironment: initial results and report of 2010 and 2011 fieldwork. Libyan Stud 42:139–149

Ehrmann W, Schmiedl G, Beuscher S, Krüger S (2017) Intensity of African Humid Periods estimated from Saharan dust fluxes. PLoS ONE 12:e0170989. https://doi.org/10.1371/journal.pone.0170989

Forman SL, Wright DK, Bloszies C (2014) Variations in water level for Lake Turkana in the past 8500 years near Mt. Porr, Kenya and the transition from the African Humid Period to Holocene aridity. Quat Sci Rev 97:84–101

Garcin Y, Vincens A, Williamson D, Buchet G, Guiot J (2007) Abrupt resumption of the African monsoon at the Younger Dryas-Holocene climatic transition. Quat Sci Rev 26:690–704

Gatto MC, Zerboni A (2015) Holocene Supra-regional environmental changes as trigger for major socio-cultural processes in Northeastern Africa and the Sahara. Afr Archaeol Rev 32:301–333

Giraudi C, Mercuri AM, Esu D (2012) Holocene palaeoclimate in the northern Sahara margin (Jefara Plain, northwestern Libya). Holocene 23:339–352

Guagnin M (2014) Patina and environment in the Wadi al-Hayat: towards a chronology for the rock art of the Central Sahara. Afr Archaeol Rev 31:407–423

Hodell DA, Nicholl JA, Bontognali TRR, Danino S, Dorador J, Dowdeswell JA, Einsle J, Kuhlmann H, Martrat B, Mleneck-Vautravers MJ, Rodríguez-Tovar FJ, Röhl U (2017) Anatomy of Heinrich Layer 1 and its role in the last deglaciation. Paleoceanography 32:284–303

Hopcroft PO, Valdes PJ, Harper AB, Beerling DJ (2017) Multi vegetation model evaluation of the Green Sahara climate regime. Geophys Res Lett 44:6804–6813

Khalidi L, Mologni C, Ménard C, Coudert L, Gabriele M, Davtian G, Cauliez J, Lesur J, Bruxelles L, Chesnaux L, Redae BE, Hainsworth E, Doubre C, Revel M, Schuster M, Zazzo A (2020) 9000 years of human lakeside adaptation in the Ethiopian Afar: fisher-foragers and the first pastoralists in the Lake Abhe basin during the African Humid Period. Quat Sci Rev 243:106459. https://doi.org/10.1016/j.quascirev.2020.106459

Lézine A-M, Zheng W, Braconnot P, Krinner G (2011) Late Holocene plant and climate evolution at Lake Yoa, northern Chad: pollen data and climate simulations. Clim past 7:1351–1362

Manning K, Timpson A (2014) The demographic response to Holocene climate change in the Sahara. Quat Sci Rev 101:28–35

Marks L, Welc F, Milecka K, Zalat A, Chen Z, Majecka A, Nitychoruk J, Salem A, Sun Q, Szymanek M, Gałecka I, Tołoczko-Pasek A (2019) Cyclonic activity over northeastern Africa at 8.5–6.7 cal kyr B.P., based on lacustrine records in the Faiyum Oasis, Egypt. Palaeogeogr Palaeoclimatol Palaeoecol 528:120–132

Matter A, Mahjou A, Neubert E, Preusser F, Schwalb A, Szidat S, Wulf G (2016) Reactivation of the Pleistocene trans-Arabian Wadi ad Dawasir fluvial system (Saudi Arabia) during the Holocene humid phase. Geomorphology 270:88–101

Ménot G, Pivot S, Bouloubassi I, Davtian N, Hennekam R, Bosch D, Ducassou E, Bard E, Migeon S, Revel M (2020) Timing and stepwise transitions of the African Humid Period from geochemical proxies in the Nile deep-sea fan sediments. Quat Sci Rev 228:106071. https://doi.org/10.1016/j.quascirev.2019.106071

Mologni C, Revel M, Blanchet C, Bosch D, Develle A-L, Orange F, Bastian L, Khalidi L, Ducassou E, Migeon S (2020) Frequency of exceptional Nile flood events as an indicator of Holocene hydro-climatic changes in the Ethiopian Highlands. Quat Sci Rev 247:106543. https://doi.org/10.1016/j.quascirev.2020.106543

Pachur H-J, Kröpelin S (1987) Wadi Howar: paleoclimatic evidence from an extinct river system in the Southeastern Sahara. Science 237:298–300

Palchan D, Torfstein A (2019) A drop in Sahara dust fluxes records the northern limits of the African Humid Period. Nat Comm 10:3803. https://doi.org/10.1038/s41467-019-11701-z

Parker A, Harris B, White K, Drake N (2008) Phytoliths as indicators of grassland dynamics during the Holocene from lake sediments in the Ubari sand sea, Fazzan Basin, Libya. Libyan Stud 39:29–40

Pausata FSR, Messori G, Zhang Q (2016) Impacts of dust reduction on the northward expansion of the African monsoon during the Green Sahara period. Earth Planet Sci Lett 434:298–307

Pausata FSR, Emanuel KA, Chiacchio M, Diro GT, Zhang Q, Sushama L, Stager JC, Donnelly JP (2017) Tropical cyclone activity enhanced by Sahara greening and reduced dust emissions during the African Humid Period. PNAS 114:6221–6226

Phelps LN, Chevalier M, Shanahan TM, Aleman JC, Courtney-Mustaphi C, Kiahtipes CA, Broennimann O, Marchant R, Shekeine J, Quick LJ, Davis BAS, Guisan A, Manning K (2020) Asymmetric response of forest and grassy biomes to climate variability across the African Humid Period: influenced by anthropogenic disturbance? Ecography 43:1118–1142

Quade J, Dente E, Armon M, Ben Dor Y, Morin E, Adam O, Enzel Y (2018) Megalakes in the Sahara? a review. Quat Res 90:253–275

Renssen H, Brovkin V, Fichefet T, Goosse H (2003) Holocene climate instability during the termination of the African Humid Period. Geophys Res Lett 30:1184. https://doi.org/10.1029/2002GL016636

Roubeix V, Chalié F (2019) New insights into the termination of the African Humid Period (5.5 ka BP) in central Ethiopia from detailed analysis of a diatom record. J Paleolimnol 61:99–110

Sha L, Ait Brahim Y, Wassenburg JA, Yin J, Peros M, Cruz FW, Cai Y, Li H, Du W, Zhang H, Edwards RL, Cheng H (2019) How far north did the African Monsoon fringe expand during the African Humid Period? insights from Southwest Moroccan speleothems. Geophys Res Lett 46:14093–14102

Shanahan TM, McKay NP, Hughen KA, Overpeck JT, Otto-Bliesner B, Heil CW, King J, Scholz CA, Peck J (2015) The time-transgressive termination of the African Humid Period. Nat Geosci 8:140–144

Stojanowski CM, Knudson KJ (2014) Changing patterns of mobility as a response to climatic deterioration and aridification in the middle Holocene Southern Sahara. Am J Phys Anthropol 154:79–93

Tachikawa K, Vidal L, Cornuault M, Garcia M, Pothin A, Sonzogni C, Bard E, Menot G, Revel M (2015) Eastern Mediterranean Sea circulation inferred from the conditions of S1 sapropel deposition. Clim Past 11:855–867

Tierney JE, Lewis SC, Cook BI, LeGrande AN, Schmidt GA (2011) Model, proxy and isotopic perspectives on the East African Humid Period. Earth Planet Sci Lett 307:103–112

Tierney JE, Pausata FSR, deMenocal PB (2017) Rainfall regimes of the Green Sahara. Sci Adv 3:e1601503. https://doi.org/10.1126/sciadv.1601503

Timm O, Köhler P, Timmermann A, Menviel L (2010) Mechanisms for the onset of the African Humid Period and Sahara greening 14.5–11 ka BP. J Clim 23:2612–2633

Truc L, Chevalier M, Favier C, Cheddadi R, Meadows ME, Scott L, Carr AS, Smith GF, Chase BM (2013) Quantification of climate change for the last 20,000 years from Wonderkrater, South Africa: implications for the long-term dynamics of the intertropical convergence zone. Palaeogeogr Palaeoclimatol Palaeoecol 386:575–587

Watrin J, Lézine A-M, Hély C (2009) Plant migration and plant communities at the time of the "Green Sahara." C R Geosci 341:656–670

Wright DK (2017) Humans as agents in the termination of the African Humid Period. Front Earth Sci 5:4. https://doi.org/10.3389/feart.2017.00004

Wu J, Liu Z, Stuut J-BW, Zhao Y, Schirone A, de Lange GJ (2017) North-African paleodrainage discharges to the central Mediterranean during the last 18,000 years: a multiproxy characterization. Quat Sci Rev 163:95–113

Zerai K (2009) Chronostratigraphy of holocene alluvial archives in the Wadi Sbeïtla basin (central Tunisia). Géomorphol: Relief, Proc, Environ, 271–286

Zielhofer C, von Suchodoletz H, Fletcher WJ, Schneider B, Dietze E, Schlegel M, Schepanski K, Weninger B, Mischke S, Mikdad A (2017) Millennial-scale fluctuations in Saharan dust supply across the decline of the African Humid Period. Quat Sci Rev 171:119–135

Human Activity and Occupation in the Sahara

16

Stefano Biagetti

Abstract

The long history of human activity and occupation in the Sahara is inextricably linked to environmental and climatic events that have occurred throughout the last 7 million years. Oscillations between wet and arid conditions marked the occupation pulses of early hominins and hominids, up to historical societies. In this chapter, the human occupation history in the Sahara is reviewed, highlighting the variety of settlement, economic, and cultural responses that arose as a consequence of changes in physiographic and ecological settings in the Sahara through time.

Keywords

Sahara · Climatic oscillations · Human responses · Adaptation · Pleistocene · Holocene

16.1 Climatic Pulses, Explorers, and Rivers of Sand

Climatic and environmental factors have always shaped the history of *Homo* and their ancestors. Yet, the Sahara is a special place, where the shifts from arid to wet and vice versa (Grant et al. 2017) have always profoundly affected different biota. Due to its diverse geomorphology and physiography, some places of the Sahara can switch from formidably challenging lands to verdant areas and back again, with implications for human occupation. On the other hand, central Saharan mountains like the Hoggar, the Tassili, the Tibesti, and the Aïr may remain suitable to human occupation even during extreme aridity (Fig. 16.1). These elevated areas act as reservoirs of natural resources, and were periodically interconnected by waterways generated by humid pulses through time (Drake et al. 2011) that facilitated the dispersals of animals and humans. The 'green corridors' periodically conveyed movements of people across the Sahara, accelerating their diffusion over a vast territory. Therefore, the interplay between climate and local geomorphological factors is key to understand the history of population pulses that occurred in the different areas of the Sahara, from earliest times to the present day. It is widely accepted that climatic shifts and fluctuations have propelled the evolution of *Homo*. Different proxies indicate that warm/cold climate fluctuations became substantial around 6 Ma, further strengthening from 3.0 to 2.8 Ma onwards (Potts 2012). The increased severity of these episodes played a major role in activating hominin and, later on, human evolution in response to increasing instability and variability in physiographic and ecological settings in the Sahara (Maslin and Trauth 2009; Carrión et al. 2011; Potts 2013).

16.2 Earlier Times (7–3 Ma): From the Miocene to the Pliocene

It is well known that human evolution mostly took place in southern and eastern Africa. Yet, two Saharan discoveries have added further pieces to the history of humankind. A nearly complete cranium of a new species, named *Sahelanthropus tchadensis* (Brunet et al. 2002), was found in 2001 in the Toros-Menalla area of the Djurab Desert of northern Chad, followed by further such material (Brunet et al. 2005). *Sahelanthropous* may be the earliest hominin evidence so far known, dating back to 7 Ma (Lebatard et al. 2008). *Sahelanthropus* may have dwelt in a woody environment although reconstruction of its precise habitat has not been fully accomplished (Vignaud

S. Biagetti (✉)
Departament d'Humanitats, Universitat Pompeu Fabra, Ramon Trias Fargas 25-27, 08005 Barcelona, Spain
e-mail: stefano.biagetti@upf.edu

J. Knight et al. (eds.), *Landscapes and Landforms of the Central Sahara*, World Geomorphological Landscapes, https://doi.org/10.1007/978-3-031-47160-5_16

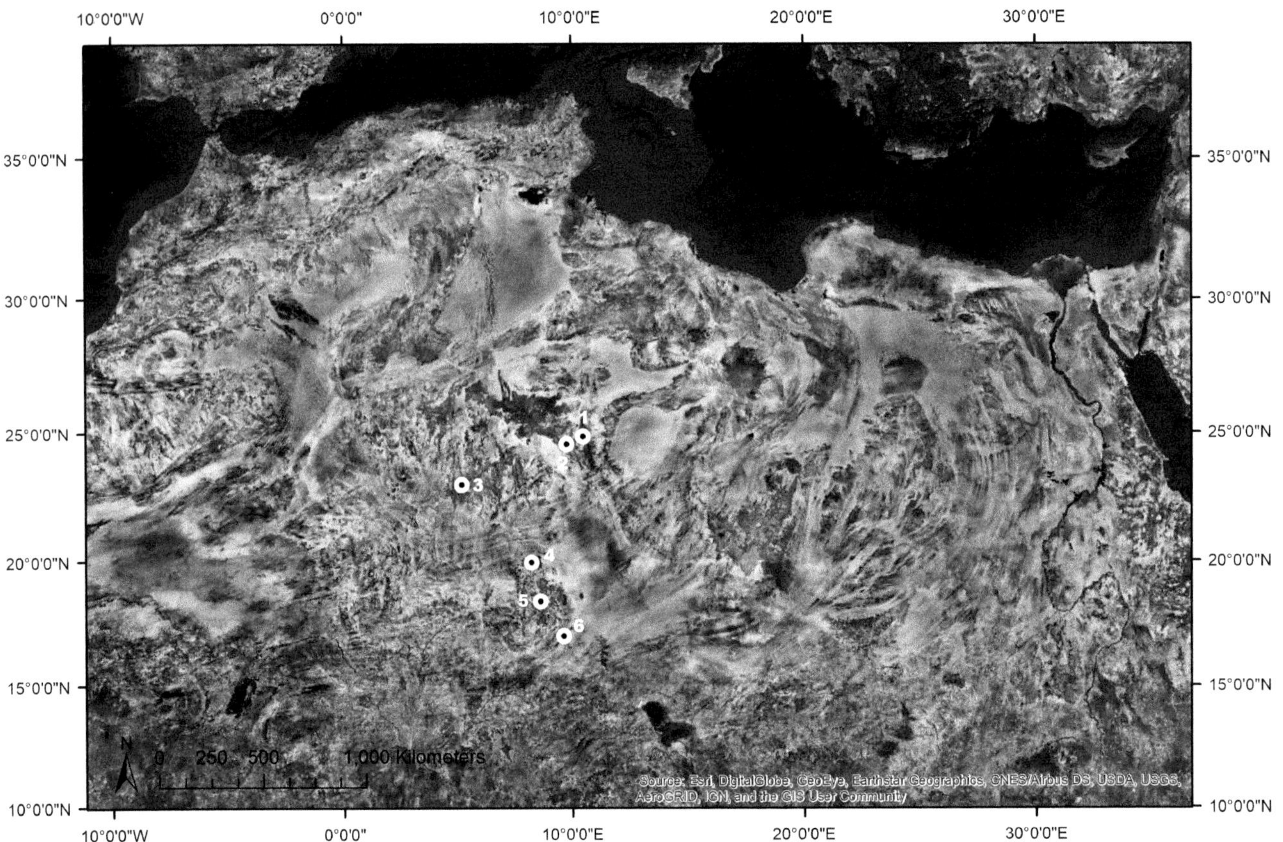

Fig. 16.1 Stratified archaeological sites in central Sahara. Key: (1) Takarkori, Uan Tabu, Uan Afuda, Ti-n-Torha, Uan Muhuggiag in the Tadrart Acacus mountains; (2) Ti-n-Hanakaten and (3) Amekni in the Tassili n'Ajjer plateau; (4) Temet, (5) Tagalagal, and (6) Gobero in the Aïr range (background image: Google Earth)

et al. 2002). *Sahelanthropus* predates by 3 million years other hominin fossils from southern and eastern Africa. The idea that our human roots lie in the Sahara is certainly fascinating, although the evidence so far recorded is still controversial (Wolpoff et al. 2006). In any case, this would be the only alleged evidence of hominin occupation in Sahara and in the North Africa in the late Miocene (Table 16.1). Prior to the *Sahelanthropus* discovery, a mandible of an Australopithecine had been discovered from the region of Bahr el Ghazal (Chad), and named thereafter *Australopithecus bahrelghazali* (Brunet et al. 1995). Dated to 3.6–3.0 Ma, this is notably the only Australopithecine discovered out of south and east Africa, resembling in various morphological characteristics *A. afarensis*. It is likely that *A. bahrelghazali* represents the results of a local adaptation to the trend towards aridification recorded in Africa, as demonstrated by the focus of this individual's diet on specific grasses that were retreating in humid ecological niches following the trend towards aridification (Macho 2015).

16.3 Into the Pleistocene (3.0–0.25 Ma): The Early Stone Age

With regard to the Pleistocene, hominin-bearing sites are absent in the Sahara. So far, no sites with direct evidence of *Homo habilis* nor *H. erectus* or associated species have been found. Such fossils have been retrieved from a limited number of North African coastal sites in northern Algeria, where the earliest evidence of human occupation in the region has been recorded and dated to ~1.8 Ma (Sahnouni et al. 2011). The southern Mediterranean coast was the arrival destination of many early human groups coming from the south, crossing the Sahara following the dense hydrographic network that existed at that time and after (Drake et al. 2011; Drake and Breeze 2016).

The Early Stone Age is characterized by two different technologies in the Sahara, the Oldowan or Mode 1, and the Acheulean or Mode 2, according to the traditional and most widespread terminology (Clark 1977). The Oldowan

Table 16.1 Synthesis of periods of human occupation in the Sahara

Stage	Cultural phase	Phases of human/hominin evolution	Period	Approximate date (before present)
–	–	*Sahelantropus tchadiensis*	Late Miocene	7–6 Ma
–	–	*Australopithecus barelghazali*	Pliocene	3.5–3.0 Ma
ESA	Oldowan, Mode 1 industries	*H. erectus*	Lower Pleistocene	1.8–0.8 Ma
ESA	Acheulean, Mode 2	*H. heidelbergensis*	Early and Middle Pleistocene	800–250 ka
MSA	Middle Stone Age (Middle Palaeolithic)	*H. sapiens?*	Middle and Upper Pleistocene	250–40 ka
MSA-Aterian	Variation of MSA	*H. sapiens*	Upper Pleistocene	110–60 ka
LSA	Later Stone Age (Upper Palaeolithic)		Upper Pleistocene–Early Holocene	40–10 ka
LSA	Epipalaeolithic/Mesolithic		Early–Middle Holocene	From 10 ka[a]
Pastoral Neolithic	Neolithic		Middle and Late Holocene	From 7.5 ka[a]
Garamantian	Historical		Late Holocene	3.0–1.3 ka
Post-Garamantian	Historical			1.3–0.9 ka
Islamic	Historical			900–300 BP
Recent	Recent historical period			300 BP to present

The transition between the Later Stone Age and Neolithic occurred at different times throughout the Sahara. *ESA* Early Stone Age, *MSA* Middle Stone Age, *LSA* Late Stone Age. [a] refers to the earliest date

in North Africa is dated from 1.8 Ma (Sahnouni et al. 2011) although stone tools of this period are extremely rare in the Sahara. Along with the recent review by Vernet (2017) for northwest Africa, evidence has been reported from the Libyan Fazzan (Lahr et al. 2011), the Massif des Richât, central Mauritania (Vernet and Naffé 2003), and the Ennedi Mountains in Chad (Monod 1976). None of those Oldowan findings has been dated, and the scarcity of evidence is generally related to the ephemeral nature (Lahr et al. 2009) of human occupation in the Sahara between 1.8 and 0.8 Ma (Table 16.1).

Being one of the longest technological traditions in Africa, the Acheulean is known from a number of sites in the Sahara. In southeast Algeria, the development of the Acheulean can be reconstructed from different archaeological findings, although in Mauritania, Libya, and Egypt the differences in Acheulean assemblages are harder to identify. Most of the evidence comes from surface collections, without radiometric control, that can be compared with the stratigraphies of Early Stone Age sites from the Maghreb (Raynal et al. 2001). Acheulean finds in the Sahara point to light human occupation pulses, following favourable environmental conditions between 800,000 and 250,000 years ago. In the absence of any direct finds from the Sahara, the association between Oldowan and *H. erectus*, and Acheulean with *H. heidelbergensis* can be traced (Table 16.1).

16.4 From Middle to Upper Pleistocene: The Middle Stone Age

The Middle Stone Age is associated in Africa with the emergence of modern humans, expanding from East Africa (Foley et al. 2013). Dated to 200,000–40,000 years ago in the Sahara, it is known from many archaeological sites in the Fazzan (Biagetti et al. 2013; Foley et al. 2013; Cancellieri et al. 2016). Nevertheless, no human remains have been found in the desert. A dense network of rivers and lakes (Drake et al. 2011; Drake and Breeze 2016) supported different episodes (Garcea 2013) of expansion through the Sahara of *H. sapiens*, following climatic pulses and shifts. Indeed, the Middle Pleistocene was a period of dramatic climatic changes, which in turn activated different 'green corridors' across the Sahara. Recent studies have focused on the complex hydrographic network that, in humid times, became green and supported the movement of humans through the desert (Drake et al. 2011). The occurrence of those favourable periods prompted the dispersal of modern humans from East Africa from as early as 300,000 years ago onwards, as highlighted by very fresh data from the Maghreb (Richter et al. 2017).

Most emblematic and better-known Saharan Middle Stone Age (Mode 3) phase is the Aterian, whose typical lithic technology appeared by 110 ka in the Maghreb. It is well documented in the Sahara (Clark 1980; Garcea

2001; Larrasoaña 2012; Biagetti et al. 2013; Scerri 2013; Cancellieri et al. 2016), where it disappears by 40 ka (Table 16.1). Unfortunately, Saharan palaeoclimatic and archaeological records are both fragmentary for that period, and the lack of human fossils leaves room for further discussion. Nevertheless, some authors suggest that the Aterian populations might have developed under arid conditions (Clark 1980; Cremaschi et al. 1998; Garcea 2012; Hawkins 2012; Larrasoaña 2012; Cancellieri et al. 2016; *contra* Scerri 2013), and that the arid Sahara was deliberately chosen for human occupation. Key data for this period come from the thick layer of Aterian deposit at the rockshelters of Uan Tabu (Garcea 2001) and Uan Afuda (di Lernia 1999), both located in the Tadrart Acacus (southwest Libya). Uncertainty in radiometric dating, however, affects the reconstruction of human dynamics in Aterian times (Drake and Breeze 2016). The end of Aterian occupation in the Sahara dates back to 40 ka, followed by a significant decrease in human occupation until the early Holocene.

16.5 The Last Green Sahara Stage and the Transition Towards the Holocene

With the exception of the Nile Valley (Wendorf et al. 1989) where relatively favourable conditions occurred between 20 and 11 ka, the Sahara shows little evidence of human activity by the end of the Pleistocene. The existence of Late Stone Age occupation in the Sahara during the late Pleistocene is considered as problematic (Cancellieri et al. 2016). Yet, with the beginning of the Holocene and the onset of wetter climatic conditions of the so-called 'Green Sahara' (DeMenocal et al. 2000; Gasse 2000), there is an increasing amount of evidence of human activity. The origin of these Saharan peoples is not yet fully understood. It is likely that newcomers arrived from the surrounding regions, following reactivation of the Saharan hydrological network from 14 ka onwards. These hunter-gatherer-fishers used a varied repertoire of stone tools (Mode 5), ranging from microlithics to (rarer) grinding equipment, along with bone and wooden tools, and composed hafted tools (Barich 2013). The large repertoire of tools likely reflects the wide spectrum of natural resources used by these communities. Conventionally divided into Epipalaeolithic and Mesolithic phases, scholars have used different labels to describe local sequences. In central Sahara, it is normally referred to as 'Early Acacus' (~9.8–8.9 ka BP) and as 'Late Acacus' (8.9–7.5 ka BP) (di Lernia and Garcea 1997; di Lernia 1998) (Table 16.1). The earlier phase was characterized by specialized hunting of Barbary sheep and rare plant processing, a typical Late Stone Age industry with microlithics and bladelets mainly in quartz (Cancellieri and di Lernia 2014). After a short arid spell, the Late Acacus phase features signs of increased sedentarism, with the collection and storage of a wide range of wild plants (di Lernia 1999). Pottery makes its appearance in the Sahara, and the same types of vessel decorations are now to be found from the Nile Valley to the central Sahara (Haaland and Haaland 2013). Ostrich eggshell was also widely used. Attempts in corralling Barbary sheep occurred in the Libyan Tadrart Acacus (di Lernia and Cremaschi 1996). Ultimately, some burial grounds have been found in the Sahara with dated from ~8 ka onwards (Barich 2013 for a review, also Biagetti and di Lernia 2013), whilst an open-air cemetery has been recorded at Gobero (Sereno et al. 2008), at the southern margins of the Sahara. Rock art is one of the most tangible signs of early Holocene occupation, recorded throughout the Sahara (see Gallinaro, this volume).

16.6 Having Herds: Milking the Green Sahara

In chronological and often physical continuity with previous Mesolithic occupants, the diffusion of domestic cattle in Saharan settlements occurred at the end of the 8th millennium. No climatic shock or dry spell prompted the advent of pastoralism in the Sahara. Domesticated stocks likely came from the Near East, and spread throughout the Sahara. The hypothesis of an African domestication of *Bos* is still controversial (see di Lernia 2013 for a summary of the debate around the issue), whilst the near-eastern origin of sheep and goats is widely accepted. In spite of regional variants and characteristics, the Saharan pastoral Neolithic tradition developed from ~8 to 3 ka (Table 16.1). Local sequences have been established, but emphasis has been traditionally put on the consistency of the archaeological evidence.

Early stages of pastoralism took place by the mid-8th millennium, and featured small societies of cattle keepers, having also some herds of ovicaprids. Some burials in rock shelters are known from the central Saharan massifs (Barich 2013; Biagetti and di Lernia 2013; di Lernia and Tafuri 2013). After the initial spread of pastoralism throughout the 7th millennium, a dry arid spell occurred and likely determined a temporary abandonment of most of the Saharan range for some centuries. In the Middle Holocene, between 6 and 5 ka, a large set of shared 'cultural' traits is found in many parts of the Sahara. Comparable funerary practises, rock art, pottery decoration, peculiar rituals known as 'cattle cult' (di Lernia et al. 2013), and settlement patterns are now widespread over large areas. Such diffusion of pastoralism occurred in a relatively short time span, and probably took the form of pulsating forays of cattle keepers looking for pasture, following the shifts between arid and wet seasons. Subsistence was based upon domestic stocks and included dairying (Dunne et al. 2012), along with widespread use of wild resources. Rock art of this period testifies to the daily life of Saharan communities, portraying herds at graze, domestic ménages, and convivial moments.

16.7 From Green to Brown: The Last Climatic Shift 5000 Years Ago

Around 5 ka, the Sahara began to turn dry. The end of the last Green Sahara phase, known as the African Humid Period (DeMenocal et al. 2000; Gasse 2000), determined a transition towards a drier environment. Saharan societies coped with the new ecological setting by progressively abandoning cattle and emphasizing small stock. Larger seasonal movements became necessary to track patchy natural resources. First attempts of cultivation took place in the central Sahara around 3000 BP (di Lernia and Merighi 2006). The landscape of Late and Final Pastoral herders (Table 16.1) began to be marked by different types of stone cairns, often burial tumuli that still dot the whole region today. The first signs of social hierarchies were reported (di Lernia and Manzi 2002) for this period. A further increase in societal complexity led to the emergence of the Garamantian kingdom. The existence of a proper Saharan state is still not fully known or acknowledged. The Garamantes ruled over a vast area, roughly corresponding to the present Fazzan, for more than 1500 years. An extended network of trade routes, the practise of agriculture in the oases and animal husbandry in the outer lands, allowed them to flourish even in hyperarid times (Mattingly 2003; Liverani 2005; Mori 2013). The use of underground irrigation system (*foggara*), the adoption of dromedaries and a set of fortified outposts were key to the long lasting rule of the Garamantes. The archaeological traces of the Garamantian are impressive and consist of necropoleii, fortified settlements and rock art, mainly distributed between southwest Libya and southeast Algeria, often visible from satellite imagery (Mattingly and Sterry 2013; Biagetti et al. 2017) (Fig. 16.2).

Even after the end of the Garamantian age, and after the spread of Islam, Saharan societies continued dwelling in the same places. Agropastoral societies lived in the better-watered areas, while herders roamed the mountains and plains where pastures and water points could be found (Biagetti 2014). A repertoire of rock art, texts carved on the rocky flanks of the wadis, scatters of material, and rare

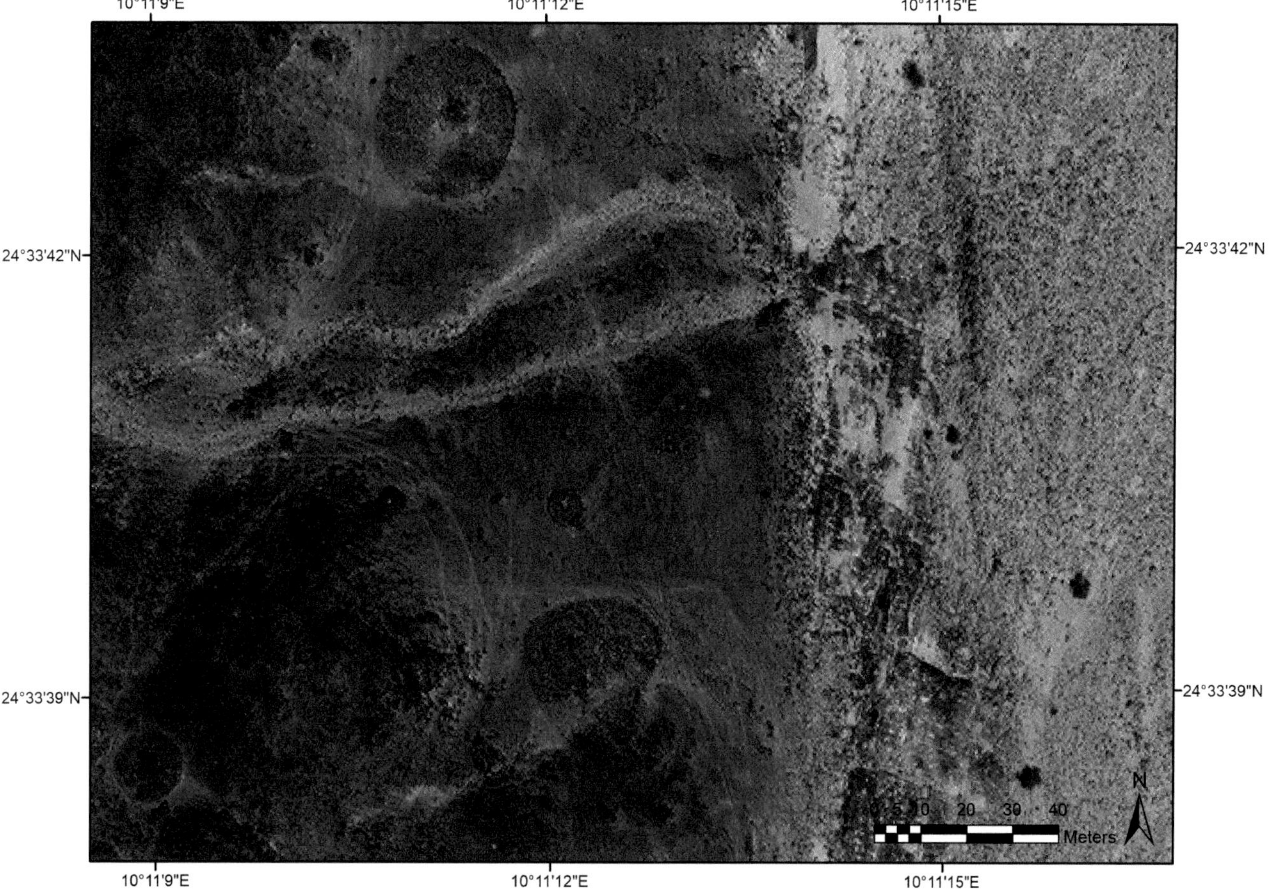

Fig. 16.2 Satellite imagery of late Neolithic tombs from southeast Algeria (*source imagery*: DigitalGlobe Foundation)

archaeological deposits is what remains of this so far poorly known period of the Sahara. Trans-Saharan trade fuelled the exchange of goods and ideas also shaped a specific Saharan identity that is at the very base of historical and current desert societies. Nowadays the Tuareg, for instance, have inherited the legacy of millennia of successful adaptation to arid and hyperarid zones. The use of a wide range of traditional knowledge is key to the resilience of current Saharan communities who are struggling to find, once more, a place in the baffling balance of power in the present Sahara.

16.8 Future Human Challenges in the Sahara

Current Saharans are part of twenty-first century states, whose borders often divide ancient bonds and shared cultural traits. The trend towards a general sedentarization of nomadic communities has taken place throughout the whole Sahara, and the oases are nowadays the foci of most permanent human occupation, along with political and economic power. Yet, relatively new technologies, like 4WD cars and mobile phones, have enhanced the capacity of current Saharan to exploit and dwell the huge 'empty spaces' at the peripheries of cities and villages. Transhumance activities have become faster, and information about rain runs quickly through smartphones. The contemporary Sahara reflects the migration and movements of people from the south, heading towards the Mediterranean through the ancient trans-Saharan trade routes established in historical times. Small communities live in all areas of the Sahara, practising mixed and adaptable strategies to adjust to and exploit climatic oscillations and different ecological niches. In spite of its current hyperarid climate, the Sahara is a lively place, hosting a repertoire of traditional knowledge to cope with unpredictable rainfall and erratic natural resources, in ways that are still largely overlooked.

16.9 Summary

Generally seen as a marginal location for the study of our human past, the Sahara is indeed a key place for the reconstruction of early hominin and human cultural trajectories. Its variable landscapes, made up of different ecological settings, allowed for past communities to exploit a variety of locales and resources to adapt to fluctuating climate. Oscillations between arid and wet environments are the hallmark of the last few million years of history in the Sahara, although the magnitude, duration, and intensity of those events were variable. The responses of early societies to environmental changes were also different. Episodes of 'Green Sahara' represented favourable moments for the whole biota. Nevertheless, the onset of arid conditions did not always determine the abandonment of Saharan regions, prompting the adoption of diverse solutions by different peoples to cope with the new ecological conditions.

References

Barich BE (2013) Hunter-gatherer-fishers of the Sahara and the Sahel 12000–4000 years ago. In: Mitchell P, Lane PJ (eds) The Oxford handbook of African archaeology. Oxford University Press, Oxford, pp 445–460

Biagetti S (2014) Ethnoarchaeology of the Kel Tadrart Tuareg. Pastoralism and Resilience in Central Sahara. SpringerBriefs in archaeology. Springer International Publishing, Cham, p 154

Biagetti S, di Lernia S (2013) Holocene deposits of Saharan rock shelters: the case of Takarkori and other sites from the Tadrart Acacus Mountains (southwest Libya). Afr Archaeol Rev 30:305–338

Biagetti S, Cancellieri E, Cremaschi M, Gauthier C, Gauthier Y, Zerboni A, Gallinaro M (2013) The 'Messak Project'. Archaeological research for cultural heritage management in SW Libya. J Afr Archaeol 11:55–74

Biagetti S, Merlo S, Adam E, Lobo A, Conesa FC, Knight J, Bekrani H, Crema ER, Alcaina-Mateos J, Madella M (2017) High and medium resolution satellite imagery to evaluate Late Holocene human-environment interactions in arid lands: a case study from the Central Sahara. Remote Sens 9:351. https://doi.org/10.3390/rs9040351

Brunet M, Beauvilain A, Coppens Y, Heintz E, Moutaye AHE, Pilbeam D (1995) The first australopithecine 2500 kilometres west of the Rift Valley (Chad). Nature 378:273–275

Brunet M, Guy F, Pilbeam D, Mackaye HT, Likius A, Ahounta D, Beauvilain A, Blondel C, Bocherens H, Boisserie J-R, De Bonis L, Coppens Y, Dejax J, Denys C, Duringer P, Eisenmann V, Fanone G, Fronty P, Geraads D, Lehmann T, Lihoreau F, Louchart A, Mahamat A, Merceron G, Mouchelin G, Otero O, Campomanes PP, Ponce de Leon M, Rage J-C, Sapanet M, Schuster M, Sudre J, Tassy P, Valentin X, Vignaud P, Viriot L, Zazzo A, Zollikofer C (2002) A new hominid from the Upper Miocene of Chad, Central Africa. Nature 418:145–151

Brunet M, Guy F, Pilbeam D, Lieberman DE, Likius A, Mackaye HT, Ponce de Leon MS, Zollikofer CPE, Vignaud P (2005) New material of the earliest hominid from the Upper Miocene of Chad. Nature 434:752–755

Cancellieri E, di Lernia S (2014) Re-entering the central Sahara at the onset of the Holocene: a territorial approach to *Early Acacus* hunter-gatherers (SW Libya). Quat Int 320:43–62

Cancellieri E, Cremaschi M, Zerboni A, di Lernia S (2016) Climate, environment, and populations dynamics in Pleistocene Sahara. In: Jones SC, Stewart BA (eds) Africa from MIS 6–2: population dynamics and palaeoenvironments. Springer, Dordrecht, pp 123–145

Carrión JS, Rose J, Stringer S (2011) Early human evolution in the Western Palaearctic: ecological scenarios. Quat Sci Rev 30:1281–1295

Clark JD (1977) World prehistory: a new perspective. Cambridge University Press, Cambridge, p 576

Clark JD (1980) Human populations and cultural adaptations in the Sahara and the Nile during prehistoric times. In: Williams MAJ, Faure H (eds) The Sahara and the Nile. Balkema, Rotterdam, pp 527–582

Cremaschi M, di Lernia S, Garcea EAA (1998) Some insights on the Aterian in the Libyan Sahara: chronology, environment, and archaeology. Afr Archaeol Rev 15:261–286

DeMenocal PB, Ortiz J, Guilderson T, Adkins J, Sarnthein M, Baker L, Yarusinsky M (2000) Abrupt onset and termination of the African Humid Period: rapid climate responses to gradual insolation forcing. Quat Sci Rev 19:347–361

di Lernia S (1998) Early Holocene pre-pastoral cultures in the Aan Afuda cave, wadi Kessan, Tadrart Acacus (Libyan Sahara). In: Cremaschi M, di Lernia S (eds) Wadi Teshuinat. Palaeoenvironment and prehistory in South-western Fezzan (Libyan Sahara). CNR Quaderni di Geodinamica Alpina e Quaternaria n 7, Firenze, pp 123–154

di Lernia S (ed) (1999) The Uan Afuda Cave: hunter-gatherers societies of Central Sahara. AZA Monographs 1. All'Insegna del Giglio, Firenze, p 272

di Lernia S (2013) The emergence and spread of herding in Northern Africa. In: Mitchell P, Lane PJ (eds) Oxford handbook of African archaeology. Oxford University Press, Oxford, pp 527–540

di Lernia S, Cremaschi M (1996) Taming barbary sheep: wild animal management by Early Holocene hunter-gatherers at Uan Afuda (Libyan Sahara). Nyame Akuma 46:43–54

di Lernia S, Garcea EAA (1997) Some remarks on Saharan terminology. Pre-pastoral archaeology from the Libyan Sahara and the Middle Nile Valley. Libya Antiqua 3:4–16

di Lernia S, Manzi G (eds) (2002) Sand, stones, and bones: the archaeology of death in the Wadi Tanezzouft Valley (5000–2000 BP). AZA Monographs 3. All'Insegna del Giglio. Firenze, p 354

di Lernia S, Merighi F (2006) Transitions in the later prehistory of the Libyan Sahara, seen from the Acacus Mountains. In: Mattingly D, McLaren S, Savage E, al-Fasatwi Y, Gadgood K (eds) The Libyan Desert: natural resources and cultural heritage. Society for Libyan studies monograph 6. Society for Libyan Studies, London, pp 111–122

di Lernia S, Tafuri MA (2013) Persistent deathplaces and mobile landmarks. The Holocene mortuary and isotopic record from Wadi Takarkori (SW Libya). J Anthropol Archaeol 32:1–15

di Lernia S, Tafuri MA, Gallinaro M, Alhaique F, Balasse M, Cavorsi L, Fullagar PD, Mercuri AM, Monaco A, Perego A, Zerboni A (2013) Inside the "African Cattle Complex": animal burials in the Holocene Central Sahara. PLoS ONE 8:e56879. https://doi.org/10.1371/journal.pone.0056879

Drake N, Breeze P (2016) Climate change and modern human occupation of the Sahara from MIS 6–2. In: Jones SC, Stewart BA (eds) Africa from MIS 6–2: population dynamics and paleoenvironments. Springer, Dordrecht, pp 103–122

Drake NA, Blench RM, Armitage SJ, Bristow CS, White KH (2011) Ancient watercourses and biogeography of the Sahara explain the peopling of the desert. Proc Nat Acad Sci 108:458–462

Dunne J, Evershed RP, Salque M, Cramp L, Bruni S, Ryan K, Biagetti S, di Lernia S (2012) First dairying in green Saharan Africa in the fifth millennium BC. Nature 486:390–394

Foley RA, Maíllo-Fernández JM, Lahr MM (2013) The Middle Stone Age of the Central Sahara: biogeographical opportunities and technological strategies in later human evolution. Quat Int 300:153–170

Garcea EAA (2001) A reconsideration of the Middle Palaeolithic/Middle Stone Age in Northern Africa after the evidence from the Libyan Sahara. In: Garcea EAA (ed) Uan Tabu in the settlement history of Libyan Sahara. AZA Monographs 2, All'Insegna del Giglio, Firenze, pp 25–50

Garcea EAA (2012) Modern human desert adaptations: a Libyan perspective on the Aterian complex. In: Hublin J-J, McPherron SP (eds) Modern origins: a North African perspective. Springer, Dordrecht, pp 127–142

Garcea EAA (2013) Hunter-gatherers of the Nile Valley and the Sahara before 12,000 years ago. In: Mitchell P, Lane PJ (eds) The Oxford handbook of African archaeology. Oxford University Press, Oxford, pp 419–430

Gasse F (2000) Hydrological changes in the African tropics since the Last Glacial Maximum. Quat Sci Rev 19:189–211

Grant KM, Rohling EJ, Westerhold T, Zabel M, Heslop D, Konijnendijk T, Lourens L (2017) A 3 million year index for North African humidity/aridity and the implication of potential pan-African humid periods. Quat Sci Rev 171:100–118

Haaland R, Haaland G (2013) Early Farming Societies along the Nile. In: Mitchell P, Lane PJ (eds) The Oxford handbook of African archaeology. Oxford University Press, Oxford, pp 541–553

Hawkins AL (2012) The Aterian of the oases of the Western Desert of Egypt: adaptation to changing climatic conditions? In: Hublin J-J, McPherron SP (eds) Modern origins: a North African perspective. Springer, Dordrecht, pp 157–175

Lahr MM, Foley R, Crivellaro F, Okumura M, Maher L, Davies T, Veldhuis D, Wilshaw A, Mattingly D (2009) DMP VI: preliminary results from 2009 fieldwork on the human prehistory of the Libyan Sahara. Libyan Stud 40:133–153

Lahr MM, Foley R. Crivellaro F, Fernandez JM, Wilshaw A, Copsey B, Rivera F, Mattingly D (2011) DMP XIV: prehistoric sites in the Wadi Barjuj, Fazzan, Libyan Sahara. Libyan Stud 42:117–138

Larrasoaña JC (2012) A northeast Saharan perspective on environmental variability in North Africa and its implications for modern humans origins. In: Hublin J-J, McPherron SP (eds) Modern origins: a North African perspective. Springer, Dordrecht, pp 19–34

Lebatard A-E, Bourlès DL, Duringer P, Jolivet M, Braucher R, Carcaillet J, Schuster M, Arnaud N, Monié P, Lihoreau F, Likius A, Mackaye HT, Vignaud P, Brunet M (2008) Cosmogenic nuclide dating of *Sahelanthropus tchadensis* and *Australopithecus bahrelghazali*: Mio-Pliocene hominids from Chad. Proc Nat Acad Sci 105:3226–3231

Liverani M (ed) (2005) Aghram Nadharif: The Barkat Oasis (Sha'abiya of Ghat, Libyan Sahara) in Garamantian Times. AZA Monographs 5. All'Insegna del Giglio, Firenze, p 520

Macho GA (2015) Pliocene hominin biogeography and ecology. J Human Evol 87:78–86

Maslin MA, Trauth MH (2009) Plio-Pleistocene East African pulsed climate variability and its influence on early human evolution. In: Grine FE, Fleagle JG, Leakey RE (eds) The first humans—origin and early evolution of the genus *Homo*. Springer, New York, pp 151–158

Mattingly DJ (ed) (2003) The archaeology of Fazzan, Vol 1: Synthesis. Society for Libyan Studies, London, p 454

Mattingly DJ, Sterry M (2013) The first towns in the central Sahara. Antiquity 87:503–518

Monod T (1976) Trois gisements de galets aménagés dans l'Adrar de Mauritanie. Provence Hist 99:87–97

Mori L (ed) (2013) Life and death of a rural village in Garamantian times. Archaeological investigations in the oasis of Fewet (Libyan Sahara). AZA Monographs 6. All'Insegna del Giglio, Firenze, p 406

Potts R (2012) Environmental and behavioral evidence pertaining to the evolution of early *Homo*. Curr Anthropol 53:S299–S317

Potts R (2013) Hominin evolution in settings of strong environmental variability. Quat Sci Rev 73:1–13

Raynal JP, Sbihi Alaoui FZ, Geraads D, Magoga L, Mohi A (2001) The earliest occupation of North-Africa: the Moroccan perspective. Quat Int 75:65–75

Richter D, Grün R, Joannes-Boyau R, Steele TE, Amani F, Rué M, Fernandes P, Raynal J-P, Geraads D, Ben-Ncer A, Hublin J-J, McPherron SP (2017) The age of the hominin fossils from Jebel Irhoud, Morocco, and the origins of the Middle Stone Age. Nature 546:293–296

Sahnouni M, van der Made J, Everett M (2011) Ecological background to Plio-Pleistocene hominin occupation in North Africa: the vertebrate faunas from Ain Boucherit, Ain Hanech and El-Kherba, and paleosol stable-carbon-isotope studies from El-Kherba, Algeria. Quat Sci Rev 30:1303–1317

Scerri EML (2013) The Aterian and its place in the North African Middle Stone Age. Quat Int 300:111–130

Sereno PC, Garcea EAA, Jousse H, Stojanowski CM, Saliège JF, Maga A, Ide OA, Knudson KJ, Mercuri AM, Stafford TW, Kaye TG, Giraudi C, Massamba N'siala I, Cocca E, Moots HM, Dutheil DB, Stivers JP (2008) Lakeside cemeteries in the Sahara: 5000 years of Holocene population and environmental change. PLoS ONE 3(8):e2995. https://doi.org/10.1371/journal.pone.0002995

Vernet R (2017) Les galets aménagés du nord-ouest de l'Afrique. Ikosim, Paris

Vernet R, Naffé BOM (2003) Dictionnaire archéologique de la Mauritanie. Université de Nouakchott, CRIAA-LERHI, p 164

Vignaud P, Duringer P, Mackaye HT, Likius A, Blondel C, Boisserie J-R, de Bonis L, Eisenmann V, Etienne M-E, Geraads D, Guy F, Lehmann T, Lihoreau F, Lopez-Martinez N, Mourer-Chauviré C, Otero O, Rage J-C, Schuster M, Viriot L, Zazzo A, Brunet M (2002) Geology and palaeontology of the Upper Miocene Toros-Menalla hominid locality, Chad. Nature 418:152–155

Wendorf F, Schild R, Close AE (1989) The prehistory of Wadi Kubbaniya, vol 2. Southern Methodist University, Dallas

Wolpoff MH, Hawks J, Senut B, Pickford M, Ahern J (2006) An ape or *the* ape: is the Toumaï TM 266 cranium a hominid? PaleoAnthropology 2006:36–50

The Evolution of Foraging and the Transition to Pastoralism in the Sahara

17

Roger Blench

Abstract

Modern humans have been crossing the Sahara as long ago as 300,000 years ago and the intermittent opening of corridors in humid periods has facilitated this human transit. Pastoralism spreads into the central Sahara, together with dairying and a striking culture of cattle necropolises, by around 7000 BP. However, it took nearly another 3500 years to spread to the Sahel, likely for ecological reasons. The chapter discusses the different elements of the pastoral package, beginning with cattle and ovicaprines and later phases of horses, donkeys, and camels. Small foraging groups still live in the Sahara, and their importance for ethnographic reconstruction is highlighted.

Keywords

Sahara · Foraging · Archaeology · Pastoralism

17.1 Introduction: Foragers and the Transition to Pastoralism in the Central Sahara

The central Sahara is one of the key zones in Africa for the evolution of modern humans. The discoveries at Jebel Irhoud in Morocco (Richter et al. 2017), linking them to comparable sites in southern Africa, show that behavioural modernity may be as much as 300,000 years old. Early modern humans had the capacity to cross the Sahara and disperse around Africa with considerable ease. At present we have little or no idea about the details of subsistence of such groups. The major Palaeolithic cultures of the Sahara (the long-lasting Aterian, then the Iberomaurusian and Capsian), which traverse the period of the introduction of both ceramics and pastoralism, transformed Saharan subsistence. This chapter discusses the transition from foraging systems, through herding up to the inception of early cereal agriculture. Some foraging cultures persisted into the twentieth century and may provide clues to the more widespread systems of the past.

17.1.1 The Aterian

The first culture with identifiable assemblages is the Aterian, a Middle Palaeolithic industry centred in North Africa and the Sahara and named after the type site of Bir el Ater, eastern Algeria. The earliest site is at Ifri n'Ammar in Morocco dating to ~145 ka (Richter et al. 2010) but most cluster in the last interglacial at ~130–116 ka (see review in Barton et al. 2009). Other Aterian sites in North Africa are described by Cremaschi et al. (1998), Hublin et al. (2012), Scerri (2013), and Dibble et al. (2013). The latest Aterian occurs around 20 ka. Evidence for the subsistence strategies of the Aterians remains tenuous. This culture is characterised by the presence of tanged or pedunculated tools (Iovita 2011) that are possibly points for arrows or projectiles such as small spears although their size does not suggest large game. Seashore and riverbank sites point to the use of aquatic resources, especially snails and other molluscs, and crabs (Lubell 2004). Exploitation of plant resources is likely and some contemporaneous sites evidence this, although not unambiguously related to the Aterian (Van Peer et al. 2003).

R. Blench (✉)
McDonald Institute for Archaeological Research, University of Cambridge, Downing Street, Cambridge CB2 3ER, UK
e-mail: rogerblench@yahoo.co.uk

J. Knight et al. (eds.), *Landscapes and Landforms of the Central Sahara*, World Geomorphological Landscapes, https://doi.org/10.1007/978-3-031-47160-5_17

17.1.2 The Iberomaurusian and Capsian

The Iberomaurusian and Capsian are successive cultures distributed along the North Africa coast from Agadir to Tunis (Fig. 17.1). The Iberomaurusian was a backed bladelet industry which appeared ~25.0–22.5 ka BP, coinciding with the Last Glacial Maximum and lasting until the early Holocene (Ferembach 1985; Irish 2000; Hogue and Barton 2016). Campmas et al. (2016) present evidence for exploitation of marine malacofauna at the Abri Alain rockshelter (Algeria) at this time.

The Capsian was a Neolithic culture centred in the Maghreb that lasted from about 10,000 to 4700 BP, almost certainly coming out of the Near East and associated with the spread of a pastoral lifestyle along the North African coast (Roubet 1979; Rahmani 2004; Mulazzani 2013; van de Loosdrecht et al. 2018). It is named after Capsa (modern Gafsa) in Tunisia. The Capsians were principally foragers, and large shell-mounds provide evidence for seashore gathering. Evidence now suggests sporadic adoption of early agriculture in Morocco. A site at Kaf Taht El-Ghar in northern Morocco has yielded remains of several types of wheat, naked barley and broad bean (*Vicia faba*) dated to 7286 ± 85 cal BP (Ballouche and Marinval 2003). The site of Ifri Oudadane (Morocco) has barley, three wheat subtypes, pea and lentil, dated broadly to the Early Neolithic B, ~7000–7500 BP (Morales et al. 2013; Zapata et al. 2013). Thus far, the Ifri Oudadane materials are the oldest cultivated plant remains recorded from Africa. In view of the gap between these dates and the next evidence for cultivation in North Africa, it is plausible that this experiment was not judged a success and the populations reverted to a pastoral/foraging lifestyle. Evidence for ceramics in North Africa is later than those recorded in the central Sahara, and the Capsians adopted cooking pots as early as 7000 BP. Some authors have proposed a link with the spread of Afroasiatic languages, but as Blench (2001) points out, the Berber languages are too closely related to reflect a Capsian expansion.

17.2 Foraging in the Central Sahara

By definition, all pre-agricultural populations were foragers; the challenge is discover the basis of subsistence in early sites. Earlier Palaeolithic cultures depended on static resources, such as shells and plant foods, suggesting that only later did game hunting play a significant part in diet. By the Holocene, it appears that communities had begun to manage wild animal populations. Prior to true domestication, the wild cattle of northeast Africa were probably managed (Brooks et al. 2007). Similarly, di Lernia (1998) argued based on hunter-gatherer settlements in the Tadrart Acacus massif (southwest Libya) that subsistence in the early Holocene (at 9000–8000 BP) was characterised by wild animal management, notably Barbary sheep (*Ammotragus lervia*). In the same period, wild cereals such as *Brachiaria*, *Sorghum,* and *Urochlea* spp. were also exploited (Tafuri et al. 2006), and that this supports 'cultivation without domestication', at least with respect to sorghum (Haaland 1995). This would imply that foragers transplanted some plants to more convenient locations, without restricting their ability to cross-breed. This would create a supply of cereals, without the morphometric changes characteristic of a true domesticate.

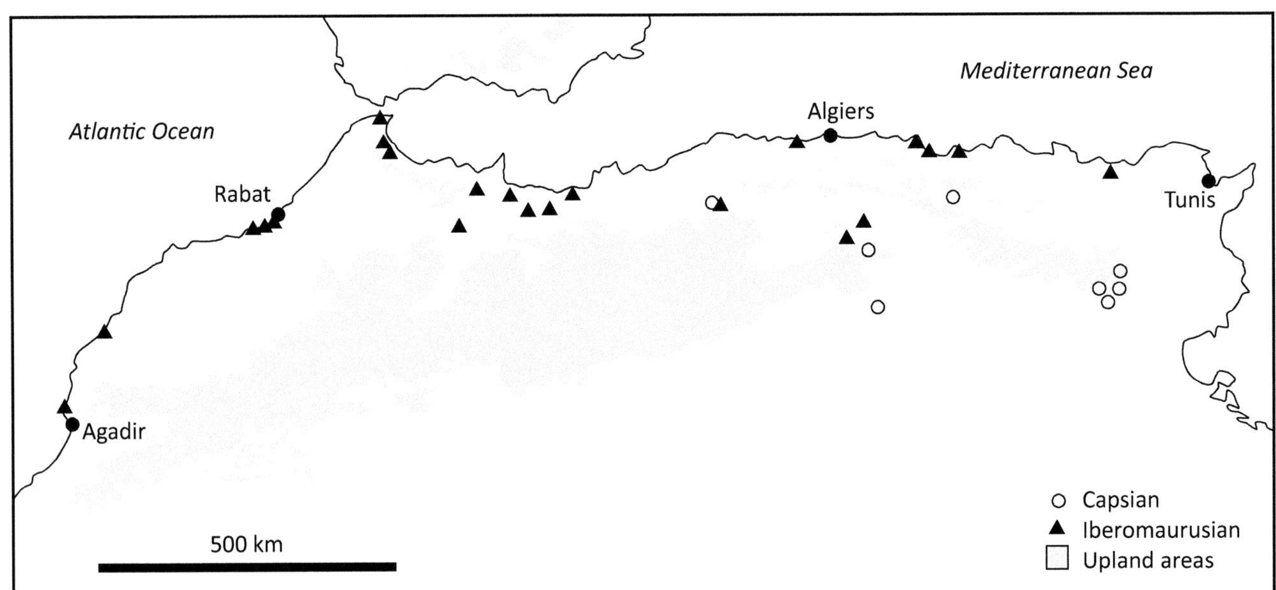

Fig. 17.1 Major Iberomaurusian and Capsian sites in the Magreb (redrawn after Barton and Bouzouggar 2013)

Table 17.1 Phases of occupation at Gobero (after Sereno et al. 2008)

Phase	Timing (years BP)	Characteristics
I	16,000–9700	General aridification, no occupation
II	9700–8200	Wet climate, Kiffian fisher-gatherer
Hiatus	8200–7200	Interruption in occupation, dry conditions make the site uninhabitable
III	7200–4200	Tenerian occupation
IV	4500–2300	Occupation ends, following return of arid conditions

A key site for understanding the interlocking strategies of foragers is Gobero in the Ténéré Desert of Niger (Sereno et al. 2008). Gobero is the oldest known cemetery in the Sahara and is the type site of the Kiffian and Tenerian cultures, dating to approximately 10 ka BP. The time frame of the site has been divided into four phases (Table 17.1).

The Kiffian and Tenerian cultures represent two distinct physical types, associated with very different subsistence patterns. Artefacts associated with Phase II occupation at Gobero include microliths, bone harpoons and hooks, dotted wavy-line pottery, and zigzag-impressed motifs. This suggests fishing-gathering, hunting of hippos, and (perhaps) Nilo-Saharan speaking. As the region aridified, the Kiffians disappeared and were replaced some 1000 years later by the Tenerians. A small amount of cattle bones were recovered, possibly suggesting a pastoralist lifestyle. Garcea (2013) strongly asserts this was the case, but an alternative interpretation is that this represents hunters trading with pastoralists located farther north in the desert.

17.3 What Species and Breeds of Livestock Are Relevant to Pastoralism?

None of the pastoral species herded today are indigenous to the Sahara and west-central Africa, with the possible exception of the ass. Wild taurine cattle were present in the Western Desert of Egypt but domestication allowed the spread of cattle into the central Sahara. The main ruminant species are cattle (both taurine (including kuri) and zeboid), sheep (wool and hair), goat, camel, horse, pony, donkey, and buffalo (known in the Nile Delta, but never spread to sub-Saharan Africa). Monogastric species including pigs, dogs, cats, and poultry cross the desert relatively late and are only marginally associated with pastoralists, or not all in the case of pigs. Blench (1993) summarised the distribution of breeds in the ethnographic record and the assumptions that can be made about past species' distributions. Within these species, there are important differences between breeds, which played specific roles in colonising particular environments.

17.3.1 Cattle

The major division within cattle is between zeboids (humped cattle of Indian origin), and taurines of northeast African and European origin. Taurines divide into humpless shorthorns (muturu), humpless longhorns (n'dama), and kuri. Humpless shorthorns are today confined to the West African coast, with a residual population in southwest Ethiopia (the Sheko). The humpless longhorns are confined to the west of West Africa, the Senegal River valley, and the highlands of Guinea Republic (Starkey 1984; Blench 1998). These are milked and also herded long distances by pastoralists. However, they do not survive well in high humidity coastal forests which strongly suggests that they represent a different wave of migration from North Africa and are probably related to the longhorn cattle of Ancient Egypt. Cattle of this variety are also represented in the rock paintings at Birnin Kudu in northern Nigeria (Fig. 17.2).

The third taurine breed is more perplexing. The kuri is a large-bodied humpless longhorn found only around Lake Chad whose exact origin is unknown (Blench 1993; Tawah and Rege 1997; Meghen et al. 2000). The kuri has distinctive, inflated, spongy horns unknown in any other breed and with a mean height of 1.5 m, and weight of some 550 kg, which make it one of the largest breeds of African cattle. Kuri are noted for their extremely variable colours and their ability to thrive in semi-aquatic conditions. Kuri cattle are kept by the Yedina people, who inhabit the islands on Lake Chad, and are notable for their resistance to insect bites and their ability to swim between islands in the lake. Unfortunately, aridification in this region has severely reduced the population of kuri and they now constitute a threatened breed.

Distinct from the taurines are the zeboids, the consequence of domestication of a wild bovid in northwest India (MacHugh et al. 1997; Chen et al. 2010). Exactly when they arrived in Africa is unknown, but it is likely this was by a sea route (the Sabaean Lane) south of the Arabian Peninsula, whence they were introduced into the Horn of Africa, and then spread westward. Early Ethiopian rock art depicts humpless longhorns, but these are displaced at

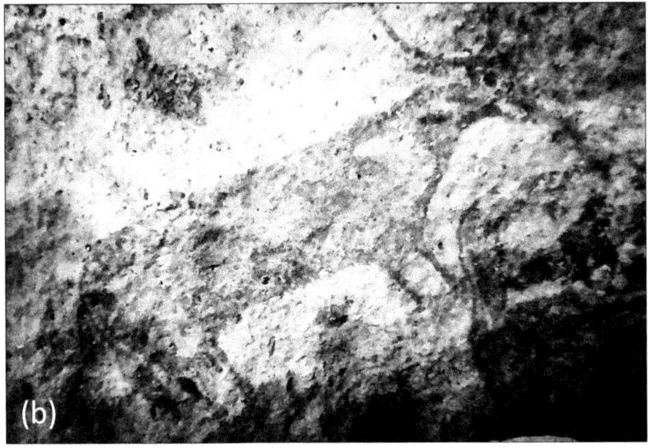

Fig. 17.2 Anthropological contexts of cattle in the Sahara. **a** Ploughing with cattle, from the burial chamber of Sennedjem, ~3200 BP (source: Creative Commons); **b** rock painting of humpless longhorn at Dutsen Mesa (source: Bradshaw Foundation)

~3500 BP. Zeboids had the disadvantage that they were non-trypanotolerant and so spread along the southern margins of the Sahara, or were crossbred with taurines to create the East African Sanga breeds (Epstein 1971). The best evidence for the spread of humped cattle is from images, terracotta models and rock art, and these indicate that the zeboids spread west rather slowly and most dates for their presence in West Africa are relatively late, except for a sixth century BC date at Zilum in Chad (Magnavita 2006). Similarly the genetic impact of indicine (i.e. zebu) cattle on indigenous cattle breeds weakens markedly farther west and is highest in eastern and southern Africa. From this it can be concluded that although the zebu was to transform Saharan pastoralism, this process took place relatively late and took a long time to become fully established.

17.3.2 Ovicaprines

Both sheep and goats were domesticated between Mesopotamia and Central Asia, and spread to the Nile Valley around 7–6 ka. Goats were domesticated in the Fertile Crescent around 11–10 ka, but it is only after 7000 BP that clearly identifiable caprine remains are identified in the African archaeological record at the eastern Sahara and Red Sea Hills (Sodmein Cave near Quseir, Egypt) (Hassan 2000). Radiocarbon dates on sheep and goat bones from archaeological sites along the North African coast (e.g. 6000 BP at Grotte Capeletti in Algeria; 6800 BP at Haua Fteah in Cyrenaica, Libya) are similar to those in the eastern Sahara, suggesting rapid dispersal of small ruminants from southwest Asia into North Africa around this time. This dispersal has been identified through the spread of Capsian culture across the region (Newman 1995) but the earliest Capsian is dated to some 3000 years before this.

The first sheep entered Africa via the Isthmus of Suez and/or the southern Sinai Peninsula, between 7500 and 7000 BP. They were likely of the thin-tailed type. Fat-tailed sheep entered through northeast Africa and the Horn of Africa. Mitochondrial DNA analysis supports a common maternal ancestral origin for all African sheep, while autosomal and Y-chromosome DNA analysis indicates a distinct genetic history for African thin-tailed and sub-Saharan fat-tailed sheep (Muigai and Hanotte 2013).

17.3.3 Camel

The one-humped dromedary is originally an Asian domesticate (Epstein 1971; Wilson 1984) although wild camels were known in North Africa in the Pleistocene. Camels are present during the Quaternary in the Maghreb but are thought to have become extinct and then reintroduced during the Graeco-Roman period (Bulliet 1990). Shaw (1979) reviews this debate and argues that the camel was present continuously in the Maghreb, but that its presence was at a low level and therefore less archaeologically visible. He argues that the camel survived from the Pleistocene and is therefore indigenous to North Africa. In the absence of archaeological evidence from more recent periods, most scholars do not accept this proposition. The camel is represented significantly in the Maghreb only in the first few centuries BC. There are camel bones at Qasr Ibrim in Egypt dated to the 1st millennium BC (Rowley-Conwy 1988). In North Africa the camel appears as a plough-animal and to carry loads (Morales Muñiz et al. 1995). In sub-Saharan West Africa, the camel is more recent: bones dating to 250–400 AD have been found in the Middle Senegal Valley.

17.3.4 Equids

The history of equids in the Sahara and sub-Saharan Africa consists of two significantly different layers. Ponies or small trypanotolerant horses were present across the Sahel by 2500 BP (Blench 1993). The earliest finds are dated to the eighth century AD at Aissa Dugjé (northern Cameroon) (MacEachern et al. 2001) and 600–1000 AD at Akumbu (Mali) (MacDonald and MacDonald 2000). These were ridden without saddles and appear to have adapted to both local vegetation and disease challenges. They were likely the progeny of horses depicted in Saharan rock art, and used to pull chariots in rock art depictions in the Tibesti. By the mediaeval period, the establishment of larger political entities and the greater volumes of trans-Saharan trade encouraged caravanners to bring large North African horses, typically barbs, across the desert (Law 1980). By the fifteenth century, these were in use in the Kingdom of Kanem around Lake Chad. Indeed some oral traditions speak of battles between the peoples of the 'small horses' and the 'large horses', presumably ponies of the mobile acephalous societies and horses of the existing kingdoms (Palmer 1928, p. 29).

The wild ass *Equus asinus africanus* is indigenous to Africa and comprises subspecies spreading from the Atlas Mountains eastwards to Nubia, down the Red Sea and as far as present-day northern Kenya (Haltenorth and Diller 1980, p. 109; Blench 2000). It is not considered likely that the wild ass occurred in sub-Saharan regions. The wild ass of the Atlas Mountains became extinct by 300 AD and is known only through depictions (Haltenorth and Diller 1980, p. 109).

The donkey may have been domesticated for its meat and milk with portage a later development. The fact that the wild ass and donkey have remained interfertile suggests there was little breeding and selection, which may reflect a management system based on seasonal corralling. Donkeys can only be distinguished from wild asses if they are shown in use; representations are not therefore evidence of domestication but only of their presence. The historical and archaeological evidence for domestic donkeys in the Maghreb is reviewed by Camps (1988). Osteological evidence for domestic donkeys is found in Egypt in the 4th millennium BC (Midant-Reynes 1992). Donkeys from the 2nd millennium BC occur at Shaqadud in the Butana grasslands of Sudan (Peters 1991). Kaache (1996) reviews the evidence for donkeys in Morocco; there are possible finds of ass bones at Dar-es-Soltane and Tangier but no certain representations in rock art.

17.4 The Evolution and Spread of Pastoralism

The main tool for determining the chronology of livestock in the Sahara is archaeology, and records of early dates remain highly provisional. Archaeozoological coverage is also patchy and some sites may not be accessible because of political instability. Distinctions between breeds are important in understanding subsistence strategies but cannot be inferred from the archaeological record. Taurine/zeboid distinctions can only be made when specific osteological features survive, and sheep cannot be reliably distinguished from goats (hence the general label ovicaprines). Equids and canines are also under-represented for taphonomic reasons. These are limitations in interpreting the evidence for the evolution and spread of pastoralism in the central Sahara region.

17.4.1 Cattle Pastoralism

The exact origins of Saharan pastoralism remain controversial, in part because it is uncertain whether the earliest finds of cattle are wild or domestic (Kuper and Reimer 2013). This has been the subject of considerable debate (Smith 1992; Chenal-Vélardé 1997; Hanotte et al. 2002; Jousse 2004; Barker 2006; Linseele 2010; Bollig et al. 2013). The earliest taurine cattle finds in North Africa are from the 9th millennium BC (Gautier 2002, 2007; Gautier and Van Neer 2009). These are all located in the Western Desert of Egypt where wild cattle were certainly endemic. However, secure evidence for domestication is present by 6000 BC (Reimer 2007). From the Nile Valley, cattle spread early into the Sahara, and the existence of cattle necropolises and evidence for dairying are indicators of their importance. Gabriel (1987) claims that more than fifty 'fireplace' sites across the Sahara can be attributed to early cattle herders. Stojanowski and Carver (2011) argue that the emergence of cattle pastoralism in the southern Sahara can be detected through localised hypoplasia of primary canine teeth.

Linseele (2013) summarises the evidence for the earliest taurine cattle finds in sub-Saharan West Africa, including 4200–2850 BP in Mali (MacDonald 1996), and 3600–3500 BP in northeast Nigeria (Linseele 2007). Figure 17.3 synthesises dates for early cattle sites in Saharan Africa and illustrates the gradual diffusion from the Nile Valley. These are very similar to dates form the Horn of Africa, at around 3500 BP (Lesur et al. 2014). Kulichová et al. (2017) argue that there is a correlation between some haplotypes found among the Fulɓe, the principal West Africa pastoralists and those found among Sahara pastoralists, suggesting a simultaneous transfer of genes and cattle.

One striking indicator of the importance of cattle in Saharan subsistence are the 'cattle necropolises' found in Algeria, Libya, and Niger (Paris 2000; Jelinek 2003; Tauveron et al. 2009; di Lernia et al. 2013). The necropolis sites at Kerma on the Upper Nile include hundreds of crania, including those with 'forward-pointing', i.e. intentionally deformed horns (Chaix and Hansen 2003). Sawchuk et al. (2018) relate Saharan mortuary practises to the spread

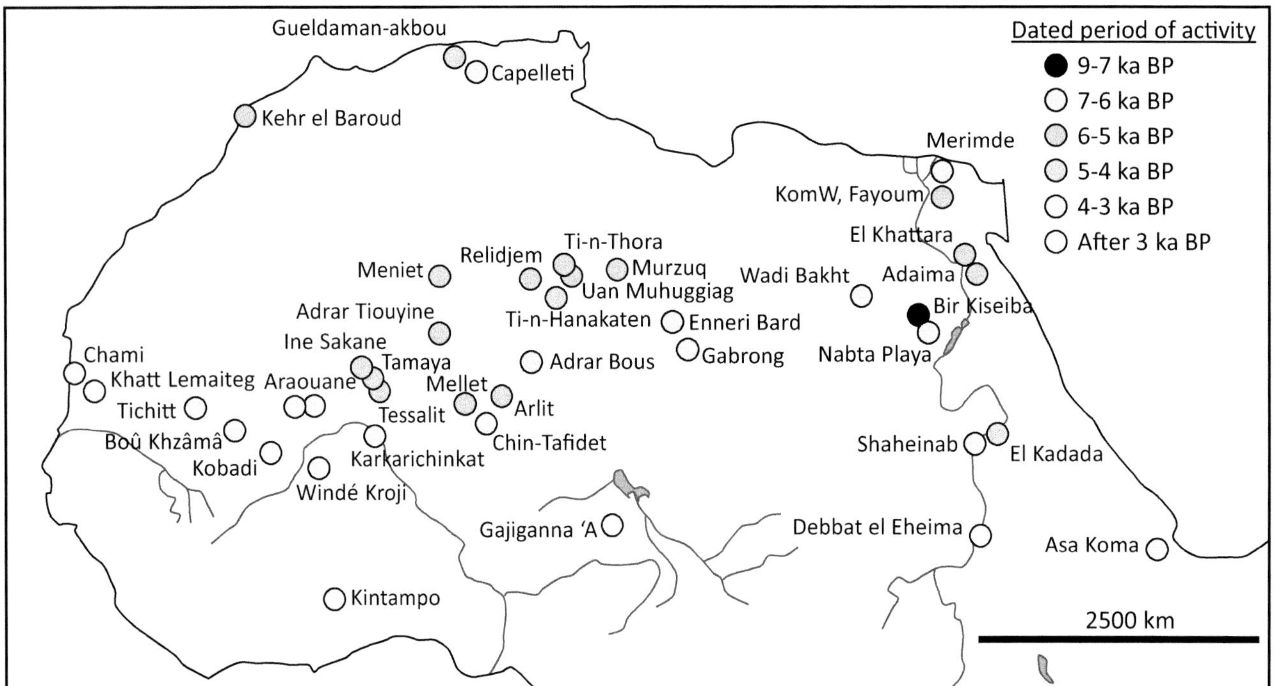

Fig. 17.3 Map of early sites for cattle in Saharan Africa (redrawn from Brooks et al. 2009)

of pastoralism into East Africa, through comparison of funerary monuments. The site of Adrar Bous in Niger is a type site for Tenerian lithic assemblages, and contains the mummified corpse of a cow, with affinities to the taurine/sanga complex, dated to 5670 ± 500 BP (Clark et al. 2008). This may be evidence for the presence of pastoralism at this period, and this may have been the source of the cattle bones at Gobero.

Pastoralism in Africa was revolutionised by the appearance of the larger zebu cattle (*Bos primigenius* f. *indicus*), domesticated in India and then reaching the Horn of Africa. Making a distinction between zebu and other cattle types is difficult from osteological remains despite their obvious phenotypic differences. Finds are sparse (Magnavita 2006) but the earliest evidence is from Mali and Senegal at possibly 800–1400 AD (MacDonald 1999; MacDonald and MacDonald 2000).

17.4.2 Ovicaprines

The traditional sheep and goats of sub-Saharan Africa have hair, not wool, and are generally not milked. The evolution of West African dwarf types in high humidity regions resulted from a long period of adaptation (Epstein 1971). As a consequence, larger Sahelian breeds were brought across the desert in a later period to fill the niche created by the southward spread of dwarf breeds. These larger breeds include the Uda'en, a breed specifically associated with pastoralists today. In the Horn of Africa, fat-tailed sheep were introduced from Arabia (Blench 2009). These cannot be detected in the archaeological record, but are represented in rock art as far south as Zimbabwe and Zambia, by 2000 BP. However, they do not spread very far west in the Sahara.

17.4.3 Camel Pastoralism

The camel was first domesticated in the Arabian peninsula at ~5000 BP (Ripinsky 1975; Compagnoni and Tosi 1978). It is occasionally represented in Egypt from the early Dynastic period, but whether as an exotic import or a working animal is disputed (Ripinsky 1985; Rowley-Conwy 1988). Brogan (1954) concludes that the camel came into general use in North Africa in the first century AD as a baggage and transport animal (Fig. 17.4). We can presume the camel was rapidly adopted by the Berbers and replaced the horse on trans-Saharan routes. Camel remains in West Africa, like those of donkey, are first documented from the Middle Senegal Valley, with a single first phalanx from the site of Siouré dated to 250–400 AD (MacDonald and MacDonald 2000, pp. 141–142).

Fig. 17.4 Anthropological contexts of camel in the Sahara. **a** Camel used for agricultural labour (source: Brogan 1954); **b** camel and donkey caravan in the central Sahara (source: Creative Commons)

17.4.4 Dairying

Earlier work argues that dairying was the key innovation which drove domestication. Chemical evidence from gas chromatography analyses of organic residues from pottery in the Takarkori rock shelter, southwest Fezzan, Libya, shows milking to go back to 7200–5800 BP (Dunne et al. 2013). However, the earliest cattle in the Sahara are unlikely to have been milked. Murdoch (1959) identifies that non-milking (taurines, humpless shorthorns) are distinct from milked n'dama (humpless longhorns) and zeboids. Ethnographic evidence suggests taurines were kept as 'royal', prestige animals, only occasionally slaughtered for meat and never milked. The absence of milking is strongly connected with trypanotolerance; to develop resistance to the trypanosomoses, cattle must have antibodies to the pathogens. Milking the cattle would weaken them and thus reduce their resistance. Another notable feature of cattle production in West-Central Africa is the absence of traction. Traditionally, neither ploughing nor carting were practised in the Sahel, although these are old technologies north of the desert. The suggestion is that work-stress related to these activities would have resulted in high cattle mortality. Today, cattle are the most important species milked, but sheep, goats, and camels are also exploited sporadically. Among the Fulɓe herders of West-Central Africa, the Uda'en herds a distinctive breed of sheep, the Uda, which has been bred for milk production. Goats are milked by the peoples in the central Sahara and North Africa but rarely in the Sahel and southwards. Arab herders of the Sahara, such as the Uled Suleiman, milk their camels and make a crumbly cheese which is sold in the markets of the region.

Historically, Sahelian nomads grew no crops, but exchanged surplus milk for cereal grains. Although pastoral nomads have lactase persistent genes, the sedentary peoples by and large cannot digest fresh milk. Studies of global lactase persistence have suggested multiple parallel evolutions, including one in West-Central Africa (Itan 2010). The usual practise is to process the fresh milk into yoghurt, butter or cheese, which avoids this problem. Although these storable dairy products are well represented in the

Table 17.2 Saharan foragers persisting into recent times

Culture	Location	Subsistence strategy	Status
Nemadi	Mauritania/Mali border	Hunting with dogs	Unknown
Imraguen	Mauritania coast	Inshore fisheries	Still active
Dawada	Libya	Shrimp harvesting	Dispersed in the 1970s
Haddad	Northern Chad	Hunting of porcupine, birds, gazelles, net-hunting	Unknown

ethnographic record, there is as yet no means to distinguish them from fresh milk in the archaeological record.

17.4.5 Sub-Saharan African Agriculture and the Diffusion of Pastoralism

A major enigma in Saharan pastoralism is that cattle occur in the Fezzan (Libya), apparently being milked, as early as 7000 BP, yet cattle, sheep, and goats only reached the Sahel of West-Central Africa around 3500 BP, some 3000 years later. This may be related to the regional genesis of agriculture, and also explain the failure to transfer traction technologies across the region. The arrival of domestic animals is in much the same time frame as the earliest evidence for cereal domestication and agricultural food production, and this may be no coincidence. Manning (2010) argued that early pastoralists domesticated cereals; this chapter suggests the process was the reverse, namely that pastoralism could succeed only because of the niche created by cereal domestication (see also Marshall and Hildebrand 2002). Unlike the Sahara, Sahelian West-Central Africa has a high incidence of pathogens that are lethal to animals in poor nutritional condition. Cereals can assist livestock, especially cattle, to overcome nutritional stress during the dry season when pasture is least available, and manure can be used as a field fertiliser.

An additional factor is the susceptibility of cattle to tsetse and trypanosomoses. When the Sahel only had low-density foragers, wild animal vectors of tsetse would have been very prevalent, especially in the gallery forest along rivers where livestock were obliged to drink. Sedentary farmers rapidly hunt out wild animals in the local area, in contrast to mobile foragers, which has the effect of creating low pathogen zones around settlements, making them attractive to livestock owners. Similar factors may explain the spread of pastoralism into East Africa where pastoral cultures were present in the Lake Turkana area as early as 5000 BP. However, pastoralism did not spread to Tanzania until ~2200 BP, and the 3000-year gap parallels that in the Sahara and the appearance of agriculture in the region.

17.5 Pastoralists and Foragers Today

Some fragmentary remnants of forager cultures of the Sahara persisted into the twentieth century, providing ethnographic insight into such lifestyles (Blench 2019). These residual foraging groups are listed in Table 17.2. It is of note that the types of foraging specialisations implied by ethnographic records have not been proposed in interpretations of the archaeological record. The Haddad (Nicolaisen 2010), for example, specialised in net-hunting for birds and digging porcupines out of their dens which strongly suggests that cultures such as the Aterian cannot have been monolithic but probably encompassed a range of subsistence strategies.

17.6 Conclusions and Wider Outlook

Africa is the core region for the evolution of modern humans, and the opening and closing of the Sahara over the last 300,000 years has provided a gateway for population flows. The alternation between hyper-aridity and river and lake systems of the Green Sahara provided a rich subsistence environment and created corridors for both humans, animals, and plants (Tierney and Pausata 2017). The Aterian dominated the region for close to 100,000 years, after which the pace of change accelerated with the Iberomaurusian and Capsian cultures, which provided the transition to the pottery-using era. Pastoralism, initially cattle and later small ruminants, transformed the forager cultures of the Sahara from around 10,000 BP although its spread into sub-Saharan Africa was slow and appears to accelerate only with the development of cereal agriculture. Elements of these subsistence systems persisted in the Sahara into the twentieth century, and only recent political insecurity in the region may have brought to an end socioeconomic and cultural systems which had lasted for millennia.

Acknowledgements Thanks to Nick Drake, Kevin Macdonald, Veerle Linseele for discussions and forwarding papers relevant to these topics.

References

Ballouche A, Marinval P (2003) Données palynologiques et carpologiques sur la domestication des plantes et l'agriculture dans le Néolithique ancien du Maroc septentrional. (site de Kaf Taht El-Ghar). Rev Archéomét 27:49–54

Barker G (2006) The agricultural revolution in prehistory: Why did foragers become farmers? Oxford University Press, Oxford, p 616

Barton RNE, Bouzouggar A, Collcutt SN, Schwenninger J-L, Clark-Balzan L (2009) OSL dating of the Aterian levels at Dar es-Soltan I (Rabat, Morocco) and implications for the dispersal of modern *Homo sapiens*. Quat Sci Rev 28:1914–1931

Barton N, Bouzouggar A (2013) Hunter-gatherers of the Maghreb 25000–6000 years ago. In: Mitchell P, Lane PJ (eds) The Oxford handbook of African archaeology. Oxford University Press, Oxford, pp 431–444

Blench RM (1993) Ethnographic and linguistic evidence for the prehistory of African ruminant livestock, horses and ponies. In: Shaw T, Sinclair P, Andah B, Okpoko A (eds) The archaeology of Africa. Routledge, London, pp 71–103

Blench RM (1998) Le N'dama et le bétail issu de croisements (Keteku) au Nigeria. In: Seignobos C, Thys É (eds) Les taurins et des Hommes, Cameroun et Nigeria. IEMVT, Maisons-Alfort, Paris, pp 293–310

Blench RM (2000) A history of donkeys and mules in Africa. In: Blench RM, MacDonald KC (eds) The origins and development of African livestock. Archaeology, genetics, linguistics and ethnography. University College Press, London, pp 339–354

Blench RM (2001) Types of language spread and their archaeological correlates: the example of Berber. Origini 23:169–190

Blench RM (2009) Was there an interchange between Cushitic pastoralists and Khoesan speakers in the prehistory of Southern Africa and how can this be detected? In: Möhlig WJG, Seidel F, Seifert M (eds) Language contact, language change and history based on language sources in Africa. Sprache und Geschichte in Afrika vol 20, SUGIA, Cologne, pp 31–49

Blench RM (2019) The linguistic prehistory of the Sahara. In: Gatto MC, Mattingly DJ, Ray N, Sterry M (eds) Burials, migration and identity in the Ancient Sahara and beyond. Cambridge University Press, Cambridge, pp 431–463

Bollig M, Schnegg M, Wotzka H-P (eds) (2013) Pastoralism in Africa: past, present and future. Berghahn Books, Oxford, p 544

Brogan O (1954) The camel in Roman Tripolitania. Papers Brit School Rome 22:126–131

Brooks N, Chiapello I, di Lernia S, Drake N, Legrand M, Moulin C, Prospero J (2007) The climate-environment-society nexus in the Sahara from prehistoric times to the present day. J North Afr Stud 10:253–292

Brooks N, Clarke J, Garfi S, Pirie A (2009) The archaeology of Western Sahara: results of environmental and archaeological reconnaissance. Antiquity 83:918–934

Bulliet RW (1990) The camel and the wheel, 2nd edn. Columbia University Press, New York, p 327

Campmas E, Chakroun A, Merzoug S (2016) Données préliminaires sur l'exploitation de la malacofaune marine par les groupes ibéromaurusiens de l'abri Alain (Oran, Algérie). PALEO, Rev d'Archéol Préhist 27:83–104

Camps G (1988) Âne. Encyclopedie Berbère 14:647–654

Chaix L, Hansen JW (2003) Cattle with "forward-pointing horns": archaeozoological and cultural aspects. In: Krzyzaniak L, Kroeper K, Kobusiewicz M (eds) Cultural markers in the later Prehistory of Northeastern Africa and recent research. Studies in African archaeology 8. Poznan Archaeological Museum, Poznan, pp 269–281

Chen S, Lin BZ, Baig M, Mitra B, Lopes RJ, Santos AM, Magee DA, Azevedo M, Tarroso P, Sasazaki S, Ostrowski S, Mahgoub O, Chaudhuri TK, Zhang Y, Costa V, Royo LJ, Goyache F, Luikart G, Boivin N, Fuller DQ, Mannen H, Bradley DG, Beja-Pereira A (2010) Zebu cattle are an exclusive legacy of the South Asia Neolithic. Molec Biol Evol 27:1–6

Chenal-Vélardé I (1997) Les Premières Traces de Boeuf Domestique en Afrique du Nord: Etat de la Recherche Centré sur les Données Archéozoologiques. Archaeozoology 9:11–40

Clark JD, Agrilla EJ, Crader DC, Galloway A, Garcea EAA, Gifford-Gonzalez D, Hall DN, Smith AB, Williams MAJ (2008) Adrar Bous: archaeology of a central Saharan granitic ring complex in Niger. Royal Museum of Central Africa, Tervuren, p 403

Compagnoni B, Tosi M (1978) The camel: its distribution and state of domestication in the Middle East during the third millennium B.C. in light of the finds from Shahr-i Sokhta. In: Meadow RH, Zeder MA (eds) Approaches to Faunal analysis in the Middle East. Peabody Museum Bulletin vol 2. Peabody Museum of Archaeology and Ethnology, New Haven, CT, pp 119–128

Cremaschi M, Di Lernia S, Garcea EAA (1998) Some insights on the Aterian in the Libyan Sahara: chronology, environment, and archaeology. Afr Archaeol Rev 15:261–286

di Lernia S (1998) Cultural control over wild animals during the early Holocene: the case of Barbary sheep in central Sahara. In: di Lernia S, Manzi G (eds) Before food production in North Africa. ABACO Edizione, Rome, pp 113–126

di Lernia S, Tafuri MA, Gallinaro M, Alhaique F, Balasse M, Cavorsi L, Fullagar PD, Mercuri AM, Monaco A, Perego A, Zerboni A (2013) Inside the "African cattle complex": animal burials in the Holocene central Sahara. PLoS ONE 8:56879. https://doi.org/10.1371/journal.pone.0056879

Dibble HL, Aldeias V, Jacobs Z, Olszewski DI, Rezek Z, Lin SC, Alvarez-Fernández E, Barshay-Szmidt CC, Hallett-Desguez E, Reed D, Reed K, Richter D, Steele TE, Skinner A, Blackwell B, Doronicheva E, El-Hajraoui M (2013) On the industrial attributions of the Aterian and Mousterian of the Maghreb. J Human Evol 64:194–210

Dunne J, Evershed RP, Cramp L, Bruni S, Biagetti S, di Lernia S (2013) The beginnings of dairying as practised by pastoralists in 'green' Saharan Africa in the 5th millennium BC. Doc Praehist 40:118–130

Epstein H (1971) The origin of the domestic animals of Africa, 2 vols. Africana Publishing Corporation, New York

Ferembach D (1985) On the origin of the Iberomaurusians (Upper Paleolithic, North Africa): a new hypothesis. J Human Evol 14:393–397

Gabriel B (1987) Palaeoecological evidence from Neolithic fireplaces in the Sahara. Afr Archaeol Rev 5:93–103

Garcea EAA (ed) (2013) Gobero: the no-return frontier. Archaeology and Landscape at the Saharo-Sahelian Borderland. Africa Magna Verlag, Frankfurt, p 312

Gautier A (2002) The evidence for the earliest livestock in North Africa: or adventures with large bovids, ovicaprids, dogs and pigs. In: Hassan FA (ed) Droughts, food and culture. Ecological change and food security in Africa's later prehistory. Springer, Boston, pp 195–207

Gautier A (2007) Animal Domestication in North Africa. In: Bollig M (ed) Aridity, Change and Conflict in Africa. Heinrich-Barth-Institut, Cologne, pp 75–89

Gautier A, Van Neer W (2009) Animal remains from predynastic sites in the Nagada region, Middle Egypt. Archaeofauna 18:27–50

Haaland R (1995) Sedentism, cultivation, and plant domestication in the Holocene Middle Nile region. J Field Archaeol 22:157–174

Hanotte O, Bradley DG, Ochieng JW, Verjee Y, Hill EW, Rege JEO (2002) African pastoralism: genetic imprints of origins and migrations. Science 296:336–339

Haltenorth T, Diller H (1980) A field guide to the mammals of Africa: including Madagascar. Harper Collins, London, p 400

Hassan FA (2000) Climate and cattle in North Africa: a first approximation. In: Blench RM, MacDonald KC (eds) The origin and development of African livestock. Archaeology, genetics, linguistics and ethnography. University College Press, London, pp 61–86

Hogue JT, Barton RNE (2016) New radiocarbon dates for the earliest Later Stone Age microlithic technology in Northwest Africa. Quat Int 413:62–75

Hublin J-J, Verna C, Bailey S, Smith T, Olejniczak A, Sbihi-Alaoui FZ, Zouak M (2012) Dental evidence from the Aterian human populations of Morocco. In: Hublin J-J, McPherron SP (eds) Vertebrate paleobiology and paleoanthropology. Springer, Netherlands, pp 189–204

Iovita R (2011) Shape variation in Aterian tanged tools and the origins of projectile technology: a morphometric perspective on stone tool function. PLoS ONE 6:e29029. https://doi.org/10.1371/journal.pone.0029029

Irish JD (2000) The Iberomaurusian enigma: North African progenitor or dead end? J Human Evol 39:393–410

Itan Y (2010) A worldwide correlation of lactase persistence phenotype and genotypes. BMC Evol Biol 10:36. https://doi.org/10.1186/1471-2148-10-36

Jelinek J (2003) Pastoralism, burials and social stratification in central Sahara. Les Cahiers de l'AARS 8:41–44

Jousse H (2004) A new contribution to the history of pastoralism in West Africa. J Afr Archaeol 2:187–201

Kaache B (1996) L'Origine des animaux domestiques au Maroc: état des connaissances. Préhist Anthropol Méditerran 5:85–92

Kulichová I, Fernandes V, Deme A, Nováčková J, Stenzl V, Novelletto A, Pereira L, Černý V (2017) Internal diversification of non-Sub-Saharan haplogroups in Sahelian populations and the spread of pastoralism beyond the Sahara. Am J Phys Anthropol 164:424–434

Kuper R, Reimer H (2013) Herders before pastoralism: prehistoric prelude in the Eastern Sahara. In: Bollig M, Schnegg M, Wotzka H-P (eds) Pastoralism in Africa: past, present, and future. Berghahn Books, Oxford, pp 31–65

Law R (1980) The horse in West African history. the role of the horse in the societies of pre-colonial West Africa. Oxford University Press, London, p 254

Lesur J, Hildebrand EA, Abawa G, Gutherz X (2014) The advent of herding in the Horn of Africa: New data from Ethiopia, Djibouti and Somaliland. Quat Int 343:148–158

Linseele V (2007) Archaeofaunal remains from the past 4000 years in Sahelian West Africa: domestic livestock, subsistence strategies and environmental changes. Archaeopress, Oxford, p 340

Linseele V (2010) Did specialized pastoralism develop differently in Africa than in the Near East? An example from the West African Sahel. J World Prehist 23:43–77

Linseele V (2013) From the first stock keepers to specialised pastoralists in the West African Savannah. In: Bollig M, Schnegg M, Wotzka H-P (eds) Pastoralism in Africa: past, present, and future. Berghahn Books, Oxford, pp 145–170

Lubell D (2004) Prehistoric edible land snails in the circum-Mediterranean: the archaeological evidence. In: Brugal J-J, Desse J (eds) Petits Animaux et Sociétés Humaines. Du Complément Alimentaire Aux Ressources Utilitaires. XXIVe rencontres internationales d'archéologie et d'histoire d'Antibes. Éditions APDCA, Antibes, pp 77–98

MacDonald KC (1996) The Windé Koroji complex: evidence for the peopling of the Eastern Inland Niger Delta (2100–500 BC). Préhist Anthropol Méditerran 5:147–165

MacDonald KC (1999) Invisible pastoralists: an inquiry into the origins of nomadic pastoralism in the West African Sahel. In: Gosden C, Hather J (eds) The prehistory of food: appetites for change. Routledge, London, pp 333–349

MacDonald KC, MacDonald RH (2000) The origins and development of domesticated animals in arid West Africa. In: Blench RM, MacDonald KC (eds) The origins and development of African livestock. Archaeology, genetics, linguistics and ethnography. University College Press, London, pp 127–162

MacEachern S, Bourges C, Reeves M (2001) Early horse remains from northern Cameroun. Antiquity 75:62–67

MacHugh DE, Shriver MD, Loftus RT, Cunningham P, Bradley DG (1997) Microsatellite DNA variation and the evolution, domestication and phylogeography of taurine and zebu cattle (*Bos taurus* and *Bos indicus*). Genetics 146:1071–1086

Magnavita C (2006) Ancient humped cattle in Africa: a view from the Chad Basin. Afr Archaeol Rev 23:55–84

Manning K (2010) A developmental history of West African agriculture. In: Allsworth-Jones P (ed) West African archaeology: new developments, new perspectives. BAR International Series 2164. Archaeopress, Oxford, pp 43–52

Marshall F, Hildebrand E (2002) Cattle before crops: the beginnings of food production in Africa. J World Prehist 16:99–143

Meghen C, MacHugh DE, Sauveroche B, Kana G, Bradley DG (2000) Characterisation of the Kuri Cattle of Lake Chad using molecular genetic techniques. In: Blench RM, MacDonald KC (eds) The origins and development of African livestock. Archaeology, genetics, linguistics and ethnography. University College Press, London, pp 259–268

Midant-Reynes B (1992) Préhistoire de l'Egypte. Des premiers hommes aux premiers pharaons. A. Colin, Paris, p 288

Morales J, Pérez-Jordà G, Peña-Chocarro L, Zapata L, Ruíz-Alonso M, López-Sáez JA, Linstädter J (2013) The origins of agriculture in North-West Africa: macro-botanical remains from Epipalaeolithic and Early Neolithic levels of Ifri Oudadane (Morocco). J Arch Sci 40:2659–2669

Morales Muñiz A, Riquelme JA, von Lettow-Vorbeck CL (1995) Dromedaries in antiquity: Iberia and beyond. Antiquity 69:368–375

Muigai AW, Hanotte O (2013) The origin of African sheep: archaeological and genetic perspectives. Afr Archaeol Rev 30:9–50

Murdoch GP (1959) Africa: its peoples and their culture history. McGraw Hill, New York, p 456

Mulazzani S (ed) (2013) Le Capsien de Hergla (Tunisie). Culture, Environnement et économie. Reports in African Archaeology 4. Magna Verlag, Frankfurt, p 439

Newman JL (1995) The peopling of Africa: a geographic interpretation. Yale University Press, New Haven, p 252

Nicolaisen I (2010) Elusive hunters: the Haddad of Kanem and the Bahr el Ghazal. Aarhus University Press, Aarhus, p 522

Palmer HR (1928) Sudanese memoirs, 2 vols. Government Printer, Lagos

Paris F (2000) African livestock remains from Saharan mortuary contexts. In: Blench RM, MacDonald KC (eds) The origins and development of African livestock. Archaeology, genetics, linguistics and ethnography. University College Press, London, pp 111–126

Peters J (1991) The faunal remains from Shaqadud. In: Marks AW, Mohammed-Ali A (eds) The late prehistory of the Eastern Sahel. The Mesolithic and Neolithic of Shaqadud, Sudan. Southern Methodist University Press, Dallas, pp 197–235

Rahmani N (2004) Technological and cultural change among the last Hunter-Gatherers of the Maghreb: the Capsian (10,000 B.P. to 6000 B.P.). J World Prehist 18:57–105

Richter D, Moser J, Nami M, Eiwanger J, Mikdad A (2010) New chronometric data from Ifri n'Ammar (Morocco) and the chronostratigraphy of the Middle Palaeolithic in the Western Maghreb. J Human Evol 59:672–679

Richter D, Grün R, Joannes-Boyau R, Steele TE, Amani F, Rué M, Fernandes P, Raynal JP, Geraads D, Ben-Ncer A, Hublin J-J (2017) The age of the hominin fossils from Jebel Irhoud, Morocco, and the origins of the Middle Stone Age. Nature 546:293–296

Reimer H (2007) When Hunters Started Herding-Pastro-Foragers and the complexity of Holocene economic change in the western desert of Egypt. In: Bollig M (ed) Aridity, change and conflict in Africa. Heinrich-Barth-Institut, Cologne, pp 105–144

Ripinsky M (1975) The camel in Ancient Arabia. Antiquity 49:295–298

Ripinsky M (1985) The camel in Dynastic Egypt. J Egypt Archaeol 71:134–141

Roubet C (1979) Économie Pastorale Préagricole en Algérie Orientale: le Néolithique de Tradition Capsienne. CNRS, Paris, p 582

Rowley-Conwy P (1988) The camel in the Nile Valley: new radiocarbon accelerator (AMS) dates from Qaşr Ibrîm. J Egypt Archaeol 74:245–248

Sawchuk EA, Goldstein ST, Grillo KM, Hildebrand EA (2018) Cemeteries on a moving frontier: mortuary practices and the spread of pastoralism from the Sahara into eastern Africa. J Anthropol Archaeol 51:187–205

Scerri EML (2013) The Aterian and its place in the North African Middle Stone Age: the Middle Palaeolithic in the Desert. Quat Int 300:111–130

Sereno PC, Garcea EAA, Jousse H, Stojanowski CM, Saliege J-F, Maga A, Ide OA, Knudson KJ, Mercuri AM, Stafford TW Jr, Kaye TG, Giraudi C, N'siala IM, Cocca E, Moots HM, Dutheil DB, Stivers JP (2008) Lakeside cemeteries in the Sahara: 5000 years of Holocene population and environmental change. PLoS ONE 3:e2995. https://doi.org/10.1371/journal.pone.0002995

Shaw BD (1979) The camel in ancient North Africa and the Sahara: history, biology and human economy. Bull Instit Fond Afrique Noire 41:663–721

Smith AB (1992) Origins and spread of pastoralism in Africa. Ann Rev Anthropol 21:125–141

Starkey PH (1984) N'Dama cattle—a productive trypanotolerant breed. World Animal Rev 50:2–15

Stojanowski CM, Carver CL (2011) Inference of emergent cattle pastoralism in the southern Sahara desert based on localized hypoplasia of the primary canine. Int J Paleopathol 1:89–97

Tafuri MA, Bentley RA, Manzi G, di Lernia S (2006) Mobility and kinship in the prehistoric Sahara: Strontium isotope analysis of Holocene human skeletons from the Acacus Mts. (southwestern Libya). J Anthropol Archaeol 25:390–402

Tauveron M, Striedter KH, Ferhat N (2009) Neolithic domestication and pastoralism in central Sahara: the cattle necropolis of Mankhor (Tadrart algérienne). In: Baumhauer R, Runge J (eds) Holocene palaeoenvironmental history of the Central Sahara. CRC Press, Boca Raton, pp 201–208

Tawah CL, Rege JEO (1997) A close look of a rare African breed—the Kuri cattle of Lake Chad Basin: Origin, distribution, production and adaptive characteristics. S Afr J Animal Sci 27:31–40

Tierney JE, Pausata FS (2017) Rainfall regimes of the Green Sahara. Sci Adv 3:e1601503. https://doi.org/10.1126/sciadv.1601503

van de Loosdrecht M, Bouzouggar A, Humphrey L, Posth C, Barton N, Aximu-Petri A, Nickel B, Nagel S, Talbi EH, El Hajraoui MA, Amzazi S, Hublin J-J, Pääbo S, Schiffels S, Meyer M, Haak W, Jeong C, Krause J (2018) Pleistocene North African genomes link Near Eastern and sub-Saharan African human populations. Science 360:548–552

Van Peer P, Fullagar R, Stokes S, Bailey RM, Moeyersons J, Steenhoudt F, Geerts A, Vanderbeken T, De Dapper M (2003) The Early to Middle Stone Age transition and the emergence of modern human behaviour at site 8-B-11, Sai Island, Sudan. J Human Evol 45:187–193

Wilson RT (1984) The camel. Longmans, London, p 223

Zapata L, López-Sáez JA, Ruiz-Alonso M, Linstädter J, Pérez-Jordà G, Morales J, Kehl M, Peña-Chocarro L (2013) Holocene environmental change and human impact in NE Morocco: palaeobotanical evidence from Ifri Oudadane. Holocene 23:1286–1296

Oases Occupation

18

Stefania Merlo

Abstract

This chapter presents a fundamental aspect of the Saharan landscape: the oasis. The oasis, contrary to popular imagination, is not an isolated and exceptional occurrence in the desert but is an artificial construct, the product of human effort, and has characterized the landscapes of the Sahara from as early as the third millennium BC (5 ka BP). Oases are supported by the perennial presence of water, which allows complex hydro-agro and pastoral production systems to develop; they are also hubs that attracted populations as a result of their available environmental resources, and historically played a crucial role as transit points of goods, people, and ideas throughout the Sahara. Framed by this perspective, this chapter presents some examples of different types of oases found in the Sahara today, and then describes the emergence and evolution of oasis occupation. The Garamantes of the Fazzan, in Libya, are discussed as an example of an ancient polity that mastered the management of oases from the first millennium BC to the first millennium AD (~3–1 ka BP).

Keywords

Oasis · Water · Irrigation systems · Urbanism · Human adaptation · Sahara · Garamantes

18.1 Introduction

There is no desert without oases. The oasis is a fundamental aspect of the Sahara, engrained in the public perception of the desert itself (Mattingly and Sterry 2020). Associated with the concept of *mirage* in the popular imagination (the perennial freshwater lake surrounded by lush green vegetation in the middle of sand dunes, often disappearing as the thirsty desert traveller approaches), it has also been used to define the geographical limits of the Sahara itself. Braudel (1982), for example, used the cultivation of the date palm in the oases as delineating the borders of the Sahara.

Whilst natural oases, defined as *loci* with the presence of water and vegetation, exist in depressions in the surface of the desert where they can move according to sand dune migration, the term more often implies human presence and human intervention. The oasis is an artificial construct, the product of human effort, as demonstrated by the gradual disappearance of oases without regular management (Benmihoub et al. 2021). Used for the first time by Herodotus in 450 BC to define the settlement of Kharga in Egypt, the term 'oasis' most probably originates from ancient Egyptian and meant 'inhabited environment' (Purdue et al. 2018). Most modern definitions, although not universally agreed, consider oases primarily as agricultural spaces, more specifically as places of production marked by the management of water resources and the mastery of hydraulic systems, but they fail to consider them as strategic nodes of political and commercial significance. It has been demonstrated that oases are also economic hubs that attracted populations as a result of available resources, and played a crucial role as transit points of goods, people, and ideas in the past (Purdue et al. 2018). They are also networks that are actively maintained and fostered today in spite of the elevated cost of sustaining oasis farming (Scheele 2010). Purdue et al. (2018, p. 27) define oases as 'long-term socio-environmental refugia (or niches) supported by perennial sources of water in otherwise harsh environments with dynamic boundaries dependent on the structure of the collectively or individually-managed complex hydro-agro and pastoral production system, integrated within a local to global economic system, and subject to decisive short-term and long-term climatic change'. Thus, oases refer to the relationship between the physical

S. Merlo (✉)
McDonald Institute for Archaeological Research, University of Cambridge, Downing Street, Cambridge CB2 3ER, UK
e-mail: sm399@cam.ac.uk

J. Knight et al. (eds.), *Landscapes and Landforms of the Central Sahara*, World Geomorphological Landscapes, https://doi.org/10.1007/978-3-031-47160-5_18

availability of water in the landscape and the ways in which these locations are affected by human activity (Amichi et al. 2020). This relationship is dynamic through space and time and continues to develop in the present. Thus in this chapter aspects of ancient and modern oases in the central Sahara are discussed.

The chapter opens with an overview of types of oases and their characteristics, followed by a brief discussion of the emergence and evolution of oases in the Sahara, and then an introduction to the Garamantes of the Fazzan, in Libya, an ancient polity that mastered the management of oases from the first millennium BC to the first millennium AD (3–1 ka BP). The chapter then offers a summary of the historical trajectory of oases to the contemporary period, highlighting in particular challenges to the preservation of Sahara oasis livelihoods.

18.2 Types of Oases and Their Characteristics

The central Sahara is characterized by high topographic variability, from salt pans below sea level to high-altitude peaks distributed along a system of mountains that can reach above 3000 m (southern Algeria's Hoggar mountains and the Emi Koussi volcano in northern Chad's Tibesti massif are amongst the most prominent elevations of today's Sahara). These mountains are interspersed with vast expanses of sand dunes and barren plateaus and hills with a mean altitude of 180–365 m (see chapters by Knight and Merlo, and Els and Knight, this volume). There is also considerable spatial variability in climate with the central Sahara plains characterized by high temperatures and intense aridity (less than 20 mm yr^{-1}) and the highlands and mountains presenting lower temperatures and higher precipitation throughout the year (e.g. mean annual temperature in the Hoggar, at the meteorological station of Assekrem is 14.2 °C and average annual rainfall 186 mm; Cherchali et al. 2022). Different types of aquifers characterize the area, ranging from alluvial, volcanic rock and sandstones, the latter including the Nubian Sandstone Aquifer System, the North Western Sahara Aquifer System, the Taoudeni basin and the Iullemmeden Aquifer System (see Knight, this volume). These water sources vary in depth to the water table and physicochemical characteristics, with consequences for the ease of water availability and the technologies used for water extraction to create and maintain oases in different locations in the Sahara (Fig. 18.1).

As desert forms are varied, so are oases. Saharan oases vary in their size, physical, and biocultural properties and function. Oases can range in size from 2 km^2 (Ourgala) to 211,980 km^2 (Beriane, located in the M'zab Valley) and in human populations from 1267 (Tamtert, Saoura Valley) (Tydecks et al. 2021) to over 250,000 inhabitants (Tamanrasset in southern Algeria) (Cherchali et al. 2022). Oasis populations are culturally diverse (e.g. 12 ethnic groups and languages were found in the study by Tydecks et al. 2021) and contribute significantly to the biodiversity of the Sahara (Brito et al. 2014).

Oases in the Sahara have been variously classified based on their history (modern *versus* ancient oases), water management (groundwater, sources, floodwater, etc.), and hydro-climatic, and geomorphological conditions (coastal,

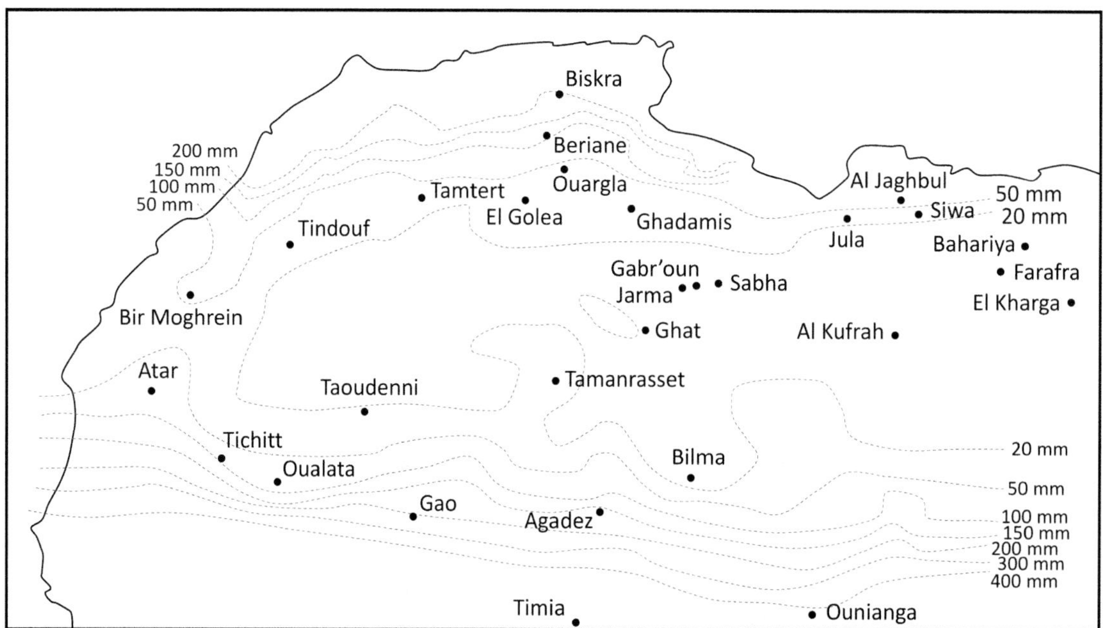

Fig. 18.1 Locations of significant oases across the Sahara (adapted from Allan 1984), including some sites mentioned in the text

plain, mountain). The functional diversity of oases makes their final classification complex (Purdue et al. 2018).

A key issue with the creation of oases in the Sahara has been accessing and managing water resources especially in the context of maintaining water supply for agricultural irrigation and avoiding salinization of the source aquifers (Khezzani et al. 2016; Khezzani and Bouchemal 2018; Purdue et al. 2018; Zegait et al. 2021). Oases have complex and highly varied hydrological conditions and as a consequence different technological answers to irrigation are often used in different areas. Wilson et al. (2020) identified several different types of water sources and irrigation technologies in the Sahara:

1. Perennial water sources and diversion canals that can be exploited by water-lifting devices such as the Archimedes screw;
2. Episodically-active streams (wadis) from which water can be diverted or stored following rainfall;
3. Natural artesian springs in footslope locations or along the margins of oases where water flow can be directed through excavated channels, wells, or along collection basins or canals;
4. Non-artesian aquifers that can be exploited by wells and boreholes, but these are more difficult to access and less reliable than artesian springs;
5. Balance wells (termed *shaduf* or *khattara*);
6. Animal-operated wells with a self-dumping bucket (termed *dalw*) for water extraction;
7. Sunken gardens (termed *bour*) where palm tree roots are able to extend down to the water table; and
8. The development of underground channel systems (termed *foggara*).

The different styles and technologies of water management employed in any one location depend on local topography, natural water availability, and water requirements for the human population and for agricultural needs. This has given rise to a complex mosaic of irrigation technologies in different parts of the Sahara and technologies that have supported the creation of a diversity of oases that evolved in different ways as a response to drivers such as population growth, expansion of agricultural area, declining water tables, and other factors (Mattingly and Sterry 2020).

18.2.1 Foggara

Foggara (plural *foggara*) refer to the subsurface drainage systems that have been excavated into rocky or sediment slopes in the Sahara in order to facilitate the withdrawal of water from the groundwater table to supply oasis communities. The foggara channels may be linear to zig-zag in shape with vertical shafts connected to the land surface, gently sloping downslope, and located mainly in footslope locations. Overviews of the Saharan foggara are provided by Dahmen and Kassab (2017), Hadidi et al. (2019), and Wilson et al. (2020). It is likely that this technology (termed differently depending on the geographic region where this system is located, for example *khettara* in Morocco, *qanat* in Iran, or *falaj* in Oman) was introduced into the Sahara region from the Middle East probably in the sixth/fifth century BC (~2.5 ka BP) (Mattingly et al. 2020a). A number of radiometric ages exist on foggara systems in the Sahara, including from luminescence dating of excavated sediments (Bailiff et al. 2019), radiocarbon dating of mud bricks on foggara walls, and archaeological evidence of human occupation of oases which is linked to enhanced availability of water provided by foggara systems (Wilson et al. 2020). These ages span a wide range from ~2.5 ka BP to within the last 200 years (Bailiff et al. 2019), suggesting that foggara systems were built, modified, and reused at different time periods, likely related to changes both in climate and cultural contexts (Mattingly et al. 2020b). Although the traditional foggara system has been gradually abandoned in most of the central Sahara since probably the end of the first millennium AD (~1 ka BP), Remini et al. (2011) state that around 250 functional foggara still exist in central Algeria; total numbers in the rest of the central Sahara are unknown.

Foggara systems are important because they represent a technological solution to limited water availability in the Sahara, and thus the development of this technology likely went in tandem with the growth of oases populations, and changes in agricultural practises with associated socioeconomic and cultural impacts (Wilson et al. 2020). The contemporary cultural impacts of foggara during their periods of operation are not well known, and this is an important area of future research (Wilson et al. 2020). However, the utilization of water resources in oases in the central Sahara was undertaken in particular by the Garamantes who developed complex urban settlements around them, spanning the earliest dated phases of foggara development (Schwartz and Ibaraki 2011; Mattingly and Sterry 2013; Wilson et al. 2020). Thus, the Garamantian cultures can be considered to particularly reflect oasis resource use and are discussed in detail in this chapter.

The online Atlas of Saharan and Arabian Oases developed by the LabOasis Foundation (http://www.laboasis.org/) provides a database of 776 traditional oases across North Africa and the Arabian Peninsula. Although not exhaustive, this is the most complete catalogue of oases in the central Sahara. The LabOasis Foundation identifies four hydromorphological oasis types (Fig. 18.2), and these are found in specific geomorphic settings with different potential for water yield for oasis agriculture. These types are:

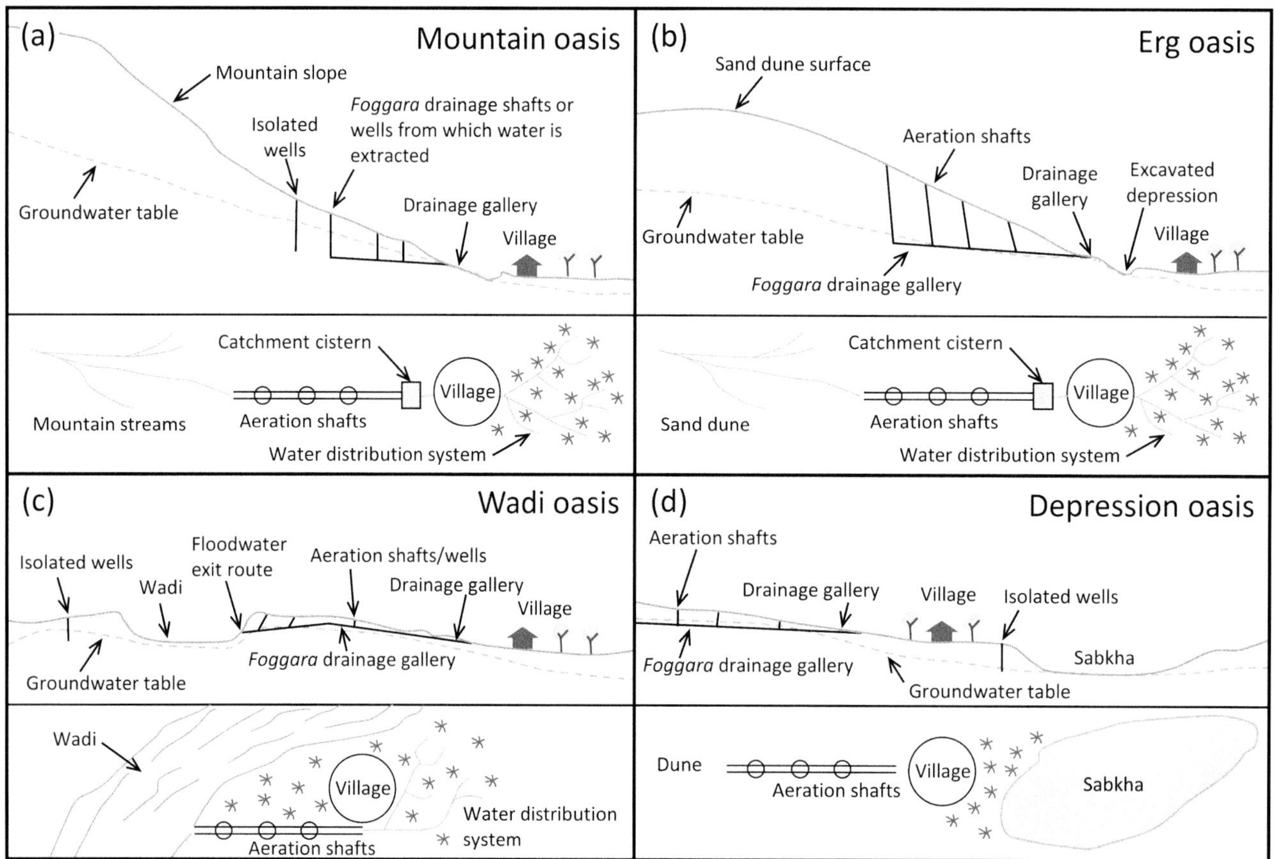

Fig. 18.2 Illustration of the four main hydrogeological and geomorphological types of oases (redrawn from the LabOasis Foundation website at laboasis.org). The upper part of the panels presents a cross-sectional view and the lower part of the panel shows a plan view

18.2.2 Mountain Oases

These are located in rocky highland areas where the water table rises in the profile and where groundwater can be extracted from this elevated water table surface (Fig. 18.2a). In these environments, groundwater sapping takes place in footslope areas (Luo et al. 1997), and this is the mechanism by which oases can be maintained (Fig. 18.3a). Shallow wells are used to extract water from phreatic aquifers at foothill locations; occasionally short-track foggara can be used for water collection. The village, palm grove, and gardens of these oases are located halfway up a slope on terraces or at the foothill at variable altitudes.

18.2.3 Wadi Oases

This is the most common type of oasis across the Sahara region. Wadi oases are associated with dry-bed river systems (*wadis*) and are therefore found in lower slope locations and adjacent to bedrock uplands where wadis are most commonly developed (see Knight, this volume) (Fig. 18.2c). Wadi systems are episodically reactivated by floodwater during rainfall events when water is able to rapidly infiltrate into the groundwater system through the coarse sediments that form the beds of most wadi channels (Megnounif et al. 2013). Water flowing through the channel can also be diverted into smaller branching channels or ponds. Foggara, when present, branch out from the riverbed and catch both residual water flows and the infiltration and micro-flows which run underground below the wadi surface. Wadi oases do not always use groundwater-maintained systems but groundwater can be captured through wells and boreholes into or in the vicinity of the coarse riverbed. The structure of agricultural systems and settlements of wadi oases is generally linear to sinuous, following the wadi channel system. Canalization supported by barriers, locks, and dams characterizes water management in these oasis type; areas of cultivation that are often terraced run parallel to the water courses.

Fig. 18.3 Examples of different oasis types. **a** Mountain oasis at Timia, Niger (*Photo*: keesleonardmaas), **b** wadi oasis at Tmisan, wadi ash-Shati, Libya (*Photo*: S. Merlo), **c** erg oasis at Umm al-Maa, Ubari sand sea, Libya (*Photo*: S. Merlo), **d** depression oasis located just outside of the central Saharan region at Siwa, Egypt (*Photo*: William John Gauthier). Photos (a) and (d) reproduced under Creative Commons licence

18.2.4 Erg Oases

This oasis type is set in the interior of enclosing sand seas (*ergs*) where oases are created in depressions between individual sand dunes, at the intersect with the groundwater table (Figs. 18.2b, and 18.3c). The oases lying within these depressions generally surround small highly saline lakes, the surface levels of which mark the approximate elevation of the water table in the valley. Creutz et al. (2016) described permanent lakes developed between linear dunes at Ounianga, Chad, that fall into the erg oasis type (Fig. 18.4). These are located in the same structural and geomorphic position as the better-known Lake Yoa which is 40 km to the northwest. Where the water table is located near to the ground surface, shallow basins or depressions called *ghout* are excavated in order to deepen the depression and thus access water more easily (Khezzani et al. 2016; Remini and Miloudi 2021). Date palm trees are planted in these depressions or just around them; this method of cultivation allows water to ascend directly by capillary action from the phreatic aquifer to the date palm roots and allows farmers to avoid the use of more labour-intensive irrigation systems (Khezzani and Bouchemal 2018). In addition to water supply from the groundwater table, water can be drained from within the surrounding sand dunes by sloping foggara, which are known to extend for up to 20 km distance. Here only a small amount of water is designed to be collected sustainably (to avoid lowering the groundwater table) and the foggara galleries collect micro-flows which drain the sands' moisture but which mainly draw condensation absorbed by the soil due to the high thermal range.

18.2.5 Depression Oases

Depression oases develop in broad, shallow natural depressions or hollows in the landscape (Fig. 18.2d). Often these depressions reflect former lake beds that were occupied

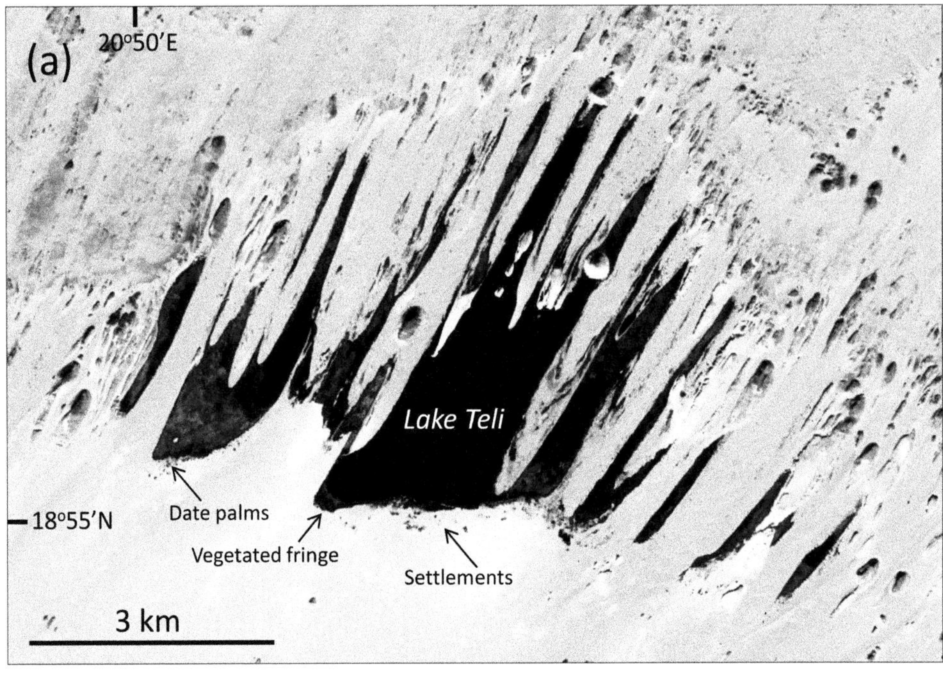

Fig. 18.4 **a** Example of the erg oasis type at Ounianga, Chad (image from Google Earth, image date 23 November 2016), **b** photo of the oasis at Ounianga (*photo*: Emilien Lebourgeois, reproduced under Creative Commons licence)

under wetter climatic conditions during the African Humid Period (see Knight, this volume). Today, these lake beds are relatively flat, may be seasonally or permanently dry under natural groundwater exfiltration conditions, and affected by salt and silt deflation. High evaporation rates from shallow water pools mean that these depressions commonly have high water salinities. The depressions are therefore termed *sabkha* (or sebkha) is a dry salt lake bed, or *chott* if they show some surface water. Several studies have examined sabkha environments in the Sahara, in particular in the Ouargla region of Algeria (Youcef and Hamdi-Aïssa 2014; Medjani et al. 2017; Zatout et al. 2020) and in southeast Tunisia (Marquer et al. 2008; Jaouadi et al. 2016). The high salinity of water within such depression oases today (>230 g L^{-1} dissolved solids, pH of 7.67; Zatout et al. 2020), or within the exfiltrating groundwater (>7000 mS m^{-1}; Idder et al. 2014), may limit their ability to provide freshwater for agricultural or domestic use. An example of adaptation to these conditions is the Siwa depression oasis in Egypt where effective water management through drainage can improve water quality and quantity (Hussein 2021). The margins of these depressions, however, may have natural freshwater springs or excavated tunnels or wells to provide water.

18.3 Emergence and Evolution of Oases in the Sahara

At various times in prehistory the Sahara has oscillated between wet and arid phases which have determined the intensity of its human occupation. The last significant wet phase was in the Early–Middle Holocene, broadly 8000–3500 BC (10.0–5.5 ka BP). During this period, known as the African Humid Period (see Knight, this volume), the wide availability of water in the form of seasonal rivers, small lakes, and a high water table supported mobile human communities of hunter-gatherers adapted to herding of domesticated animals, primarily cattle (Mattingly and Sterry 2020). In the Late Holocene from ~3500 BC (5.5 ka BP) there was a significant step-change in climate with the start of the modern hyperarid phase of the Sahara. The rate of desertification at the end of the African Humid Period is a controversial topic, with some arguing for rapid drying due to nonlinear responses to monsoon patterns, and others for gradual change from savanna to desert over a few thousand years (see Kuper and Kröpelin 2006; Shanahan et al. 2015; Drake et al. 2018). After 3000 BC (5 ka BP), minor fluctuations in rainfall did not fundamentally change the reality of the Sahara as a full desert environment (Mattingly and Sterry 2020). It is against this background that significant cultural and economic changes, well documented by recent hydrogeological and archaeological research (e.g. Cremaschi and di Lernia 2001; Dachy et al. 2018; Mattingly et al. 2020a), took place in the central Sahara. These changes reflect different adaptive responses of human populations to hyperaridity and they include the replacement of cattle with sheep, goat, and camel pastoralism; the emergence of arable agriculture; the depopulation of large parts of the Sahara and the beginning of more permanent urban settlements; alongside continued mobility and the short and long distance movements of goods and people as part of trade and migration networks (Cremaschi and di Lernia 2001; Scheele 2010; di Lernia 2013; Mattingly 2017; Sterry and Mattingly 2020; Fenwick et al. 2021). Different in nature and scale and not mutually exclusive, these responses are all directly or indirectly linked to the presence and availability of water, and are expressed by a diverse range of strategies undertaken by humans for water access and utilization. The creation of agricultural oases, characterized by highly organized harvesting of water and complex irrigation systems, is therefore only one of several ways in which humans responded to hyperaridity in the desert, albeit the best-known image of a Saharan oasis in the popular imagination (Khalidi et al. 2018; Purdue et al. 2018).

The earliest evidence for oasis occupation in the Sahara at large dates back to the third millennium BC (5 ka BP) and comes from the Western Desert of Egypt (at oasis sites of Fayyum, Kharga, Dakhala, Farfara and Siwa) (Mattingly et al. 2020a). In the Fayyum, the early development of oases is linked to augmentation of the exploitation of Nile floodwaters via a canalized irrigation system, whilst other oases, away from the Nile, saw the development and expansion of a variety of desert irrigation systems, some of which are still used today (Mattingly et al. 2020a; Wilson et al. 2020). Clearly, complex rules of water management were necessary for the durability of these systems (Hussein 2021), that are often referred to as 'waterscapes' (Purdue et al. 2018). Many of these eastern Saharan oases also have other elements in common, including the presence of specific agricultural crop types (cereals, dates, figs, grapes, olives) and agricultural practises, access to Roman pottery and other material cultures, and a tendency to evolve into urbanized landscapes as a result of the ability of oases to sustain higher population densities (Mattingly et al. 2020b). These early examples of oasis agriculture and urban settlements became established in the Hellenistic and Roman periods (~300 BC to 300 AD). Despite a number of lacunae, and a particular paucity of evidence for the second millennium BC (4 ka BP), well documented and dated evidence shows that sedentary oasis communities were established much earlier and across much broader areas than has been previously recognized, including in the central Sahara such as the Fazzan, around the 1st millennium BC (3 ka BP) (Sterry and Mattingly 2020). Moreover, many of the key Saharan Islamic sites (e.g. Sijilmasa, Ghadames, Zawila, Timbuktu, and Gao) developed in areas with good evidence for pre-Islamic urbanism (Nixon 2020). Through time, some of the early oases were abandoned whereas others grew in size and importance; a number of them were created at strategic locations in the historic period, along the axes of the trans-Saharan trade (Ross 2010; Scheele 2010; Nixon 2020).

18.4 The People of the Oases: The Garamantes of Fazzan

The Garamantes are archaeologically the best-known indigenous peoples of the ancient Sahara, as a result of long-term investigations in and around their capital at Garama (the modern Jarma in the wadi al-Ajal valley in Fazzan, southwest Libya) (Daniels 1970; Mattingly 2003, 2007, 2010, 2013), in the wadi Tanezzuft, near the present oasis of Ghat (Liverani 2006; Mori 2013), in the northern Tadrart Akakus (Biagetti and di Lernia 2008) and more recently in the areas of Murzuq and Shati (Edwards 2001; Merlo et al. 2013; Mattingly et al. 2015) (Fig. 18.5). This research has demonstrated that from the early first millennium BC (~3 ka BP) the Garamantes had a significant focus on sedentary oasis cultivation, with numerous permanent villages created along a series of three parallel linear depressions that define the landscape of Fazzan, which are the wadi ash-Shati, wadi al-Ajal, and the Murzuq Basin (Mattingly et al. 2020b). Four major settlement phases occurred in the Fazzan: the Early Garamantian (1000–500 BC; 3.0–2.5 ka BP), Proto-Garamantian (500–1 BC; 2.5–2.0 ka BP), Classic Garamantian (AD 1–300; 2.0–1.7 ka BP), and Late Garamantian (AD 300–700; 1.7–1.3 ka BP) (Mattingly et al. 2020b).

The characteristics of Garamantian occupation of Saharan oases have been summarized by Mattingly and Sterry (2020) as follows:

Fig. 18.5 Map of the Garamantian heartland area in southwest Libya. Background imagery sourced from Bing Maps (accessed 22 May 2023), Microsoft product screen shot(s) reprinted with permission from Microsoft Corporation, this image from Earthstar Geographics SIO. International borders sourced from the World Bank, available from https://datacatalog.worldbank.org/search/dataset/0038272 on a CC-BY 4.0 licence. Locations of wadis and mountain ranges partly adapted from Google Earth place name data. Map drawn by Ed Burnett

- A concentration of populations around water sources or potential hydraulic resources, evidenced by the appearance of new settlement forms such as hillforts or dense funerary landscapes;
- The adoption of a developed agricultural package, with affinities with the oases of the Western Egyptian Desert;
- The association with people riding horses and driving chariots and slightly later also the camel, as evidenced in rock art scenes;
- The movement of ideas and technologies (the use of fire, irrigation including foggara, spinning, and weaving, a written Libyan script);
- The construction of distinctive styles of fortified structures (*qsur*); and
- Evidence of trade contacts and networks.

Occupation deposits from the promontory of Zinkekra, to the southwest of Jarma, attest to the first agricultural activities in the central Garamantian region dating around 800 BC (2.8 ka BP) (Pelling 2008). Archaeobotanical evidence indicates that irrigated agriculture was introduced soon after desiccation of the springs, in the early part of the first millennium BC (3 ka BP) (van der Veen 1992, 1995) presumably in response to a lack of water caused by a fall in the position of the water table, and probably consisting of a system of wells used to tap into the sunken water table near the settlement. Archaeological evidence suggests that foggara, which became the trademark irrigation system of the Garamantes in the region, were probably introduced by the final few centuries BC, and definitely before the fourth century AD (1.6 ka BP) and required the organization of large-scale labour to construct and maintain them (Wilson and Mattingly 2003). This roughly corresponds with archaeobotanical evidence for the intensification and diversification of agriculture involving the introduction of a farming system that utilized both winter (wheat and barley) and notably summer (pearl millet, sorghum and cotton) crops (Pelling 2008). Pulses were also present whilst date (*Phoenix dactylifera*) were the most popular fruit, followed by grape and fig (Pelling 2008). Foggara would have also allowed the spread of agriculture into areas adjacent to oases. In combination these developments likely led to an increase in agricultural production and the rise of the Garamantes as a major political power in the central Sahara linked to increasing socioeconomic and trading status (Mattingly et al. 2020a). Equally, it is hypothesized that the Garamantian population, urban development, and irrigated agriculture declined substantially in the second half of the first millennium AD due in great part, although not exclusively, to foggara falling out of use as a result of declining water tables (Wilson et al. 2020).

Settlements developed to a greater or lesser extent throughout the Fazzan oases, where they are numbered in the hundreds, during the Garamantian period. They initially occupied promontories and high ground areas to then descend to the slopes and eventually to lower ground areas, within the oases. These settlements were varied in size and form and ranged from isolated villages, to fortified outposts to substantial villages and a few sites of urban scale developments (including but not limited to the capital Garama) (Mattingly et al. 2020a). Some of the most spectacular evidence of pre-Islamic funerary archaeology in the Sahara relates to the Garamantes. Garamantian cemeteries are characterized by dense concentrations of simple, complex and at times monumental tomb types and are notable for the abundance of Roman ceramics and glass (di Lernia and Manzi 2002; Mattingly et al. 2019). During the Garamantian period, considerable quantities of goods were circulating in the Sahara; many imports from the Mediterranean are found in the Garamantian heartlands alongside materials from the Saharan and trans-Saharan areas (Mattingly 2017). The ability of the Garamantes to exploit oasis resources meant that they were able to sustain populations throughout the Sahara, and thus can explain their importance to the initial evolution and subsequent growth of trans-Saharan trade networks (Mattingly 2017). Salt production was also carried out in situ in oases (most probably in the Ubari sand sea in southwest Libya), in particular for the extraction of natron and resources used for metallurgy. Both trade contacts and military conflicts existed between Mediterranean populations and the Garamantes, with a particular mention by the Roman historian Plinius of the Roman proconsul L. Cornelius Balbus marching into the Fazzan and defeating the Garamantes (Mattingly 2003). The Garamantian polity was nevertheless never occupied or provincialized by the Romans (Mattingly 2003). In the Fazzan there is also evidence for other large-scale pre-Islamic societies within oases, alongside Saharan pastoral groups (Mattingly et al. 2020b; Fenwick et al. 2021), indicating that the Garamantes were not the only cultures and peoples making use of Saharan oases in the last 2500 years. It is important to note that the Garamantian settlement pattern did not follow the same distributions found in later periods and attested in colonial period reports. For example, areas that are recorded as highly populated in the 1930s, like the wadi ash-Shati, have yielded limited evidence of Garamantian occupation, whereas the traditional heartlands of the Garamantes such as the wadi al-Ajal have been more sparsely populated in modern times as compared to the wadi ash-Shati (Mattingly et al. 2020b). While varying levels of archaeological preservation and visibility cannot be disregarded, the evidence for shifting settlements in the Fazzan demonstrates that oases are not static environments and that settlements found today may not correspond with those that were highly populated in the past (Mattingly et al. 2020b).

18.5 Discussion

Saharan oases have a long and rich history of human occupation because they provide key water resources, and therefore migration and trade stopping points, in areas that are otherwise dry, inhospitable, and difficult to cross (Mattingly et al. 2015, 2020a; Sterry and Mattingly 2020). Although hundreds of past and present oases are known from the central Sahara, detailed dating studies that combine palaeoenvironmental, archaeological, and historical data are limited (Marquer et al. 2008; Brauneck et al. 2013; Mattingly et al. 2015; Zerboni et al. 2018). Likewise, foggara systems have not all been systematically mapped across the Sahara. Therefore, most oasis systems of all types in the Sahara have not been fully investigated with respect to their hydrogeology, geomorphology, environmental evolution, and patterns of past and present human occupation. Archaeological studies, where available, however, show multiple phases of oasis occupation and resource use, highlighting the complex interplay between climate, water-use technologies, socioeconomic and cultural contexts, and external factors such as trans-Saharan trade systems. Understanding the relationships between such factors requires an integrated approach that compares evidence from different oases across the region. The edited volume by Mattingly and Sterry (2020) offers a first insight into this type of approach.

Both the physical and human systems of oases are undergoing rapid contemporary changes. In particular, changing water supply systems have influenced patterns of agricultural expansion and inhabited areas movement and/or abandonment (Gonçalvès et al. 2020). Traditional water management techniques have been steadily abandoned to be supplanted, via technological innovation and financial investment, with mechanically-powered water wells and sprinkler irrigation systems that extend water supplies to thousands of hectares around oases (Wilson et al. 2020). Groundwater extraction for agricultural development has characterized state enterprises in all countries in the Sahara with additional injection of private capital in some cases (Sims 2014). For example in Egypt, 75% of all desert reclamation (for agriculture and urban projects) has relied on private investors and corporate modes of farming (Sims 2014). Egypt's development projects in the desert since the 1950s have included water well fields in various oases as part of the New Valley Project in the Western Desert. Thousands of hectares have been reclaimed for irrigation in oases such as Kharga and Dakhla and more recently, since the 1990s, farther south near the border with Sudan (East and West Oweinat) (Sims 2014). The state has also had a role in the development of groundwater resources as part of campaigns to settle nomadic pastoralists or isolated communities in rural areas linking this in, in some cases, to wider programmes to ensure rural livelihoods (Sims 2014). This has been the case of resettlement of the Dawada communities from the oases of Gabr'Aun, Mandara an Trouna in the Ubari sand sea to new settlements in the wadi al-Ajal in Libya (Merlo 2007).

In many areas, over-exploitation of freshwater resources has lowered the groundwater table and therefore deeper aquifers are now being tapped (Ramdani et al. 2009; Gonçalvès et al. 2020). Several studies have shown that water over-exploitation leads to problems with water quality and salinization, and that governance frameworks are currently insufficient to manage such impacts or to reconcile the needs of different water users (e.g. Bellal et al. 2016; Khezzani and Bouchemal 2018; Gonçalvès et al. 2020; Zegait et al. 2021).

18.6 Summary

The oases of the Sahara have undoubtedly played different roles from the early periods of human occupation to the present, as refugia, places of exchange, production, and mobility. They have undergone development and crises, they have grown and shrunk: they remain the home of at least 2 million people. Transformation of oases is still ongoing, influenced by social, political, and environmental factors. Conflict, climate change and urban migration have led to the abandonment of many of the world's oases (Purdue et al. 2018), including in the Sahara. In other cases, over-exploitation of available resources is leading to the disappearance of fossil water reservoirs and consequent damage to desert ecosystems (Gonçalvès et al. 2020). A holistic understanding of the oasis in its evolution is needed not only to fully represent and understand the landscapes of the Sahara in the past and in the present but also to guarantee its future sustainable trajectory of development.

References

Allan JA (1984) Oases. In: Cloudsley-Thompson JL (ed) Sahara desert. Permagon Press, Oxford, pp 325–333

Amichi F, Bouarfa S, Kuper M, Caron P (2020) From oasis archipelago to pioneering Eldorado in Algeria's Sahara. Irrig Drain 69(Suppl. 1):168–176

Bailiff IK, Jankowski N, Gerrard CM, Gutiérrez A, Wilkinson KN (2019) Luminescence dating of sediment mounds associated with shaft and gallery irrigation systems. J Arid Env 165:34–45

Bellal S-A, Hadeid M, Ghodbani T, Dari O (2016) Accès à l'eau souterraine et transformations de l'espace oasien: le cas d'Adrar (Sahara du Sud-ouest algérien). Cahiers Géogr Québec 60:29–56

Benmihoub A, Akli S, Benabid N (2021) Enjeux, pratiques et contraintes pour une mise en valeur agroécologique des terres au Sahara. Cas d'un périmètre péri-oasien dans la Vallée du M'Zab (Algérie). New Medit 2:79–96

Biagetti S, di Lernia S (2008) Combining intensive field survey and digital technologies: new data on the Garamantian castles of Wadi Awiss, Acacus Mts. Libyan Sahara. J Afr Archaeol 6:57–85

Braudel F (1982) La Méditerranée et le Monde Méditerranéen à l'Époque de Philippe II, 5th edn. Colin, Paris

Brauneck J, Mees F, Baumhauer R (2013) A record of early to middle Holocene environmental change inferred from lake deposits beneath a sabkha sequence in the Central Sahara (Seggedim, NE Niger). J Paleolimnol 49:605–618

Brito JC, Godinho R, Martínez-Freiría F, Pleguezuelos JM, Rebelo H, Santos X, Vale CG, Velo-Antón G, Boratyński Z, Carvalho SB, Ferreira S, Gonçalves DV, Silva TL, Tarroso P, Campos JC, Leite JV, Nogueira J, Álvares F, Sillero N, Sow AS, Fahd S, Crochet P-A, Carranza S (2014) Unravelling biodiversity, evolution and threats to conservation in the Sahara-Sahel. Biol Rev 89:215–231

Cherchali MEH, Liégeois JP, Mesbah M, Daas N, Amrous K, Ouarezki SA (2022) Central Hoggar groundwaters and the role of shear zones: $^{87}Sr/^{86}Sr$, $\delta^{18}O$, $\delta^{2}H$ and ^{14}C isotopes, geochemistry and water-rock interactions. Appl Geochem 137:105179. https://doi.org/10.1016/j.apgeochem.2021.105179

Cremaschi M, di Lernia S (2001) Environment and settlements in the Mid-Holocene palaeo-oasis of Wadi Tanezzuft (Libyan Sahara). Antiquity 75:815–825

Creutz M, Van Bocxlaer B, Abderamane M, Verschuren D (2016) Recent environmental history of the desert oasis lakes at Ounianga Serir, Chad. J Paleolimnol 55:167–183

Dachy T, Briois F, Marchand S, Minotti M, Lesur J, Wuttmann M (2018) Living in an Egyptian oasis: reconstruction of the Holocene archaeological sequence in Kharga. Afr Archaeol Rev 35:531–566

Dahmen A, Kassab T (2017) Studying the origin of the foggara in the Western Algerian Sahara: an overview for the advanced search. Water Sci Tech: Water Supply 17:1268–1277

Daniels CM (1970) The Garamantes of Southern Libya. Oleander, London, p p47

di Lernia S (2013) The emergence and spread of herding in Northern Africa: a critical reappraisal. In: Mitchell P, Lane P (eds) Oxford handbook of African archaeology. Oxford University Press, Oxford, pp 527–540

di Lernia S, Manzi G (2002) Sand, stones, and bones. The archaeology of death in the Wadi Tanezzuft Valley (5000–2000 BP). Arid zone archaeology monographs, vol 3. Università degli studi di Roma "La Sapienza", Rome, p 354

Drake NA, Lem RE, Armitage SJ, Breeze P, Francke J, El-Hawat AS, Salem MJ, Hounslow MW, White K (2018) Reconstructing palaeoclimate and hydrological fluctuations in the Fezzan Basin (southern Libya) since 130 ka: A catchment-based approach. Quat Sci Rev 200:376–394

Edwards D (2001) Archaeology in the southern Fazzan and prospects for future research. Libyan Stud 32:49–66

Fenwick C, Sterry M, Mattingly DJ, Rayne L, Bokbot Y (2021) A Medieval boom in the north-west Sahara: evolving oasis landscapes in the Wadi Draa, Morocco (c.700–1500 AD). J Islamic Archaeol 8:139–165

Gonçalvès J, Deschamps P, Hamelin B, Vallet-Coulomb C, Petersen J, Chekireb A (2020) Revisiting recharge and sustainability of the North-Western Sahara aquifers. Reg Env Ch 20:47. https://doi.org/10.1007/s10113-020-01627-4

Hadidi A, Yaichi M, Saba D, Habi M (2019) An overview of the traditional irrigation system (foggaras) in Taouat Region. AIP Conf Proc 2123:0200081–0200088

Hussein MS (2021) An "out-of-the-depression" drainage solution to the land degradation problem in Siwa Oasis. Egypt. Arab J Geosci 14:740. https://doi.org/10.1007/s12517-021-07100-8

Idder T, Idder A, Tankari Dan-Badjo A, Benzida A, Merabet S, Negais H, Serraye A (2014) Les oasis du Sahara algérien, entre excédents hydriques et salinité. L'exemple de l'oasis de Ouargla. Rev Sci l'Eau / J Water Sci 27:155–164

Jaouadi S, Lebreton V, Bout-Roumazeilles V, Siani G, Lakhdar R, Boussoffara R, Dezileau L, Kallel N, Mannai-Tayech B, Combourieu-Nebout N (2016) Environmental changes, climate and anthropogenic impact in south-east Tunisia during the last 8 kyr. Clim Past 12:1339–1359

Khalidi L, Cauliez J, Bon F, Bruxelles L, Gratuze B, Lesur J, Ménard C, Gutherz X, Crassard R, Keall E (2018) Late prehistoric oasitic niches along the Southern Red Sea (Yemen and Horn of Africa). In: Purdue L, Charbonnier J, Khalidi L (eds) From refugia to oases: living in arid environments from prehistoric times to the present day. Des refuges aux oasis: vivre en milieu aride de la préhistorie à aujourd' hui. XXXVIIIe. Éditions APDCA, Antibes, pp 71–99

Khezzani B, Bouchemal S (2018) Variations in groundwater levels and quality due to agricultural over-exploitation in an arid environment: the phreatic aquifer of the Souf oasis (Algerian Sahara). Env Earth Sci 77:142. https://doi.org/10.1007/s12665-018-7329-2

Khezzani B, Bouchemal S, Halis Y (2016) Some agricultural techniques to cope with the fluctuation of the groundwater level in arid environments: Case of the Souf Oasis (Algerian Sahara). J Aridland Agric 2:26–30

Kuper R, Kröpelin S (2006) Climate-controlled Holocene occupation in the Sahara: motor of Africa's evolution. Science 313:803–807

Liverani M (ed) (2006) Aghram Nadharif. The Barkat Oasis (Sha'abiya of Ghat, Libyan Sahara) in Garamantian Times. All'Insegna del Giglio, Firenze, p 520

Luo W, Arvidson RE, Sultan M, Becker R, Crombie MC, Sturchio N, El Alfy Z (1997) Ground-water sapping processes, Western Desert, Egypt. GSA Bull 109:43–62

Marquer L, Pomel S, Abichou A, Schulz E, Kaniewski D, Van Campo E (2008) Late Holocene high resolution palaeoclimatic reconstruction inferred from Sebkha Mhabeul, southeast Tunisia. Quat Res 70:240–250

Mattingly DJ (2003) The archaeology of Fazzan, vol 1. Society for Libyan Studies, London, Synthesis, p p426

Mattingly DJ (2007) The archaeology of Fazzan, vol 2, Site Gazetteer, Pottery and other Survey Finds. Society for Libyan Studies, London, p 520

Mattingly DJ (2010) The archaeology of Fazzan, vol 3, Excavations carried out by C.M. Daniels. Society for Libyan Studies, London, p 641

Mattingly DJ (2013) The archaeology of Fazzan, vol 4, Survey and excavations at Old Jarma (Ancient Garama) carried out by C. M. Daniels (1962–69) and the Fazzan Project (1997–2001). Society for Libyan Studies, London, p 640

Mattingly D (2017) The Garamantes and the origins of Saharan Trade: state of the field and future agendas. In: Mattingly D, Leitch V, Duckworth C, Cuénod A, Sterry M, Cole F (eds) Trade in the Ancient Sahara and beyond. Cambridge University Press, Cambridge, pp 1–52

Mattingly DJ, Sterry M (2013) The first towns in the Central Sahara. Antiquity 87:503–518

Mattingly DJ, Sterry M (2020) Introduction to the themes of sedentarisation, urbanisation and state formation in the Ancient Sahara and beyond. In: Mattingly DJ, Sterry M (eds) Urbanisation and state formation in the Ancient Sahara and beyond. Cambridge University Press, Cambridge, pp 3–50

Mattingly DJ, Sterry MJ, Edwards DN (2015) The origins and development of Zuwīla, Libyan Sahara: an archaeological and historical overview of an ancient oasis town and caravan centre. Azania: Archaeol Res Afr 50:27–75

Mattingly D, Sterry M, Ray N (2019) Dying to be Garamantian: burial, migration and identity in Fazzan. In: Gatto M, Mattingly D, Ray N, Sterry M (eds) Burials, migration and identity in the ancient Sahara and beyond, Trans-Saharan archaeology. Cambridge University Press, Cambridge, pp 53–107

Mattingly DJ, Sterry M, Rayne L, Al-Haddad M (2020a) Pre-Islamic oasis settlements in the Eastern Sahara. In: Mattingly DJ, Sterry M (eds) Urbanisation and state formation in the Ancient Sahara and beyond, Trans-Saharan archaeology. Cambridge University Press, Cambridge, pp 112–146

Mattingly DJ, Merlo S, Mori L, Sterry M (2020b) Garamantian oasis settlements in Fazzan. In: Mattingly DJ, Sterry M (eds) Urbanisation and state formation in the Ancient Sahara and beyond. Cambridge University Press, Cambridge, pp 53–111

Medjani F, Aissani B, Labar S, Djidel M, Ducrot D, Masse A, Hamilton CM-L (2017) Identifying saline wetlands in an arid desert climate using Landsat remote sensing imagery. Application on Ouargla Basin, southeastern Algeria. Arab J Geosci 10:176. https://doi.org/10.1007/s12517-017-2956-6

Megnounif A, Terfous A, Ouillon S (2013) A graphical method to study suspended sediment dynamics during flood events in the Wadi Sebdou, NW Algeria (1973–2004). J Hydrol 497:24–36

Merlo S (2007) Survey of historic settlements in the Wadi ash-Shati and the Dawada lake villages. Libyan Stud 38:115–156

Merlo S, Hakenbeck S, Balbo AL (2013) Desert migrations Project XVIII: the archaeology of the northern Fazzan. A preliminary report. Libyan Stud 44:141–161

Mori L (ed) (2013) Life and death of a rural village in Garamantian times: Archaeological investigations in the oasis of Fewet (Libyan Sahara). AZA Monographs 6, All'insegna del Giglio, Firenze, p 405

Nixon S (2020) The Sahara. In: Walker BJ, Insoll T, Fenwick C (eds) The Oxford handbook of Islamic archaeology. Oxford University Press, Oxford, pp 286–309

Pelling R (2008) Garamantian agriculture: the plant remains from Jarma, Fazzan. Libyan Stud 39:41–71

Purdue L, Charbonnier J, Khalidi L (2018) Introduction. Living in arid environments from prehistoric times to the present day: approaches to the study of refugia and oases. In: Purdue L, Charbonnier J, Khalidi L (eds) From refugia to oases: living in arid environments from prehistoric times to the present day. Des refuges aux oasis: vivre en milieu aride de la préhistorie à aujourd' hui. XXXVIIIe. Éditions APDCA, Antibes, pp 9–32

Ramdani M, Elkhiati N, Flower RJ (2009) Lakes of Africa: north of Sahara. In: Likens GE (ed) Encyclopedia of inland waters. Academic Press, London, pp 544–554

Remini B, Miloudi A (2021) Souf (Algeria), the revolution of crater palm groves (ghouts). Larhyss J 47:161–188

Remini B, Achour B, Albergel J (2011) Timimoun's foggara (Algeria): an heritage in danger. Arab J Geosci 4:495–506

Ross E (2010) A historical geography of the Trans-Saharan trade. In: Krätli S, Lydon G (eds) The Trans-Saharan book trade. Manuscript culture, Arabic literacy and intellectual history in Muslim Africa. Brill, Leiden, pp 1–34

Scheele J (2010) Traders, saints, and irrigation: reflections on Saharan connectivity. J Afr Hist 51:281–300

Schwartz FW, Ibaraki M (2011) Groundwater: a resource in decline. Elements 7:175–179

Shanahan TM, McKay NP, Hughen KA, Overpeck JT, Otto-Bliesner B, Heil CW, King J, Scholz CA, Peck J (2015) The time-transgressive termination of the African Humid Period. Nat Geosci 8:140–144

Sims D (2014) Egypt's desert dreams: development or disaster. American University Press, Cairo, p p436

Sterry M, Mattingly DJ (2020) Discussion: sedentarisation and urbanisation in the Sahara. In: Mattingly DJ, Sterry M (eds) Urbanisation and state formation in the Ancient Sahara and beyond. Cambridge University Press, Cambridge, pp 330–356

Tydecks L, Jeschke J, Zarfl C, Boehning-gaese K, Schütt B, Bremerich V, Tockner K (2021) Oases in the Sahara desert—linking biological and cultural diversity. Preprint available at https://doi.org/10.32942/osf.io/hxqw9

van der Veen M (1992) Botanical evidence for Garamantian agriculture in Fezzan, southern Libya. Rev Palaeobot Palynol 73:315–327

van der Veen M (1995) Ancient agriculture in Libya: a review of the evidence. Acta Palaeobot 35:85–98

Wilson A, Mattingly DJ (2003) Irrigation technologies: foggaras, wells and field systems. In: Mattingly DJ (ed) The archaeology of Fazzan, vol 1. Synthesis. Society for Libyan Studies, London, pp 235–278

Wilson A, Mattingly DJ, Sterry M (2020) The diffusion of irrigation technologies in the Sahara in antiquity. Settlement, trade and migration. In: Duckworth CN, Cuénod A, Mattingly DJ (eds) Mobile technologies in the Ancient Sahara and beyond. Cambridge University Press, Cambridge, pp 68–114

Youcef F, Hamdi-Aïssa B (2014) Paleoenvironmental reconstruction from Palaeolake sediments in the area of Ouargla (Northern Sahara of Algeria). Arid Land Res Manag 28:129–146

Zatout M, Kadri MM, Hacini M, Hamzaoui AH, M'nif A (2020) Origin and signs of disturbance of hyperarid climate wetland ecosystem: case of Chott Ain Beida, Algerian Sahara. Env Earth Sci 79:170. https://doi.org/10.1007/s12665-020-08906-7

Zegait R, Bensaha H, Addoun T (2021) Water management and the agricultural development constraints in the Algerian Sahara: case of the M'Zab Valley. J Water Land Develop 50:173–179

Zerboni A, Bernasconi A, Gatto MC, Ottomano C, Cremaschi M, Mori L (2018) Building on an oasis in Garamantian times: geoarchaeological investigation on mud architectural elements from the excavation of Fewet (Central Sahara, SW Libya). J Arid Env 157:149–167

Desert City Landscapes: At the Crossroads of Urbanization, Agricultural Intensification, and Trans-Saharan Connections

19

Oliver Pliez

Abstract

The Sahara is not a desert wasteland but a location of thriving historical and contemporary towns and cities. Their recent developments reflect both revitalization of commerce along traditional trans-Saharan trade routes as well as state-planned interventions such as agricultural development around contemporary cities and routeways. The variable successes and failures of such recent interventions shows the fragility of Saharan towns and cities to resource exploitation, but also their resilience to sustain certain types of economic activities despite external intervention. These contradictory factors resulted in the making of today's urban landscapes in the central Sahara.

Keywords

Urbanization · Traders · Agricultural development · Settlements · Trans-Sahara

19.1 Introduction: The Urban Sahara

The Sahara has undergone considerable changes in the past decades. The desert has long been regarded as a peripheral area because it is sparsely populated, marginalized after the decline of trans-Saharan trade in the second half of the nineteenth century, and situated far from the economic and political dynamics that have involved the region since independence. The Sahara has always been an open space where many religious or ethnic minorities have found refuge over the centuries. In the Middle Ages, trans-Saharan trade routes crossed the entire desert, from north to south between the Mediterranean and the Sahel, but also from west to east where long routes were used by traders and pilgrims to Port Sudan and Mecca on the Red Sea (Fig. 19.1).

Monod (1968) distinguished four meridian axes (1, 3, 5, 7) that bypass the most inhospitable regions of the Sahara (2, 4, 6). These axes propose a 'geographical division of the Saharan world', identified by their main populations of Saharan peoples: the western Moorish zone, from Morocco to Senegal; the central Tuareg zone, from the Algerian Sahara to the loop of the Niger River; and the eastern Tubu zone, from Libya to Lake Chad and the Waddaï in present day Chad. The Nile Valley is generally excluded from the 'intra-Saharan' model, where roads and tracks cross vast empty areas to reach the oases and towns that lie along the way. These traffic routes may disappear and then be reborn along the same route depending on political and economic circumstances. This organization is no longer conspicuous today, after colonization and competition from maritime traffic damaged it during the nineteenth and twentieth centuries, but it is still partly topical since it has determined the location of many cities and roads in the central Sahara historically and to the present day. Nevertheless, the desert is a key space for North African states because of its vastness, its geopolitical and strategic importance, and the richness of its subsoil in minerals, hydrocarbons and deeply-located but non-renewable aquifers.

One of the most significant properties of desert regions is that the Sahara is urban. Until recently, cities were considered anomalies in desert landscapes, even if one often distinguishes the 'old Saharan city' which is part of the history of the desert, from the 'city in the Sahara', which is the product of the deliberate design and plan of the public authorities (Côte 2002). It is considered that the 'old Saharan city' has been able to evolve in concert with the surrounding environment, whereas designed and planned cities have been stigmatized as an unsustainable model, imposed by actors outside the desert, such as traders, industrialists or

O. Pliez (✉)
Université Paul-Valéry, Montpellier 3, Site Saint-Charles 1, 34090 Montpellier, France
e-mail: Olivier.PLIEZ@cnrs.fr

J. Knight et al. (eds.), *Landscapes and Landforms of the Central Sahara*, World Geomorphological Landscapes, https://doi.org/10.1007/978-3-031-47160-5_19

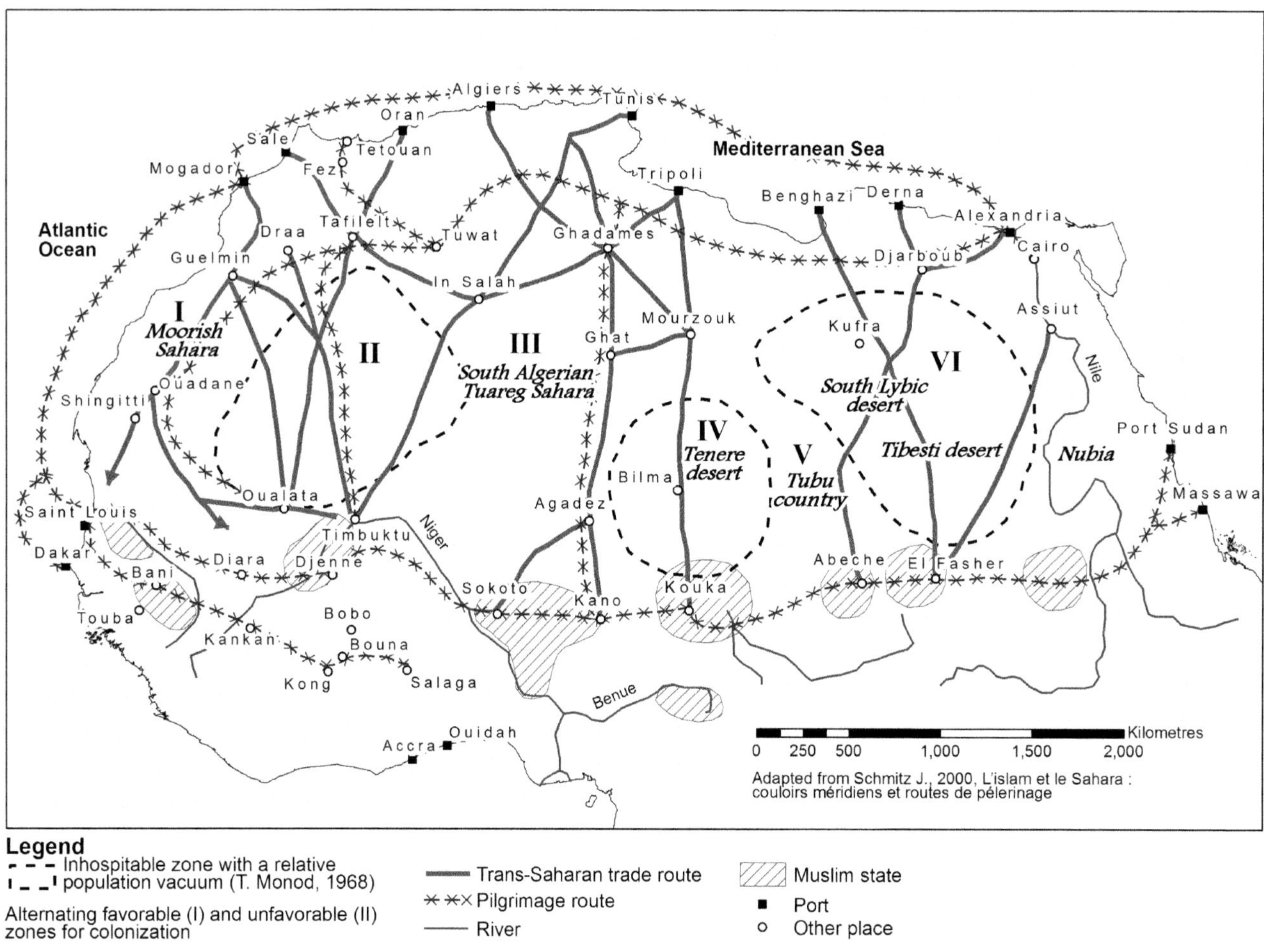

Fig. 19.1 A global view of trans-Saharan networks before colonization (reproduced with permission from Dixon et al. 2019)

the state (Pliez 2011). This distinction, however, has become less valid, since desert cities irrespective of their origins are now facing a lack of governance in urban planning and environmental problems that can also be found in other areas globally (Bendjelid et al. 2018). These problems include silting up of water courses, deficiencies of or excess water, pollution, and waste management.

The vast majority of Saharan populations now live in conurbations that can be defined as groups of cities (or agglomerations) close to each other, each city remaining individualized. The situation is paradoxical since, during the first half of the twentieth century, the desert was essentially a space without cities: the prestigious names of Agadez, Ghadames or Ghardaïa refer to a trans-Saharan trade then in full decline. The population of the Sahara stagnated until the 1960s, rising from 1.7 to 2 million inhabitants between 1948 and 1966, then increasing rapidly during the following decades: around 1995, the Sahara had 5 million inhabitants, excluding the desert margins (Bisson 2003). Over the same period, the Saharan portions of the Sahelian states, from Mauritania to Sudan, experienced at best a stagnation of population and more frequently a decline in urbanization. The contributors to this change are multiple and not unique to the Sahara: relocation of land-demanding installations (new cities), development of mining and power engineering facilities, development of transport infrastructure, development of means of pumping fresh water, and desalination technologies that provide water in areas where it was previously not available (Ezcurra 2006). These changes have resulted, over the last 40 years in particular, in a process of urbanization unprecedented in its magnitude, particularly in the Saharan regions of Morocco and Egypt, with less impact in those of Mauritania and Sudan (OECD 2014). In the mid-1990s, there were 53 cities in the Saharan Maghreb but only 8 in the Sahelian Sahara. In Morocco, Algeria and Libya, several agglomerations now exceed 100,000 inhabitants and urbanization rates in the desert are higher than national averages, at 70% in the Algerian Sahara, 80% in Libya, and 92% in Western Sahara (Côte 2012).

Saharan urbanization adopts two dominant spatial forms: on the one hand, administrative cities with 50,000–150,000 inhabitants and, on the other, numerous towns of 5000–15,000 inhabitants. Both of these forms are produced by either sedentarization of pastoral peoples since the 1960s (Bataillon et al. 1963) or more recently by urban demographic expansion along oasis valleys, often influenced by two main factors that sometimes combine: exploitation of mineral or hydrocarbon wealth, and territorial control, where the city is a strong marker of boundaries of the national territory.

19.2 Urban Landscapes: From the Oasis to the City

It would be difficult to describe a 'typical landscape' of the urbanized Sahara, as its landscapes are not uniform. At a small scale the Sahara is a vast desert covering some 8 million km^2 and dominated by very low annual rainfall (less than 100 mm/year) (Ezcurra 2006). However, this viewpoint of a monotonous desert does not describe its many nuances. For example, the Sahara contains a diversity of biomes, including a northern Mediterranean fringe, tropical, continental environments in the south, a central arid core, and variability determined by altitude, latitude, and proximity to ocean water sources (Le Houérou 1990).

Despite the diversity of situations, each city or town urban centre provides a burst of vegetation, which is linked to the presence of oases, as vividly described in the novel *Cities of Salt* by Abdulrahman Munif (1984): "It is Wadi Al-Oyoune. Suddenly, in the midst of the hard and stubborn desert, this green spot springs out, as if it had sprung from the belly of the earth or fallen from the sky. It is different from all that surrounds it, or, more exactly, there is no connection between it and the rest ..." (translated from the original French by O. Pliez). However, when entering the city, the disenchantment is quickly noticeable: "The vast agglomeration of camel-skin tents ... has given way to precarious constructions, topped with white domes ... There are straight avenues, sidewalks, a few meagre gardens. The mystical city has become a military garrison and a trading centre ..." (Le Clézio and Le Clézio 1997, translated from the original French by O. Pliez).

Urbanization in the oasis environment remained 'invisible' for a long time (Popp 1993). It materialized through the expansion of villages as a result of demographic growth and nomadic settlement, in particular in the latter part of the twentieth century. The standardization of urban landscapes, generally led by the state as a function of centralized urban planning, resulted in a break with traditional growth patterns, through the imposition of modern urban design and development of infrastructure, and the introduction of urban models designed for other environments. However, other less visible ancient functions of Saharan settlements are still present, in particular those representing the transit of people, goods, knowledge, and culture, since an oasis is still a place of passage in the trans-Saharan road network.

19.3 Cities at Different Scales and Historical Links

Whilst at a local scale a settlement is dependent on the proximity of water, this is rarely a determinant factor for population settlement. In fact, Despois (1946) asserted that settlements have developed their own water sources through wells and springs, rather than making use of the distribution of pre-existing water sources in determining settlement locations. Thus, water sources and agricultural fields have been deliberately created close to settlements. The viewpoint that villages serve merely a productive role and that the oasis is merely an island in the midst of the sand sea is an erroneous perception.

At a smaller scale, the presence of oases is explained by their role in a circulation network of people, goods, knowledge, and culture. Despois (1946) pointed out the need to change the scale of analysis in order to understand Saharan urbanisation. In studying the Fezzan region (Libya) after the Second World War, one of the most important regions of the central Sahara, he pointed out that it represents both a crossroads of historic trade routes, and also of modern aviation routes. It remains a rural area but also a nodal point. But the circulatory function and inhabited places of the central Sahara are part of a spatial organization that arises from the varied actors who traverse these roads, whether Saharan or not. The locations of settlements on the axes of the trans-Saharan trade routes can help explain why certain regions are in crisis or can be revitalized.

It is therefore useful to reason at two scales to understand the variations of the inhabited Sahara. The local scale corresponds to the place, in this case the oasis, which allows for a description of the landscape and the practises of those who live there. At the scale of the Sahara itself, the location of places and cities within the homogeneous desert can allow one to understand why and how places are linked to one another and found in regions more or less hostile to human life. As a result, although population densities of the Sahara are very low (on average less than 5 inhabitants per km^2; Fig. 19.2), an increasing part of the Saharan landscape is affected by multiple factors (natural, economic, political) for which urban landscapes, built over a long period of time, are the most visible expression.

19.4 Agricultural Revitalization Driven by Cities: The Example of Fezzan (Libya)

The rural crisis is one of the most emblematic elements of the decline of Fezzan after the Second World War. Despois (1946) insisted on the 'miserable aspect' of the farms, although land development is the main activity of the inhabitants of this region of the central Sahara. Half a century later, in the 1990s, agriculture in the Libyan Sahara has been experiencing a revival, linked to a number of favourable conditions including off-season agricultural production, privatization, urbanization, and development of the service sector that have contributed to a rapid increase in the demand for fruit, vegetables, and meat (Pliez 2003). Governments have played a key role in improving the quality of road networks, increasing incentives (drilling, subsidies, etc.), and sometimes subsidizing the price of gasoline. These incentives contrast with a chronic shortage of labour. Thus, the goal of 're-greening' the desert is often accompanied by the need to attract agricultural workers from abroad where, half a century earlier, rural emigration seemed an unstoppable trend.

Everywhere, there is evidence for a profound remodelling of landscapes (Côte 2012), with the juxtaposition of state fenced areas and private plots planted with orchards, grass, and vegetable crops, which are sold on the national market. It is around major cities that changes are most dramatic, for example in Sebha (Libya), an agglomeration of 150,000 inhabitants whose peri-urban area is the subject of speculative investments in and development of agricultural land. New farms are generally located along roads, to produce crops geared to local demand and for larger coastal settlements (Fig. 19.3). The gap between the area occupied by the palm groves and that of the new gardens is large, commensurate with the vast consumer market constituted by the city. These are dozens of individual farms

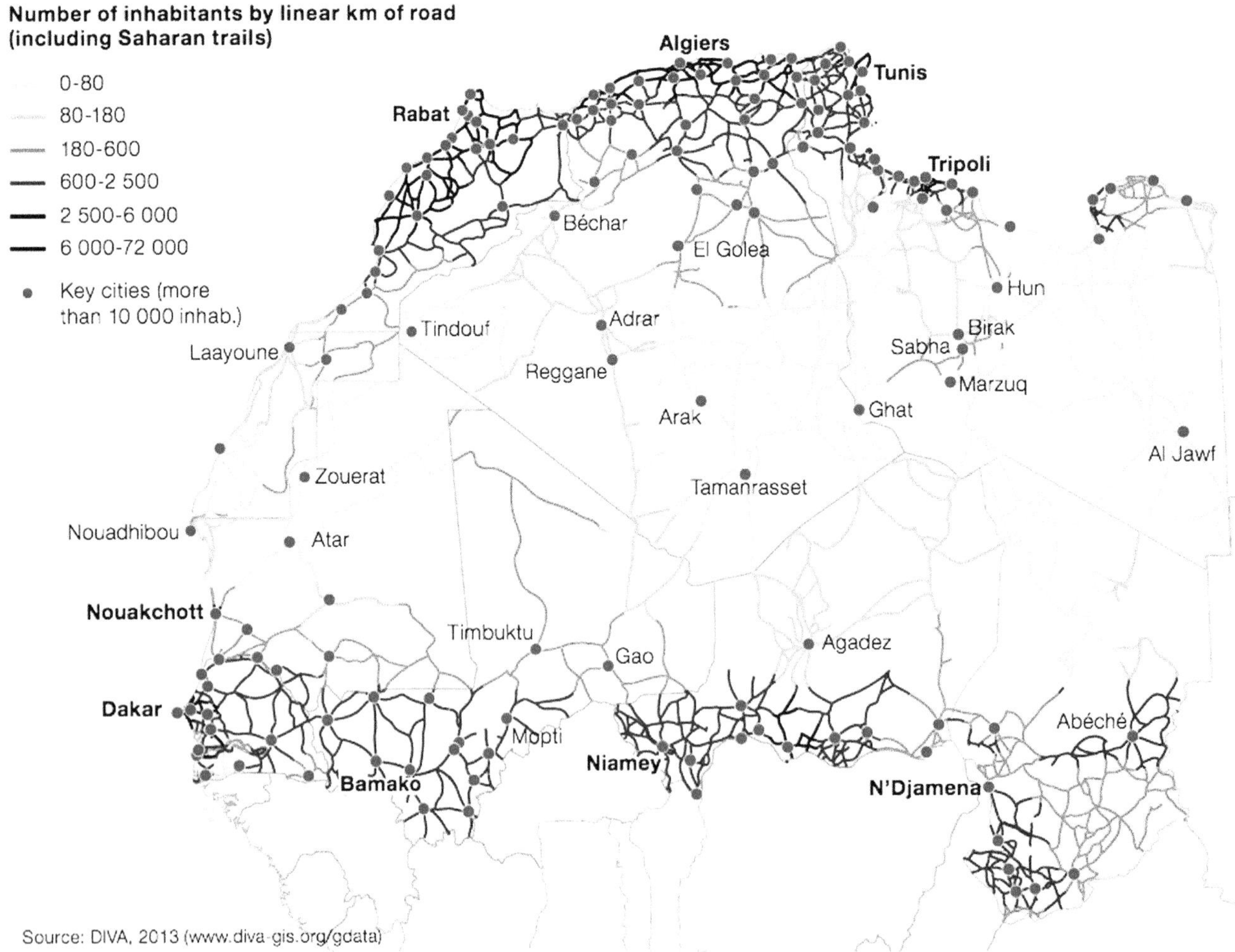

Fig. 19.2 Number of inhabitants by linear km of road (including Saharan trails) (*source* OECD 2014)

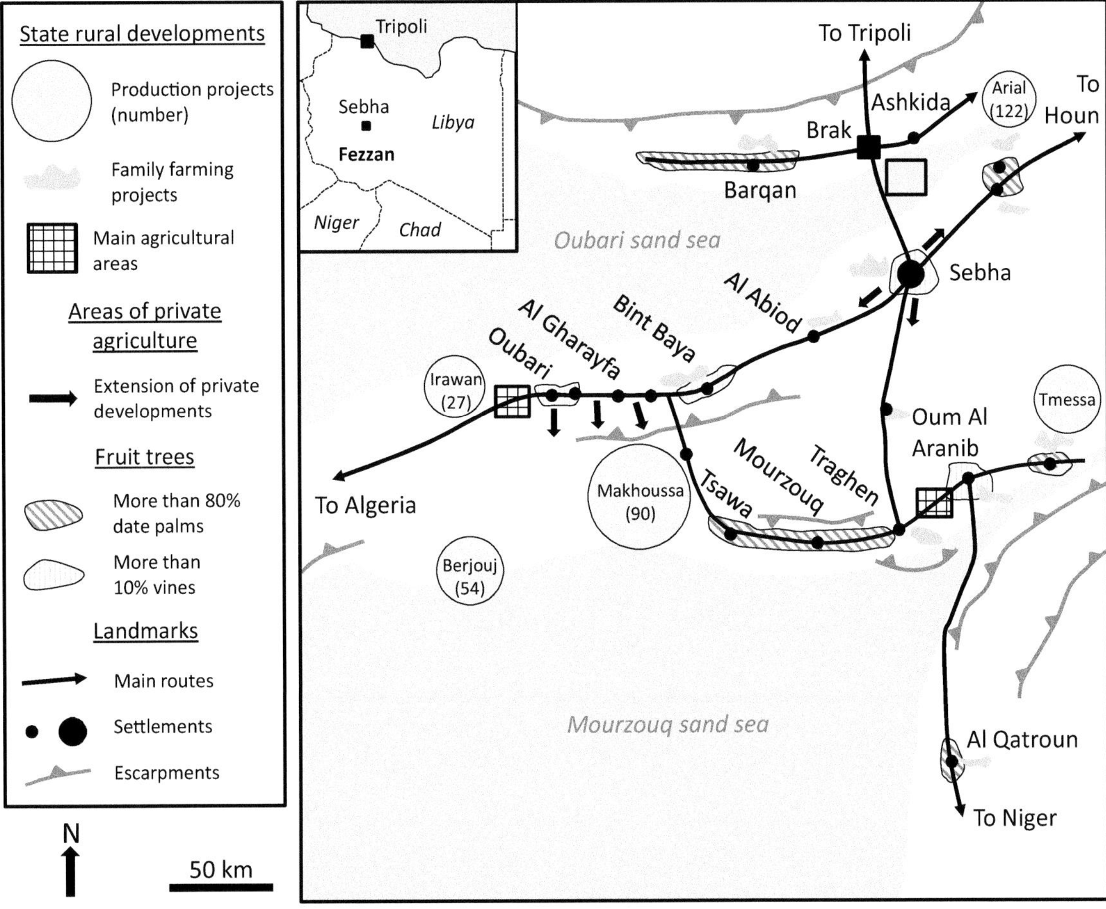

Fig. 19.3 Agricultural development near Sebha, Libya (adapted from Pliez 2004)

producing fruit and vegetables for Libya as a whole, which operate thanks to the presence of migrants, transient, or seasonal. This shows the role of sub-Saharans, Egyptians, and Asians participating in the rural revitalization of the Sahara. Their presence is both important in the context of the city and also diffuse in the context of the Sahara, and provides a contested viewpoint of the meaning of the 'the city of natives'. The participation of migrants in the preservation of the oasis heritage is therefore very real: they invest and renovate villages (*ksour*) and provide cheap labour necessary for the maintenance of gardens and irrigation systems.

19.5 Kufra (Libya): A Crossroad or an Oasis Without Agriculture?

Kufra's profile is rather unusual compared to the common perceptions of the Saharan oases as green havens dotting the desert. Located in the heart of the Libyan Sahara, the town has a hyperarid climate, with a rainfall of less than 50 mm per year. The nearest settlement is hundreds of kilometres away. Downslope, Kufra has access to abundant water resources, and water management and equity of access has never been an issue in this oasis community. Likewise, the gap between oasis gardeners and nomad cattle herders has lost its relevance, since Kufra's gardens have seldom been improved, while pastoralism is almost extinct, due to the lack of availability of pasture land. On the other hand, Kufra is the epitome of the stopover town, described by Davis (1983, 1987). Isolated from both the Sahel and the Mediterranean, the oasis has thrived from strong relations with these outer regions, whilst other towns in the desert have declined. Here, economic activities have focused on its role as a key trading post, historically of both goods and people. During the second half of the nineteenth century, in times of intensified colonization in North Africa, trading in goods and slaves reached its peak in this area along the trans-Saharan road towards the Mediterranean coast at Benghazi (Libya). This meant the Wadi Kufra was an important stop between both sides of the desert. It remained a crossroads until 1932, when Italian colonists settled in the oasis and precipitated the town's economic decline.

Kufra's renaissance came after Libyan independence (1951) when the oasis town became the emblem of a green, modern desert city that was developed based on a drive by the new state to develop national food self-sufficiency. Two

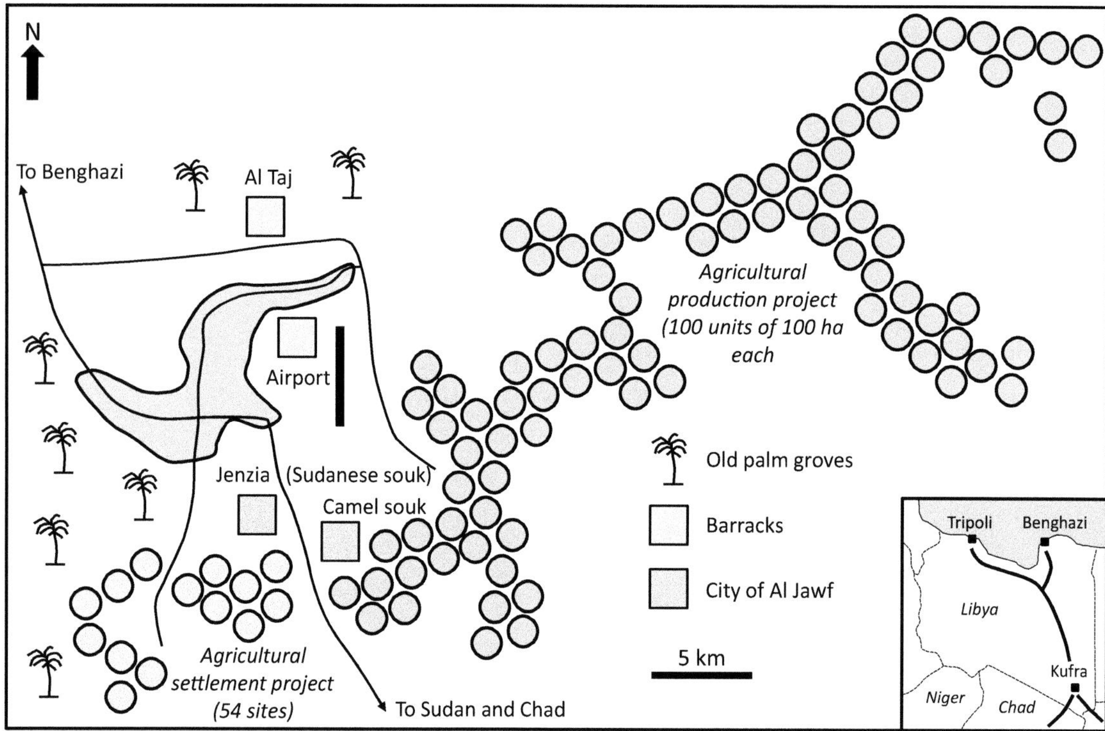

Fig. 19.4 Map of the Kufra region, Libya (redrawn after Pliez 2004)

'modern oases' were developed (Fig. 19.4). The first was made up of 100 circular irrigation units of 100 hectares each, where wheat, barley, and berseem (Egyptian clover) were grown. The second, known as the 'population project', was based on development of 54 hamlets of 16 farms built around a main square. The 16 farms were arranged in a hexagonal pattern totalling 5600 hectares, and focused around water production plants, whose resources were used at home and in the fields. The project was initially meant to house the whole population of the Kufra oasis, of between 5000 and 7000 inhabitants. However, only 14 hamlets out of 54 were granted with farmlands. After the failure of the artificial oasis, a more ambitious scheme was designed in order to pump and transfer fossil water (groundwater) to the coast. The government plan was that a united Libya would emerge from this great artificial river, with the country fertilized in its entirety from subsoil groundwater. This plan never materialized. Thus, by the 2000s, Kufra looked like an oasis without farms, where dilapidated gardens and palm groves were in stark contrast to other similar regions in the Sahara. The town also looked like a military wasteland, in which many barracks closed down after the end of the war against Chad. Today, through its landscapes, population, and economy, Kufra presents itself as a town built by the state, albeit shaped by those involved in trans-Saharan exchanges more than from local agricultural activities.

19.6 Conclusions

The Sahara is not merely a place where either migrants on their way to Europe or terrorist groups may roam free. It also underwent considerable changes within the last half-century, such as massive urbanization (Pliez 2011). Many built-up areas, which were created or developed by the central state at the crossroads of major trans-Saharan routes, have experienced spectacular growth rates for decades, as they received at different times farmers fleeing from their oases, nomads, refugees, and economic migrants. In all cases, the increase in urbanization and migrant flows has been a significant factor in bringing about social and spatial change. In the 1990s to the 2000s, declining state subsidies and expanding mobility in the Sahara characteristically created ambivalent settlements which are at the same time marginal towns (according to the state) and connected towns (according to transnational mobility operators). These towns are now in crisis following the demise of centralized state power across many different Saharan countries and the wars that have shaken the Sahara since the early 2000s. Yet the survival of individual towns and cities may lie in the long term resilience of trans-Saharan mobility networks.

References

Bataillon C, Bisson J, Capot-Rey R (1963) Nomades et nomadisme au Sahara. UNESCO, Paris, p 195

Bendjelid A, Brûlé J-C, Fontaine J, Kouzmine Y, Ormaux S (2018) Introduction. Mutations Sahariennes Les Cahiers d'EMAM. https://doi.org/10.4000/emam.1476

Bisson J (2003) Mythes et réalités d'un désert convoité : le Sahara. L'Harmattan, Paris, p 480

Côte M (2002) De quelques villes nouvelle au Sahara. Méditerranée 99:71–76

Côte M (2012) Signatures sahariennes: Terroirs et territoires vus du ciel. Presses Universitaires de Provence, p 308

Davis J (1983) La structure sociale de Koufra. Annuaire de l'Afrique du Nord 22:545–564

Davis J (1987) Libyan politics: tribe and revolution. Tauris, London, p 310

Despois J (1946) Mission scientifique du Fezzân, 1944–1945. Institut de recherches sahariennes de l'Université d'Alger, III Géographie humaine, p 260

Dixon J, Garrity D, Boffa J-M, Williams TO, Amede T, Auricht C, Lott R, Mburathi G (eds) (2019) Farming systems and food security in Africa. Priorities for science and policy under global change. Routledge, London, p 674

Ezcurra E (ed) (2006) Global deserts outlook. UNEP, Nairobi, p 164

Le Clézio J-MG, Le Clézio J (1997) Gens des nuages. Stock, Paris, p 151

Le Houérou H-N (1990) Définition et limites bioclimatiques du Sahara. Science et Changements Planétaires/Sécheresse 1:246–259

Monod T (1968) Les bases d'une division géographique du domaine saharien. Bulletin de l'IFAN 30:269–288

Munif A (1984) Cities of salt. Random House, London, p 627

OECD (2014) An Atlas of the Sahara-Sahel. Geography, economics and security. OECD, Paris, p 19

Pliez O (2003) Villes du Sahara: Urbanisation et urbanité dans le Fezzan libyen. CNRS Éditions, Paris, p 199

Pliez O (2004) La fin de l'État demiurge? Les nouvelles facettes de l'urbain dans le Sahara libyen. Autrepart 31:59–74

Pliez O (2011) Les cités du desert: des villes sahariennes aux saharatowns. Presses Universitaires du Mirail, Toulouse, p 164

Popp H (1993) Une modernisation «invisible». Changements économiques et sociaux à l'oasis de Figuig. In: Espace et société dans les oasis marocaines, Série Colloques 6. Meknès, Université Moulay Ismail, pp 97–106

Trade, Mobility and Migration

20

Judith Scheele

Abstract

The Sahara has long been seen, by Arab and European geographers alike, as a predominantly 'empty' space crossed by more or less fixed routes. Based on historical material, secondary sources and original ethnographic fieldwork in the central Sahara, this chapter argues that it is better to think of the region as a space of internal connectivity that underpins all lifeways in the region, both sedentary and nomadic. This internal connectivity subtly shapes its landscape, through regional trade, pastoral or labour migration and political ambition and domination.

Keywords

Pastoralism · Connectivity · Itineraries · Oases · Slavery

20.1 Introduction

In the twelfth century, the North African geographer al-Idrīsī described the central Sahara in the following terms:

> The majority of these lands are continuous uninhabited deserts, wild solitudes and rough, barren hills devoid of vegetation and with very little water… In these destitute lands one may meet nomads who move about over the country, pasturing their livestock in places near and far. They are not attached to any place and have no station in any land, but spend their time traveling, constantly on the move (quoted in Hopkins and Levtzion 1981, p. 119).

Al-Idrīsī contrasted this image with that of prosperous Saharan trading towns, built to rival cities elsewhere in the Islamic world. He thereby implicitly opposed two types of mobility, one—of local 'nomads' – erratic and of no consequence, the other—of urban traders—civilisational and transformative, but essentially 'foreign' to the area. In such a view, trade and migration were productive of civilised space in a way that pastoral mobility was not.

Much progress has been made in the last few decades to recognise the peculiar archaeological footprint left by nomadic pastoralists on the landscape. 'Mobility' nonetheless remains 'an elusive phenomenon and is not easily studied or detected by archaeologists' (Honeychurch and Makarewicz 2016, p. 347). This is partly due to the continued prevalence of explanatory models that, just like al-Idrīsī, relegate mobile herders to the passive margins of 'civilisation', conceived according predominantly sedentary models (Makarewicz 2013, p. 162). Oasis-dwellers, apt to leave a more lasting imprint on the landscape, are partly exempt from this image of marginality and impermanence, at the cost of being counted as purely sedentary. Yet there can be little doubt that in the central Sahara, for most of history, mobile herders were economically, socially and politically dominant, well able to 'enclose' (Honeychurch 2014, p. 309) and manage settlements according to their own priorities; and that thus oases—and the trans-Saharan trade they helped to channel—need to be understood as an integral part of the mobile socioeconomic regimes that encompassed them. The aim of this chapter is to outline these mobile socioeconomic regimes and to point to the many ways in which they contributed to the shaping of the central Saharan landscape.

20.2 Trans-Saharan Trade

The longstanding Arab and European interest in (trans-) Saharan caravan routes partakes in the conception of the Sahara as essentially empty and untouched by human occupation: as a barrier that needs to be crossed (McDougall 2005). Arab geographers like al-Idrīsī approached the Sahara mostly in terms of trading towns, ideally inhabited by Arab merchants, and the routes that connected them. European colonial observers and geographers, similarly

J. Scheele (✉)
EHESS–Centre Norbert Elias, Centre de la Vieille Charité, 13002 Marseille, France
e-mail: judith.scheele@ehess.fr

J. Knight et al. (eds.), *Landscapes and Landforms of the Central Sahara*, World Geomorphological Landscapes, https://doi.org/10.1007/978-3-031-47160-5_20

obsessed with 'trans-Saharan trade', followed their lead. Yet we have to be wary of this emphasis on routes. Routes exist only if they are used. They create places where they intersect, but these places may be reached in different ways and shift or disappear when mobility lapses. Prior to the construction of paved roads in the 1940s, itineraries in the Sahara were never the same: they changed from caravaneer to caravaneer, season to season, dry periods to wet periods and above all political circumstances which meant that people who could claim different allegiances and protection travelled in different ways. Both places and itineraries were social and political projects as much as geographical fixtures.

Furthermore, given the exclusive reliance on livestock for Saharan transport until recently, single fixed routes were ecologically untenable. Camels need pasture in order to cross distances of the kind involved here; a substantial caravan depleted resources in a way that made it unwise for others to follow suit immediately. Navigation in the Sahara was not so much a matter of knowing one 'route', but rather of knowing how to interpret the signs that indicate the best direction to take at a particular moment of time and with the amount and state of animals at hand. As, in the mid-nineteenth century, French colonial officers started to move towards the Sahara from what is now Algeria, they had to learn this lesson the hard way. Not only did all their attempts to reroute or 'capture' trans-Saharan trade fail miserably, but their establishment of a 'supply route' from northern Algeria to the French colonial outposts in the Sahara led to hecatombs of camels, while those who survived refused to continue, as 'numerous dead camels from the preceding convoys were lying on the road … smelling very bad'.[1] Things only improved as the French army command decided to delegate transport to pastoral nomads or to imitate pre-existing regional patterns of exchange.

Historically, then, trans-Saharan trade depended on a pre-existing infrastructure of regional connectivity that successful trans-Saharan traders could use to their own advantage. This infrastructure was social and political as much as it was technical. Due to variations in pasture and the nature of the terrain, camels were trained as regional specialists, finding it difficult to cope with unfamiliar pasture and ground; individual camels did not cross the Sahara in a straight line, but tended to cover a particular stage. Moreover, different people were able to extend their protection over travellers in different parts of the Sahara and at different times. Primarily of urban background, traders had to hire camels and their care-takers against a fee, which represented their value but also in many cases a kind of 'protection money' paid to locally dominant pastoralists. Traders had to change their suppliers-cum-protectors from one region to the next, with oases serving as relay points.

20.3 Itineraries, Not Routes

We would thus do better to start, not with the assumption of caravan routes, but with local historical perceptions of space, as far as we can piece them together from later accounts. 'From a Tuareg point of view', notes Claudot-Hawad (1990, p. 214), 'territory is defined through the elements that make it viable, i.e. permanent water points, rational ways of exploiting pasture, and patterns of commercial exchange'. Place matters less than relations between places: 'connections produce places, and not vice versa… centrality moves through the desert according to flows animated by social networks' (Retaillé and Walther 2013, p. 608). In other words, places and routes change with circumstances, but the underlying fact of connectivity remains.

Geographical and environmental constraints shaped movement in some ways through the existence of impassable stretches like ergs and mountain ranges, large distances that created the need for relay posts on the way and the availability of water and pasture. The latter, however, was and is notoriously changeable, while the quality of pasture is not given, but the result of careful evaluation by herders in collaboration with their animals. As it is based on judgement (of the herder and their animals), it might change not merely from one season or year to the next, but also from one caravaneer or herder to another, one particular breed of livestock to the next (Krätli and Schareika 2010, p. 614). Biomass becomes pasture when it is accessible to livestock and their herders, which in turn depends on the social ties they can invoke, and on the proximity of water and the kind of animals herded: goat or cattle pastures are the not the same as camel pastures, and both imply different forms of mobility and management, which in many Saharan societies overlapped with gender divisions and status distinctions.

The development and maintenance of an infrastructure of mobility, although often invisible to the unpractised eye, transformed the Saharan landscape in subtle ways. Soils rich in salt attracted livestock from far and wide, over and beyond their initial carrying capacity (McDougall 1990, p. 233). Humans and animals in turn created open-air mines to facilitate use; particular soils might even be carried over distance, while natron and salt was and is traded transregionally (Fig. 20.1). By digging a well, pastures become accessible; if the well is too productive or permanent, it might lead to overgrazing and destroy good pasture,

[1] Archives nationales d'outre-mer, Aix-en-Provence, 10H22, Barthel, 'In Salah et l'archipel touatien', 15/04/1902; and 22H33, 'Rapport du Lieutenant Chourreu, chef du 6ᵉ convoi de ravitaillement d'Igli, sur les conditions dans lesquelles le convoi a été effectué', 29/12/1900.

Fig. 20.1 **a** 'Trou du natron': natron deposit, Tibesti, northern Chad, January 2012; **b** natron mine in Faya, northern Chad, March 2012 (*photographs* Judith Scheele)

as might also happen in the vicinity of permanent water sources. By carefully maintaining particular plant species that are of use for pastoralists and suppressing others, the flora and fauna of the area is transformed over time; trampling by large concentrations of animals also changes soil structure.

It is important to remember here that virtually all cultivated plants and herded animals in the Sahara have not been domesticated locally, but were imported from elsewhere. The most stereotypical images of the Saharan landscape—oases of date palms and camel herds—are thus in themselves the result of immigration and adaptation of the local landscape to the needs and abilities of immigrant species and vice versa. Meanwhile, their rather remarkable genetic uniformity across the region—at least where camels are concerned (Almathen et al. 2016)—is an indicator of continued mobility.

20.4 Mobility and the Making of Place

Despite this fluidity, a general conceptual division emerges, between those parts of the Sahara that are potential dwelling places and those that are not, variously called *serir*, *tanzerouft*, *ténéré* or *awi* in different Saharan languages. Monod (1968) hence suggested a vertical division of the Sahara into three areas of interaction divided by truly arid zones. Each of these areas of interaction contained different climatic and ecological zones and was dominated by one linguistic group: Arabic speakers in the west, Tamacheq speakers in the centre and Tedaga/Dazagada speakers in the east. Ironically, the areas of the Sahara where movement corresponds most closely to our notion of 'routes' were few areas in between, those that were recognised, from an internal point of view, as 'empty', largely devoid of water and pasture, and that hence could not be dwelled in (and thereby transformed), but needed to be crossed as quickly as possible.

Elsewhere, mobility within and across the Sahara creates not so much routes as overlapping areas of heightened connectivity linked to each other through points of exchange. These could take different forms. Pastoral economies create surpluses that need to be stored (Makarewicz 2013, p. 166), in fortified silos for instance, or else invested, in irrigated or non-irrigated agriculture or arboriculture (Fig. 20.2). Mobility relies on far-flung social networks that might periodically coalesce at places of pilgrimage and worship, such as *zawāyā* (sing. *zāwiya*, fortified religious centre, Fig. 20.2c) or saints' tombs. Mobility thus created fixed points that transformed the landscape, but that did not necessarily imply a fully sedentary lifestyle. In southern Morocco, collective storehouses (*hisn*, *agadir* or *tighremt*) count among the oldest permanent buildings. They durably marked the landscape, but they had few permanent residents and relied on a widely dispersed nomadic hinterland. Saharan settlement even today might have a highly fluctuating population, turning from sleepy hamlets into large nomadic camps at the time of annual religious and commercial fairs (as in Reggane in southern Algeria) or the date harvest (as in Faya in northern Chad).

Even permanently irrigated oases were rarely self-sufficient and participated, both economically and socially, in the mobile systems that surrounded them. Pastoral nomads owned gardens and houses much as permanent oasis dwellers owned livestock entrusted to their nomadic partners. Oases can thus be seen, alongside salt mines, as the most visible imprint of mobile economies on the Saharan landscape. As Retaillé (1998, pp. 73, 77) notes, 'within nomadic societies, most people are sedentary', as these societies should be defined not statistically, by 'the apparent mobility, more or less complete, of people and their dwellings', but politically,

Fig. 20.2 **a** Irrigated horticulture: gardens in Kusan, southwestern Algeria, September 2008; **b** non-irrigated date plantations, Kirdimi, northern Chad, August 2012; **c** *Zāwiya* in Reggane, southwestern Algeria, site of a large annual pilgrimage, October 2008 (*photographs* Judith Scheele)

by the 'mobility of the violent exercise of power' that subordinated settlement to the priorities of mobile societies and economies. In such a context, Honeychurch and Makarewicz (2016, p. 348) are surely right that 'sedentism might have been as much a part of the mobile repertoire as was actual movement', although people were not necessarily free to choose their own place within this broader 'repertoire'.

20.5 Migration as Geography

While mobility in the central Sahara was just a fact of life, most Saharan groups today remember their history as a story of migration, as an arrival from elsewhere following a known itinerary. These stories cover both sometimes extravagant claims to far-flung origins, and the gradual process by which pastoral groups and their associated settlements moved from one zone to another, by successfully claiming or losing rights of access to pasture and other forms of resources. 'Migrations' of this kind might imply both real movement of people or a shift in categories, when people on the ground redefine themselves according to new cultural, linguistic or religious models and references or as a result of political changes.

These migrations have been explained as responses to ecological pressures (e.g. Webb 1995). They are, however, also always political—people move because they can or because other areas are blocked to them—and they might often be accompanied by a shift in livelihood, from camel to cattle husbandry for instance. Cleaveland (2002) thus describes the case of the population of the Mauritanian town of Walāta, whose inhabitants arrived, from the fourteenth century, in town in an act of *hijrah*, renouncing their former lifestyle and taking up religious scholarship, trade and cattle (rather than camel) raising. For the eighteenth and nineteenth centuries, Grémont (2010) traces the slow progression of the Tuareg Iwellemedan (who are now based on the border between Mali and Niger) eastward, while accruing political clout and collecting retainers and clients. In the central Sahel, Fulani cattle herders, with putative origins in Senegambia, are still slowly moving towards the Nile Valley.

These migrations were always also expressions of political ambitions and status. This is reflected in regional historiographies that tend to trace prestigious origins to external sources, in the form of genealogies. These genealogies double up as geographical descriptions, in which the putative ancestors' itineraries across the Sahara are marked in the landscape through the discovery and naming of water sources, the establishment of silos, oases, *zawāyā* or tombs. These historic itineraries continue to shape subsequent social ties and alliances, through marriage, access rights, investment and settlement; although they might be fictional, they thereby accrue reality over time. Genealogical narratives and their spatial markers both differentiate and homogenise the social and physical landscape. They mark the landscape through highly individual names and spatial markers; yet one *zāwiya*, well or ancestor's tomb looks much like another, and genealogical names recur throughout.

20.6 Slavery as Migration

Most labour used to build and maintain oasis settlement and their labour-intensive irrigation canals, to guard fortified silos and to staff complex herding systems, was imported from the south, through slavery, the most prevalent form of forced migration in the region. This means on the one hand that we need to revise common images of Saharans as quintessentially white and pastoralists. On the other, we can consider the central Sahara as a landscape of slavery (Marshall 2015), although this is expressed here in different ways from the West African coast or the US American south.

The importance of slavery and the particular kind of mobility (raiding and trading) and immobility (slave bondage) it gave rise to fully participate in what Honeychurch (2014, pp. 291–292) has identified as the defining characteristics of pastoral lifeways: namely, their inherent flexibility. Baier and Lovejoy (1977) describe how, in what they call the Tuareg 'desert-side economy', servile labour relations made it possible, for elites, to shed people and hence responsibility in times of drought, when slaves were freed and generally migrated south. In wetter periods, elites could acquire new slaves or invoke past ties of dependency, bringing people back into the desert. The resulting cyclical pattern of migration brought into the Sahara those people who, through their work, transformed it most visibly.

The importance of slavery and other forms of raiding—instantiations of Retaillé's 'mobile power', cited above—also shaped local conceptions and uses of landscape. By subordinate groups, the landscape was viewed not only in terms of access to strategic resources but also, and perhaps primarily, in terms of the degree of protection it afforded against raiders. In the Nigerien Ader, on the southern edge of the central Sahara, presentations of the landscape never omit references to hiding places, and villages were periodically displaced for similar reasons (Rossi 2015, p. 82). In the western and central Sahara, the most commonly used term to refer to oases is *qsar* (plural *qsūr*) or 'fortress', indicating an emphasis on safe storage and defence rather than on agricultural production. Mountainous regions have always attracted refugees of all kinds, settling in places chosen for their defensive potential. In a world marked by connectivity, inaccessibility in itself becomes an asset.

20.7 Conclusions

The impacts of trade, migration and mobility on central Saharan landscapes are thus best thought about, not in terms of routes and staging posts, but rather as the result of a more general and diffuse connectivity which was a necessary condition for all forms of life, even in their most 'sedentary' and hence archaeologically visible guise. The archaeological understanding of mobility has over the last two decades made great progress, both technically and conceptually. In the central Sahara, however, it is not quite enough to try to avoid thinking of mobile lifeways through sedentary stereotypes; we also need to be able to perceive sedentism from a mobile point of view. Mobile power, based on the primacy of access to resources rather than their stable ownership, 'enclosed' sedentism and created a fluid landscape, one thought about and used in terms of strategic points rather than territorial cover and coherence. It makes sense in such a context to understand mobility as a political choice, as a 'primary benefit' of pastoralism, as Honeychurch (2014, p. 290) suggests for western Eurasia, rather than just as an adaptation to marginal environments. Indeed, in recorded history, the most powerful mobile groups in the central Sahara tended to relinquish direct control over livestock to their clients, as herds would in fact impede their freedom of movement.

More recently, these relations of dominations have changed in favour of more sedentary polities, and the 'footprint' of mobile people on central Sahara landscapes has taken on new political significance. With regards to the brutal repression of the Tuareg rebellion in northern Mali in the 1990s, Randall and Giuffrida (2006, p. 457) thus note that 'one reason why it was so easy for the state army and other Malians to literally clear the landscape of Tuareg and Maures was because they were nomads', and their disappearance left no visible traces on the ground. Aware of this, many nomadic pastoralists today have started to plant millet fields and vegetable gardens even in places where these are not viable and to build houses, even though they rarely live in them. At a time when archaeologists are growing increasingly sophisticated in detected the traces of mobility on the landscape, for many pastoralists, marking the landscape has become a political priority.

References

Almathen F, Charruau P, Mohandesan E, Mwacharo JM, Orozco-terWengel P, Pitt D, Abdussamad AM, Uerpmann M, Uerpmann H-P, De Cupere B, Magee P, Alnaqeeb MA, Salim B, Raziq A, Dessie T, Abdelhadi OM, Banabazi MH, Al-Eknah M, Walzer C, Faye B, Hofreiter M, Peters J, Hanotte O, Burger PA (2016) Ancient and modern DNA reveal dynamics of domestication and cross-continental dispersal of the dromedary. Proc Nat Acad Sci 113:6707–6712

Baier S, Lovejoy PE (1977) The Tuareg of the Central Sudan: gradations in servility at the desert edge (Niger and Nigeria). In: Kopytoff I, Miers S (eds) Slavery in Africa; historical and anthropological perspectives. University of Wisconsin Press, Madison, pp 391–411

Claudot-Hawad H (1990) Nomades et État: l'impensé juridique. Droit et Société 15:211–222

Cleaveland T (2002) Becoming Walata: a history of Saharan social formation and transformation. Heinemann, Portsmouth, p 232
Grémont C (2010) Les Touaregs Iwellemmedan (1647–1896). Un ensemble politique de la boucle du Niger. Karthala, Paris, p 552
Honeychurch W (2014) Alternative complexities: the archaeology of pastoral nomadic states. J Archaeol Res 22:277–326
Honeychurch W, Makarewicz CA (2016) The archaeology of pastoral nomadism. Ann Rev Anthropol 45:341–359
Hopkins JFP, Levtzion N (eds) (1981) Corpus of early Arabic sources for West African history. Cambridge University Press, Cambridge, p 492
Krätli S, Schareika N (2010) Living off uncertainty: the intelligent animal production of dryland pastoralists. Europ J Develop Res 22:605–622
Makarewicz CA (2013) A pastoralist manifesto: breaking stereotypes and re-conceptualizing pastoralism in the Near Eastern Neolithic. Levant 45:159–174
Marshall LW (ed) (2015) The archaeology of slavery: a comparative approach to captivity and coercion. Southern Illinois University Press, Carbondale, p 426
McDougall EA (1990) Salts of the Western Sahara: myths, mysteries, and historical significance. Int J Afr Hist Stud 23:231–257
McDougall EA (2005) Conceptualising the Sahara: the world of nineteenth-century Beyrouk commerce. J North Afr Stud 10:369–386
Monod T (1968) Les bases d'une division géographique du domaine Saharien. Bull l'IFAN B 30:269–288
Randall S, Giuffrida A (2006) Forced migration, sedentarization and social change: Malian Kel Tamasheq. In: Chatty D (ed) Nomadic Societies in the Middle East and North Africa: Entering the 21st century. Brill, Leiden, pp 426–466
Retaillé D (1998) L'espace Nomade. Rev Géogr Lyon 73:71–81
Retaillé D, Walther O (2013) L'actualité sahélo-saharienne au Mali: une invitation à penser l'espace mobile. Ann Géogr 694:595–618
Rossi B (2015) From slavery to aid. Politics, labour, and ecology in the Nigerien Sahel, 1800–2000. Cambridge University Press, Cambridge, p 257
Webb JLA (1995) Desert frontier: ecological and economic change along the Western Sahel, 1600–1850. University of Wisconsin Press, Madison, p 208

Warfare in the Central Sahara

21

Hermann Häusler

Abstract

This contribution focuses on terrain evaluation of the central Sahara for military purposes in particular on reconnaissance in Libya during World War II. The deployments of the German "Sonderkommando Dora" to Libya in 1942 and of "Wehrgeologenstelle 12" to Rommel's Africa Corps in 1941–1943 are examples that illustrate the importance of a sound knowledge of geology and the interpretation of geomorphology of the central Saharan landscape during military expeditions and reconnaissance. Cross-country movement and supply of drinking water quality from groundwater are of long-standing interest for support of civil and military actions in the central Sahara.

Keywords

Military geography · Military geology · Terrain evaluation · World War II · Sonderkommando Dora · Long Range Desert Group

21.1 Introduction

Textbooks for the traditional "terrain and tactics" school of military geography often emphasise physical, cultural and political–military geography and terrain analyses and also cover military operations other than war (Palka and Galgano 2005). Terrain evaluation was and still remains the focal point of military geology or military geoscience. More recently, it is the proceedings of international conferences on military geosciences—and not textbooks, as reviewed by Häusler (2009)—that provide insights into the historical aspects of military geosciences. The comments on German and British terrain analyses of the central Sahara discussed in this chapter refer to tactics and military missions during World War II. These comments, however, do not address what Rachel Woodward has termed "critical military geography"—the geographies of militarism and military landscapes (Woodward 2014). Expeditions crossing the central Sahara date back to the first half of the nineteenth century. Later, German, Italian, French and British expeditions led to geographic and geologic investigations of the western, central and eastern Sahara. The 19th-century exploration and mapping of northern Africa demonstrated the traditional period of expedition cartography (e.g. Harold 2005) which was followed by topographic mapping in the twentieth century, which is when expedition maps became primary sources for military and cartographic intelligence (Török 2012).

21.2 Terrain Evaluation of the Central Sahara During World War II

Except for logistic support from British colonies in West Africa via French Equatorial Africa towards Sudan and Egypt (Schmidl 2013), Allied troop military operations in Africa during World War II largely took place in the coastal areas of Libya, Egypt and Tunisia. During a relatively short period—from the Italian attack on Egypt in late 1940 to the end of Axis resistance in May 1943—the German "Wehrgeologenstelle 12" evaluated the terrain for Rommel's warfare in North Africa and prepared specialist maps of Libya at regional scales. At that time, the British Long Range Desert Group (LRDG) had patrols in southern Egypt bordering Anglo-Egyptian Sudan and in southern Libya bordering French West Africa and French Equatorial Africa. LRDG patrols fought in Rommel's "backyard" (Timpson 2000), whereas the German "Sonderkommando

H. Häusler (✉)
Department of Environmental Geosciences, Universität Wein, Althanstrasse 14, 1090 Vienna, Austria
e-mail: hermann.haeusler@univie.ac.at

J. Knight et al. (eds.), *Landscapes and Landforms of the Central Sahara*, World Geomorphological Landscapes, https://doi.org/10.1007/978-3-031-47160-5_21

Dora" had patrols reconnoitring the Libyan Sahara southward to French Equatorial Africa. While Winters et al. (1998) described more general conditions of warfare in the Western Desert, Major General Alfred Toppe summed up German experiences in African desert warfare with many details on water supply and trafficability. These details were provided by the military geologist Dr Sigismund Kienow, who had joined the military geology team of Rommel's Africa Corps (Toppe 1991).

21.2.1 Terrain Evaluation by "Wehrgeologenstelle 12" of Rommel's Africa Corps

From April 1941 to May 1943, "Wehrgeologenstelle 12" supported the German Africa Corps. Initially consisting of eight persons (two senior geologists, one junior geologist, one technician, one draughtsman, one typist and two drivers), "Wehrgeologenstelle 12" was temporarily provided with four vehicles for reconnaissance and groundwater investigations. The military geology fieldwork and written expertises of "Wehrgeologenstelle 12" comprised preparation of water supply in coastal areas and deserts, landscape classification for motorised tank divisions and fortification of positions (Häusler 2003). Off-road trafficability maps at scales of 1:100,000 and 1:400,000 were based on the geology team's expeditions using wheeled vehicles and reconnaissance flights.

These hand-coloured "Befahrbarkeitskarten" informed Rommel's troops on landscape classes relevant to the movement of motorised divisions. The landscapes were predominantly classified by geomorphology (e.g. mountainous landscapes, scarps, hummocky areas, sand plain, dunes, clay depressions) and vegetation (e.g. shrub cover). In May 1942, "Wehrgeologenstelle 12" prepared trafficability maps at 1:400,000 scale for several topographic sheets in northern Libya and 1:500,000 scale for the Egyptian sheets Matrûh and Cairo. In addition, a general trafficability map of Libya at a scale of 1:3,000,000 was prepared. Figure 21.1 depicts a section of this Libyan map that had been prepared to provide general information on the interpretation of landscapes. This trafficability map ("Befahrbarkeitskarte") was compiled by "Wehrgeologenstelle 12" ("bearbeitet von Wehrgeologen-Stelle 12") that in March 1942 was deployed to the engineer officer of the army headquarters of Panzerarmee Africa ("beim Armee-Pionierführer des Panzer-Armee-Oberkommandos Afrika"). The legend of this generalised trafficability map was simple and consisted of four classes: good trafficability of plain serir ("Gut befahrbare, ebene Kieswüste"), moderate to bad

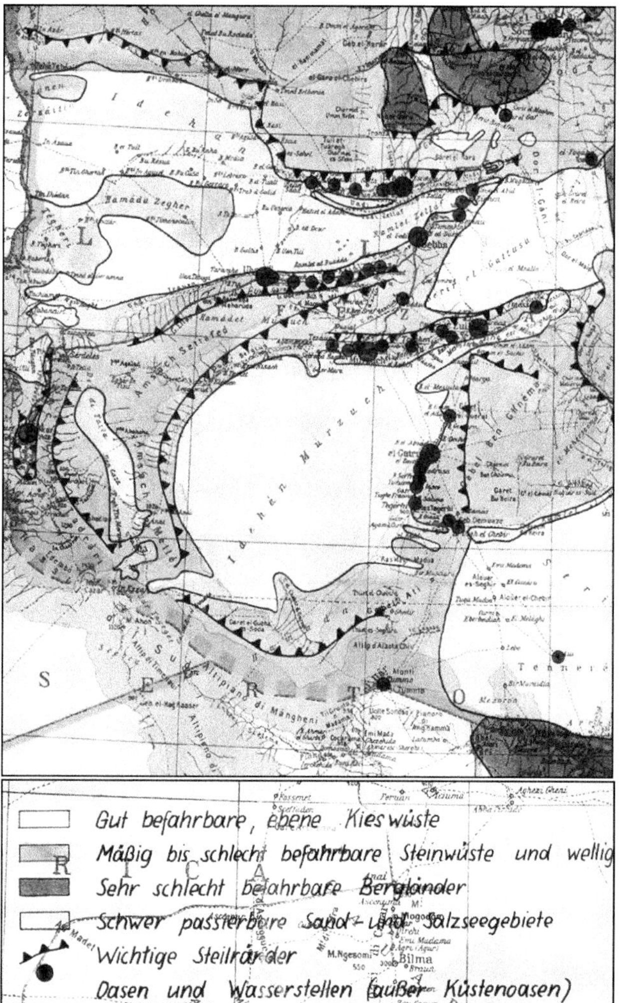

Fig. 21.1 Southwestern section of the German "Befahrbarkeitskarte" 1:3,000,000 scale map of Libya with classification of cross-country movement, enhanced scarps and important oases, compiled in March 1942 by Rommel's "Wehrgeologenstelle 12". Legend and military abbreviations are explained in the text (author's collection)

trafficability of stone desert ("Mäßig bis schlecht befahrbare Steinwüste"), very bad trafficability of mountainous landscape ("Sehr schlecht passierbare Bergländer") and difficult trafficability of sand-covered areas or salty marshes ("Schwer passierbare Sand- und Salzseegebiete"). Linear topographic elements such as steeper slopes were highlighted but not classified as passable or non-passable. In addition, important oases and water wells ("Oasen und Wasserstellen") were marked.

In summing up, the German off-road trafficability map of Libya 1:3,000,000 scale identified geomorphologic features such as major scarps, hilly areas, dune fields, gravelly plains and stone deserts which offered a better impression of landscapes for military commands than other published topographic maps of the same scale available at the time. In contrast, very detailed military geoscientific maps of southwestern Libya at scales of 1:25,000 up to 1:200,000 were produced by a special unit of the German Army High Command, the "Sonderkommando Dora", which are explained in more detail below.

21.2.2 Field Reconnaissance of the Libyan Sahara by "Sonderkommando Dora"

In March 1942, the German Counter-Intelligence Service launched "Sonderkommando Dora" (Häusler 2011). It consisted of a military group of about 100 personnel and a scientific unit of about 10 military geoscientists, including geographers, cartographers, geologists, astronomers, meteorologists and road specialists. The main mission of this command was a military geoscientific one, and although the command was fit for action, combat had to be avoided in all circumstances. Two patrol teams were fully motorised and supported by aeroplanes for reconnaissance flights. They were active in May–December 1942 and were ordered to perform cartographic mapping of expedition routes at 1:200,000 scale, providing military geology maps including description of natural obstacles, water supply, descriptions of trafficability and concluding main traffic routes (Fig. 21.2).

The Hon airport radio station had contact with Berlin headquarters as well as with smaller radio stations of "Sonderkommando Dora's" technical teams. The commanding officer and geographer Dr Otto Schulz-Kampfhenkel, a reserve lieutenant of the German Air Force, was the head of the military geoscientific staff (Häusler 2011). The explanations of subsurface geology and of tectonics relevant for military geology follow the observations of the German geologists Dr Georg Knetsch and Dr Friedrich-Karl Mixius, who took part in the counter-intelligence campaign of the German Abwehr in Libya (Knetsch and Mixius 1950). Weathering of claystone and marlstone under desert conditions rapidly formed powdery soils, which mostly are cemented and therefore protected by a hard layer. Huge dune complexes were formed in Egypt (Great Sand Sea) and in southwest Libya such as Idehan Murzuk and Idehan Ubari hamada (high rocky plateau, often with scarps). Occasionally, landscapes of serir (gravel desert) and graret, large depressions with several dm-thick powdery soil cemented by a thin layer of coarse sand or salt (fesh-fesh), prevail in the central Sahara. Sabkha, depressions filled with fine sand and enriched with salt, occur in addition to single dunes and larger dune complexes. These saltpans were either terminal lakes occasionally charged by flowing water or owe their existence to groundwater evaporation through capillary ascent.

The southernmost area investigated by "Sonderkommando Dora" was the volcanic crater-oasis Wau en-Namus close to the Serir Tibesti bordering French Equatorial Africa. Topography of this oasis was geodetically mapped, and geomorphology was interpreted from the geologic point of view (Fig. 21.3). The schematic block diagram of Fig. 21.4 depicts the geologic formations of southern Libya and northern Chad consisting of a crystalline basement overlain by the Nubian Sandstone Formation and Tertiary limestone and marl intercalated with salt beds. Very young volcanic explosions rose through this sequence forming a 140 m deep caldera 3.2 km in diameter. The flanks of the volcanic craters consisted of either salty mud flows or thin layers of volcanic ash. Where the shallow aquifer consisted of evaporated beds, the groundwater was salty (Knetsch and Mixius 1950). As a result, German classification of landscapes in terms of trafficability comprised:

- Serir: firm and good trafficability, no obstacles (Fig. 21.5, top panel).
- Dunes: good trafficability; delay by detours and/or soft patches of sand (Fig. 21.5, lower panel).
- Hamada, hills and marl: reconnaissance essential before movement (Fig. 21.6).

In short, the geologists of "Sonderkommando Dora" interpreted the geomorphology of the published topographic maps at a scale of 1:200,000 and mapped natural obstacles such as dunes and hamada, water wells, magnetic deviation and main traffic routes (Häusler 2011).

21.2.3 Terrain Evaluation by the British Long Range Desert Group (LRDG)

The LRDG was formed specifically to carry out deep penetration, covert reconnaissance patrols and intelligence missions from behind Italian lines and to improve maps of the desert. In its first year, this group was organised and led by Major (later Brigadier) Ralph Alger Bagnold, a Royal Engineers signals officer, who had lived and travelled extensively in the desert in the years between the two world wars (Bagnold 1931; Underwood et al. 2002; Goudie 2008). A renowned desert explorer and esteemed geomorphologist (Forsyth 2017), Bagnold taught the men of the LRDG to identify and avoid most areas of dry quicksand and when they did become stuck, to use sand tracks and sand

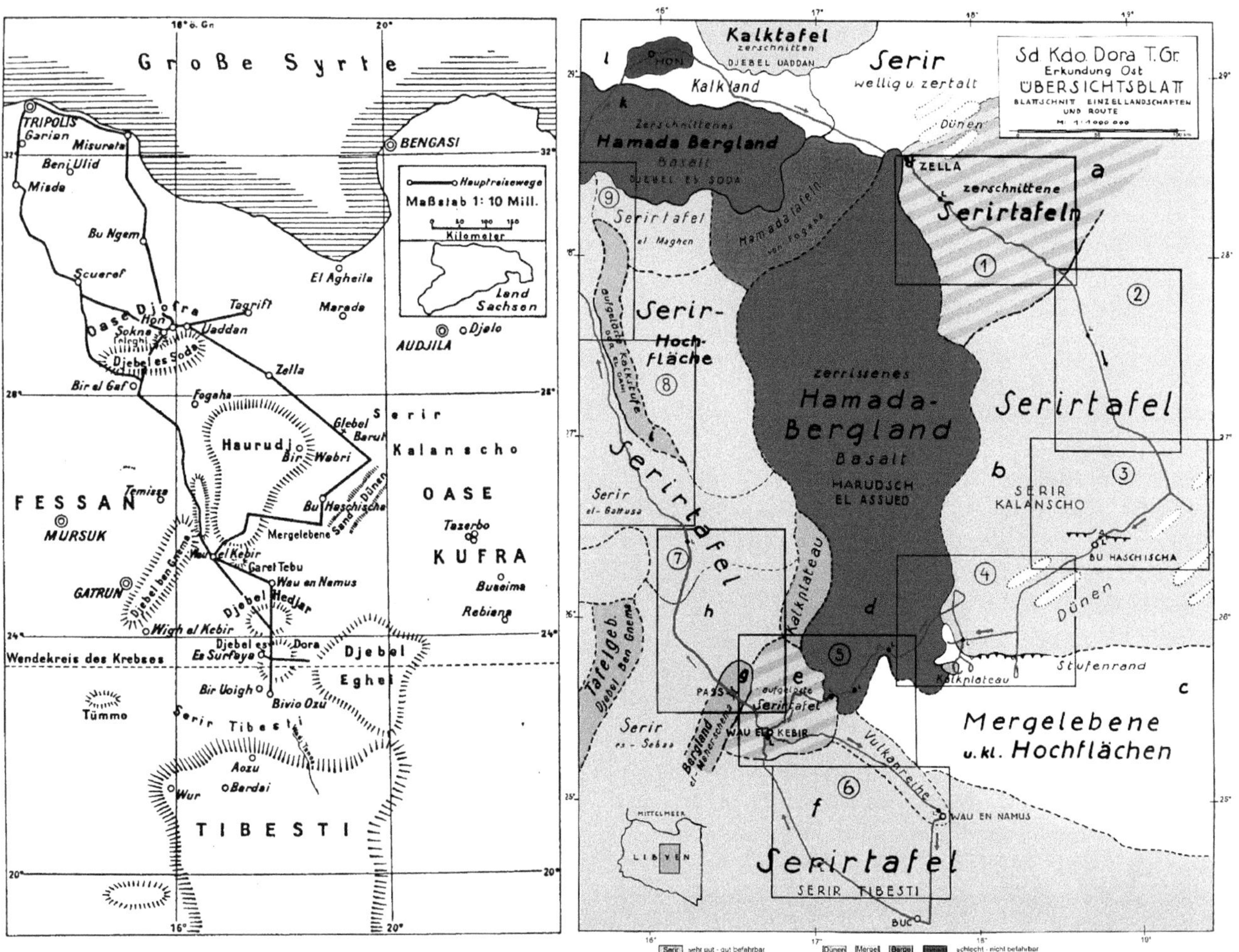

Fig. 21.2 Map of "Sonderkommando Dora's" reconnaissance routes in 1942. Route of Technical Reconnaissance Group East, east of Haruj, and route of Technical Reconnaissance Group West, west of Haruj (Left: Richter 1999, courtesy of Michal Farin, Belleville Publishers). Geologic-geomorphological map of central Libya depicting hamada (red), dunes (yellow), serir (grey) and hills and mountains (dark grey) (Right: Häusler 2011, courtesy of Redaktion Truppendienst)

mats to extract their trucks and move on (Underwood et al. 2002). All patrols had to report on both the features and the nature of the going their route covered, and it was mainly as a result of the information gained by the network of their journeys that accurate maps were eventually made of the interior of the Libyan Desert. As a result, colour-coded going maps denoting the state of the ground surface were compiled by intelligence officers at General Headquarters (Middle East) in Cairo. Timpson (2000, p. 24) reported: "Yellow was the best going, hard and smooth, like sand sheet. Dark brown was quite good, gravel, undulating with small wadi-beds and some camel-thorn; occasional dried-up mid pans were excellent, but did not last long. The purple areas were the worst, very rocky like west of Mekili. It did not include big escarpments or chaotic dune formations, since it was not a proper map of topographical features". For printing these off-road trafficability maps, the colours were slightly changed and landing grounds were added based on reconnaissance by the A-Squadron of the LRDG. Figure 21.7 depicts the legend of a British special map of Libya, sheet Derna 1:500,000 scale, with "goings" overprint dated October 1941. These going maps were classified as "most secret, not to be published".

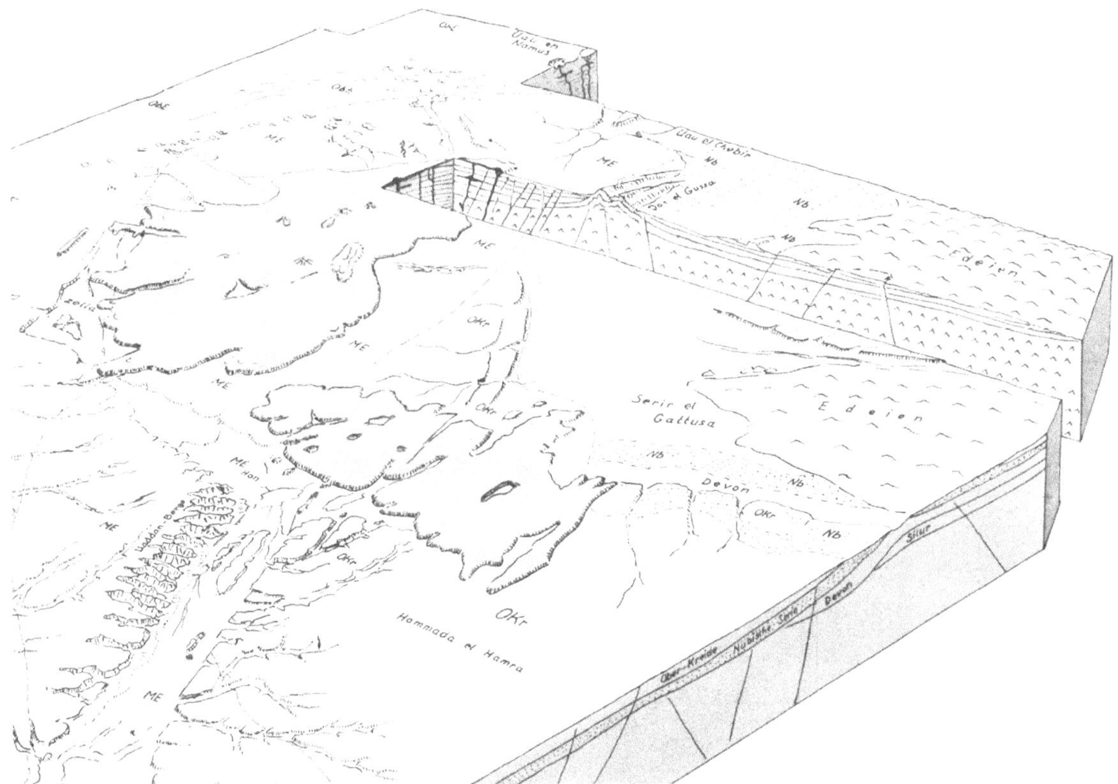

Fig. 21.3 Detail of the geologic block diagram of the Libyan Sahara depicting a major dry valley south of the Big Sirte, the tectonic Djofra Graben structure. The widespread young basaltic layer of Djebel Sauda and Djebel Haruj unconformably overlies older deposits. View in a southern direction. Abbreviation of geologic formations: *OKr* Upper Cretaceous, *Nb* Nubian Formation, *ME* Middle Eocene, *ObE* Upper Eocene (Knetsch and Mixius 1950; courtesy of German Geological Society)

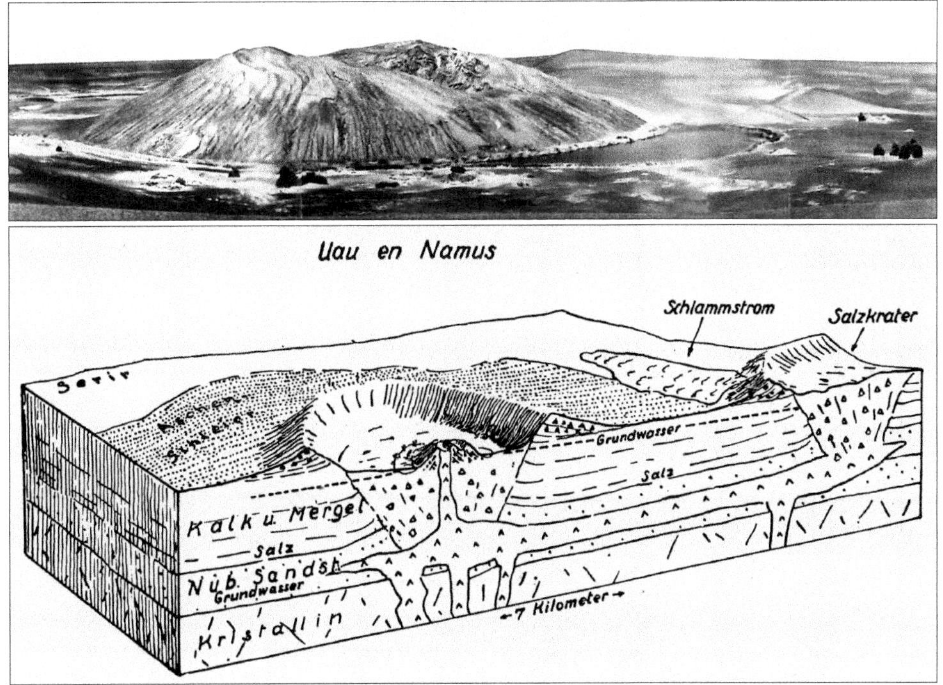

Fig. 21.4 Oblique aerial photograph of the oasis Wau en-Namus (Pillewizer and Richter 1957, above) and geologic block diagram of this volcanic crater (below; Knetsch and Mixius 1950; courtesy of German Geological Society). Meaning of German explanations: "Schlammstrom" (mud flow), "Salzkrater" (salt crater), "Kristallin" (crystalline basement), "Nub. Sandst." (Nubian Sandstone), "Kalk und Mergel, Salz" (limestone and marl, salt), "Grundwasser" (ground water) and "Aschenschleier" (thin cover of volcanic ash)

Fig. 21.5 Cross-country movement in the Libyan Sahara. Above: Good trafficability on hard sand of the Serir Kalancho, east of Djebel Haruj (1942) (Richter 1999, courtesy of Michal Farin, Belleville Publishers). Below: Bad trafficability on soft sand (Generalstab des Heeres 1940)

21.3 Discussion

Armed forces nowadays provide highly detailed fact sheets, which also refer to geomorphology, economy and infrastructure, to humanitarian and peace keeping missions. In contrast to the military geoscientific expeditions of armed forces in North Africa during World War II described above, these regional fact sheets and remote sensing data often serve civil and military missions. Additional reconnaissance at a local scale is only necessary to evaluate such databases or to update their actuality. However, the assessment of local cross-country movement for humanitarian missions during wet seasons or of local drinking water supply for refugee camps for example still needs special military geoscientific information. As a consequence, modern geoinformation on the countries of the Sahara and the Sahel zone no longer supports warfare but rather EU stabilisation missions instead. The information upon which modern fact sheets are based includes but is not limited to worldwide databases provided by CIA world fact books; internet databases of for example the World Health Organization; UN organisations; the US Geological Survey; the US Army Corps of Engineers (TEC, Topographic Engineering Centre), relief and development agencies; and articles in internet encyclopaedias. International cooperation is also needed to cope with causes and consequences of desertification with regard to climate change.

Fig. 21.6 Cross-country movement in the Libyan Sahara. Above: Bad trafficability on the marly dust plain northeast of Wau en-Namus. Below: Bad trafficability in the Black Haruj a stone desert covered by basalt blocks (Richter 1999; courtesy of Michal Farin, Belleville Publishers)

21.4 Summary

From examples of terrain evaluation for German and British military campaigns in the central Sahara during World War II, it is concluded that the geomorphologic interpretation of the landscape and evaluation of groundwater bearing formations effectively supported missions of armies that were hardly familiar with the desert environment. In the new millennium, international security politics has fundamentally changed in Africa, but military missions—this time with UN mandate—face enormous environmental problems in countries of the Sahara and the Sahel zone, a future challenge not only for military geoscientists.

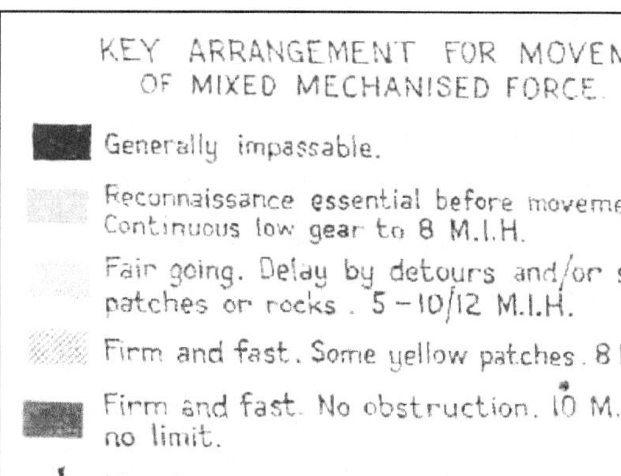

Fig. 21.7 Legend of the British going map at 1:500,000 scale, sheet Derna, printed by the South African Geodetic Survey in late 1941 (Häusler 2003)

References

Bagnold RA (1931) Journeys in the Libyan Desert 1929 and 1930: a paper read to the evening meeting of the society on 20 April 1931. Geogr J 78:13–33

Forsyth I (2017) Piracy on the high sands: covert military mobilities in the Libyan desert, 1940–1943. J Hist Geogr 58:61–70

Generalstab des Heeres (1940) Militärgeographische Beschreibung von Nordost-Afrika. Abteilung für Kriegskarten und Vermessungswesen (IV. Mil.-Geo.), Berlin, p 138

Goudie A (2008) Wheels across the desert. Exploration of the Libyan Desert by motorcar 1916–1942. The Society for Libyan Studies, Silphium Books, London, p 205

Harold J (2005) Deserts, cars, maps and names. eSharp. University of Glasgow, 4, pp 1–16

Häusler H (2003) Wehrgeologie im nordafrikanischen Wüstenkrieg (1941–1943). MILGEO 13, Bundesministerium für Landesverteidigung, Wien, p 135

Häusler H (2009) Report on national and international military geo-conferences held from 1994 to 2007. MILGEO 30E, Institute for Military Geography, Ministry of Defence and Sports, Vienna, p 70

Häusler H (2011) The "Forschungsstaffel z.b.V.", a special geoscientific unit of the German counter military intelligence service during the Second World War. In: Häusler H, Mang R (eds) International handbook military geography, vol 2. Proceedings of the 8th International Conference on Military Geosciences, 15–19 June 2009. Redaktion Truppendienst, Vienna, pp 276–286

Knetsch G, Mixius FK (1950) Beobachtungen in der libyschen Sahara. Geol Rundsch 38:40–59

Palka EJ, Galgano FA (2005) Military geography from peace to war. McGraw Hill, New York, p 496

Pillewizer W, Richter N (1957) Beschreibung und Kartenaufnahme der Krateroase Wau en-Namus in der Zentralen Sahara. Petermanns Geogr Mitt, Erg-H 264:303–320

Richter NB (1999) Unvergessliche Sahara, 2nd edn. Belleville, München, p 239

Schmidl EA (2013) Die "Takoradi Air Route"—eine strategisch bedeutsame Nachschubroute der Alliierten im Zweiten Weltkrieg. Österr Milit Z 6:640–653

Timpson A (2000) In Rommel's backyard. A memoir of the Long Range Desert Group. Leo Cooper, Barnsley, South Yorkshire, p 182

Toppe A (1991) Desert warfare: German experiences in World War II. Combat Studies Institute (U.S. Army Command and General Staff College, Fort Leavenworth), Kansas, p 103

Török ZG (2012) From expedition cartography to topographic mapping: Italian military maps of the Southern Libyan Desert from the 1930s. In: Beineke D, Heunecke O, Horst T, Kleim UGF (eds) Festschrift für Univ.-Prof. Dr.-Ing. Kurt Brunner anlässlich des Ausscheidens aus dem aktiven Dienst. Schriftenreihe des Instituts für Geodäsie der Universität der Bundeswehr München 87, München, pp 259–273

Underwood JR Jr, Giegengack RF (2002) Piracy on the high desert: the Long-Range Desert Group 1940–1943. In: Doyle P, Bennett MR (eds) Fields of battle: terrain in military history. Kluwer, Dordrecht, pp 311–324

Winters HA, Galloway GE, Reynolds WJ, Rhyne DW (1998) Battling the elements: weather and terrain in the conduct of war. John Hopkins University Press, Baltimore, Maryland, p 317

Woodward R (2014) Military landscapes: agendas and approaches for future research. Progr Human Geogr 38:40–61

22 Central Saharan Rock Art Landscapes

Marina Gallinaro

Abstract

Rock art is a ubiquitous feature of the central Sahara region. Thousands of pictograms (paintings) and petroglyphs (engravings) depicting isolated or clustered motifs—and often creating complex multi-layered palimpsests—are located all over the massifs and plateaus. The rock art represents residual, fragile and visual traces of the symbolic worlds of the diverse groups that inhabited and crossed the region over at least ten millennia and through significant climate shifts. It is a challenging heritage naturally embedded in the environment that had an active role in the construction of cultural landscapes and identities of the region.

Keywords

Central Sahara · Holocene · Rock art · Cultural landscape · Round heads · Pastoralism

22.1 Introduction

Rock art is one of the most extraordinary signs that humans left impressed on the landscape. Zoomorphic, anthropomorphic and geometric motifs—in the form of pictograms or petroglyphs—occur all over the world. Africa presents a significant number of rock art sites with noteworthy clusters of outstanding universal value, now partially recognised by the international community. To date, thirteen African sites are included in the UNESCO World Heritage Site list despite the wealth of African rock art of exceptional cultural, historical and aesthetic value (Table 22.1). Three of these sites are in the central Sahara region: the Tassili n'Ajjer (Algeria), the Tadrart Acacus (Libya) and the Ennedi (Chad) (Fig. 22.1).

If we examine closely the characteristics of the African rock art sites in the World Heritage Site list, a clear spatial pattern emerges: all the classifications correspond to clusters of sites that coincide with an entire massif or plateau (or a portion of it) rather than single, isolated caves or rock shelters. The definition of some of the World Heritage rock art sites as mixed properties (natural and cultural) and, above all, the classification of sites as cultural landscapes (e.g. Fowler 2003) recognises the close connection between rock art and the environment (Table 22.1). Indeed, rock art provides useful insights into the technical skills of its creators and their knowledge of the surrounding landscape and its resources. Furthermore, it constitutes a unique visual archive of the cultural, social and symbolic worlds of past human societies that cannot be separated from its location and the surrounding environment. This is especially true for the Sahara region, where rock art has been produced by diverse human groups that have inhabited the area during the last 10,000 years, if not longer, and through severe climate fluctuations and environmental changes with both 'green' and arid phases. Individual rock art examples all show a certain represented subject matter and 'scenes'. Here, a 'scene' refers to a set or cluster of stylistically similar motifs in spatial proximity to each other, from which an observer can infer an interaction (see Davidson and Nowell 2021). Rock art also has a specific locale which often corresponds to distinctive features in the landscape and can reflect land use under different climatic and environmental conditions, diverse human subsistence strategies and other ritual or symbolic values. In the latter case, it is difficult to use an informed approach to ethnographic or ethnohistoric interpretation, as rock art practice does not appear to have contemporary or historical continuity (e.g. Taçon and Chippindale 1998).

M. Gallinaro (✉)
Dipartimento di Scienze dell'Antichità, Università Degli Studi di Roma "La Sapienza", 00185 Roma, Italy
e-mail: marina.gallinaro@uniroma1.it

J. Knight et al. (eds.), *Landscapes and Landforms of the Central Sahara*, World Geomorphological Landscapes, https://doi.org/10.1007/978-3-031-47160-5_22

Table 22.1 List of the African rock art sites inscribed in the UNESCO World Heritage Site list

# on Fig. 22.1	Name	Date of inscription	Status	Criteria	Category	Country
1	Tassili n'Ajjer	1982		i, iii, vii, viii	Mixed	Algeria
2	Rock Art Sites of Tadrart Acacus	1985	Endangered (since 2016)	iii	Cultural	Libya
3	Ennedi Massif: Natural and Cultural Landscape	2016	Cultural landscape	iii, vii, ix	Mixed	Chad
4	Cliff of Bandiagara (Land of the Dogons)	1989		v, vii	Mixed	Mali
5	The Gedeo Cultural Landscape	2023	Cultural landscape	iii, v	Cultural	Ethiopia
6	Ecosystem and Relict Cultural Landscape of Lopé-Okanda	2007	Cultural landscape	iii, iv, ix, x	Mixed	Gabon
7	Kondoa Rock Art Sites	2006		iii, vi	Cultural	Tanzania
8	Chongoni Rock Art Area	2006		iii, vi	Cultural	Malawi
9	Tsodilo	2001		i, iii, vi	Cultural	Botswana
10	Twyfelfontein or /Ui-//aes	2007		iii, v	Cultural	Namibia
11	Matobo Hills	2003		iii, v, vi	Cultural	Zimbabwe
12	Mapungubwe Cultural Landscape	2003	Cultural landscape	ii, iii, iv, v	Cultural	South Africa
13	Maloti-Drakensberg Park	2000		i, iii, vii, x	Mixed	Lesotho, South Africa
*	Aïr and Ténéré Natural Reserves	1991	Endangered (since 1992)	vii, ix, x	Natural	Niger

* The region is inscribed as a natural property; however, a relevant concentration of rock art is present in the Aïr massif and quoted in the official WHS list description

This contribution offers an overview of the distribution and principal characters of rock art in the Saharan massifs and outlines the strict correspondence between rock art and the landscape. Three case studies from the central Sahara—the Tassili n'Ajjer, the Tadrart Acacus and the Messak—are discussed herein as examples of major rock art landscapes dating to the early, middle and late Holocene occupations, respectively.

22.2 Rock Art Regions

All the massifs and plateaus that constitute the geological frame of the central Sahara include rock art. The archaeological investigation of the Saharan region has been (and still is) greatly conditioned by political circumstances of the modern countries that share the area, and this has resulted in a patchy and heterogeneous documentation of rock art sites. Furthermore, the exploration of rock art in the region often followed a path independent from archaeological research. Since the first discoveries of rock art in the nineteenth century, it was reported and recorded by small groups of explorers, travellers, rock art enthusiasts, amateurs and archaeologists employing different recording methods and often experiencing higher field mobility than the systematic and demanding archaeological research (in terms of time, logistics and bureaucratic responsibilities), even in the same region. Although many areas of the central Sahara are still underexplored, often rock art is better known than any other traces of ancient occupation (ICOMOS 2007).

It is hard to synthesise such rich and heterogeneous data, as demonstrated by the corpus of papers and books devoted to each of the subregions of the central Sahara (Table 22.2). The few existing syntheses are now quickly superseded by new research (e.g. Muzzolini 1995; Jelinek 2004). While some general distribution patterns can be identified based on the incidence of unusual motifs, stylistic traits or artistic 'schools', the scattering of data and the frequent lack of robust archaeological classifications severely hinder in-depth analyses. In addition, any attempt at defining a comparative sequence of this impressive Saharan heritage is hampered by heated scholarly debates over rock art chronologies and styles (e.g. see Hachid 2016; di Lernia 2018). Indeed, rock art is mostly dated on a stylistic basis, highlighting the incomplete status of archaeological records.

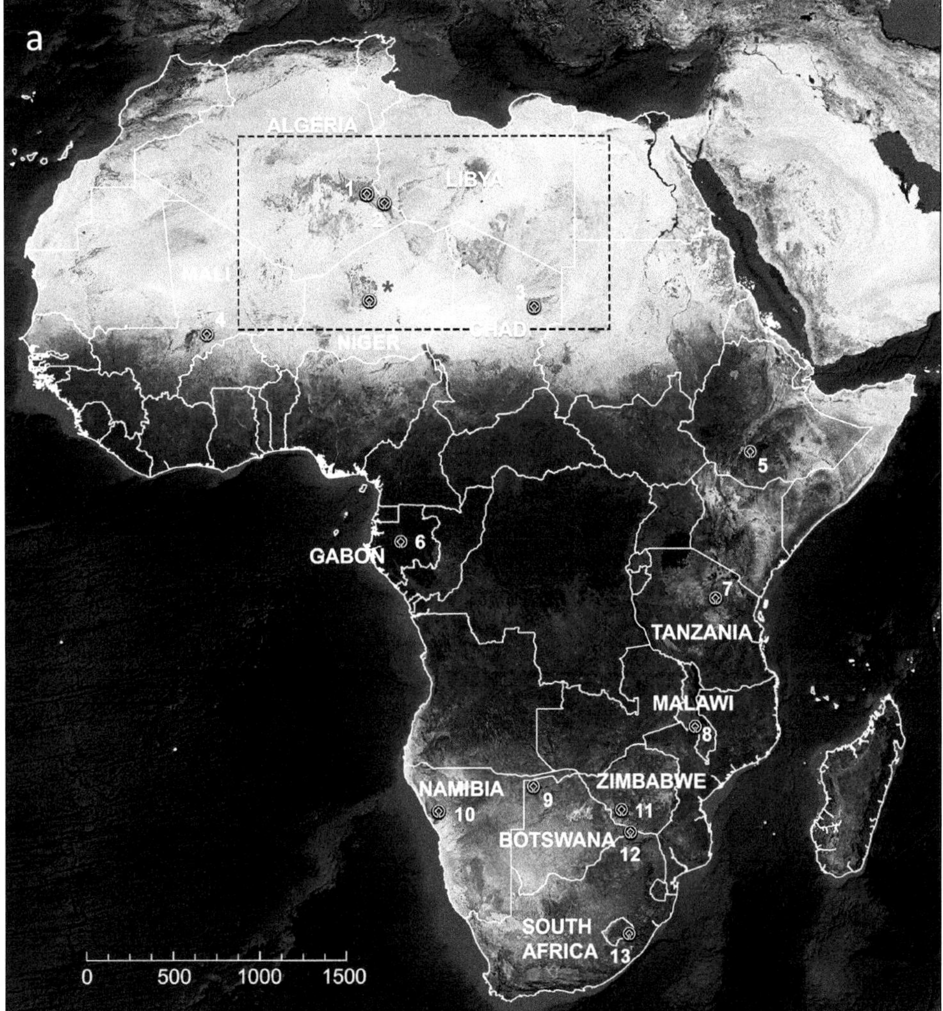

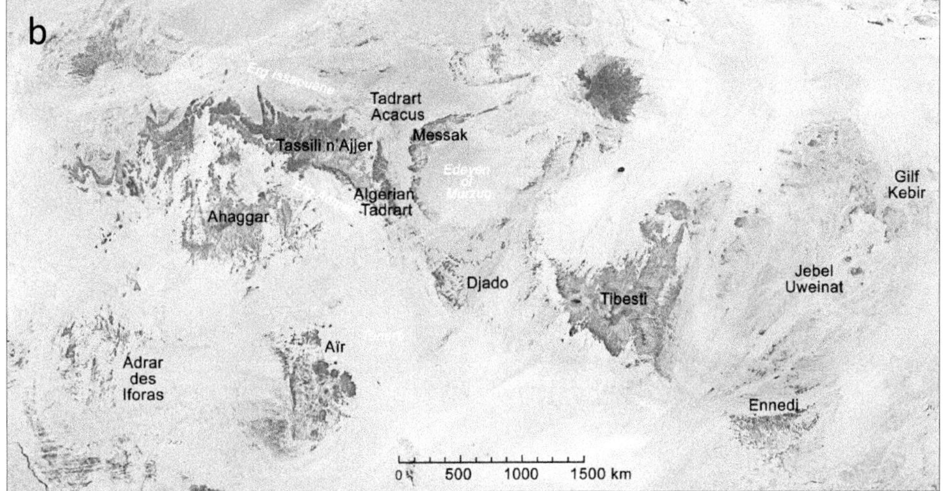

Fig. 22.1 **a** African rock art sites inscribed in the UNESCO World Heritage list, and the red star shows the Aïr and Ténéré natural reserves, **b** detail of the Saharan massifs mentioned in the text with the principal concentrations of rock art

Table 22.2 List of the Saharan massifs with rock art, including the principal techniques and styles

Region	Bubaline/Wild Fauna		Round heads		Pastoral/Bovidian		Horse/Bitriangular		Libyan Warriors		Camel		Tifinagh	
	Pictograms	Petroglyphs	Pictograms	Petroglyphs	Pictograms	Petroglyphs	Pictograms	Petroglyphs	Pictograms	Petroglyphs	Pictograms	Petroglyphs	Pictograms	Petroglyphs
Tassilin'Ajjer	–	+	++	(+)	++	++	++	+	(+)	+	+	+	+	+
Ahaggar (or Hoggar)	–	+	–	–	+	+	+	(+)	–	+	++	++	(+)	+
Algerian Tadrart	–	+	++	(+)	+	+	+	(+)	–	–	+	+	(+)	+
Tadrart Acacus	–	+	++	+	++	++	++	+	–	–	+	++	+	++
Messak	–	+	–	–	–	++	+	+	–	–	+	++	–	++
Djado	–	+	++	(+)	+	+	–	–	–	+	–	–	(+)	+
Adrar des Iforas	–	+	–	–	–	+	–	–	–	++	–	–	–	+
Aïr	–	(+)		–	–	+	–	–	–	++	–	–	–	+
Ennedi	–	+	(+)	–	+	+	+	+	–	–	+	+	–	–
Tibesti	–	–	–	–	+	+	–	–	–	–	+	+	–	(+)

Symbols: – absent, (+) uncertain/contested presence, + presence, ++ abundant presence

22.3 Central Sahara Rock Art Landscapes: Three Case Studies

22.3.1 The Round Heads of the Tassili n'Ajjer and Tadrart Acacus

Although the chronology of Saharan rock art is still debated, scholars generally agree that the so-called Round Heads style is one of the earliest evidences of rock art in the central Sahara. The term 'style' refers to 'a pattern of distinctive features of artistic expression or execution, distinctive of a specific person, people, school, or time period' (IFRAO Glossary available at http://www.ifrao.com/rock-art-glossary/; IFRAO 2000). The Round Heads style is generally agreed as early Holocene in age (see Hachid 2016; Tauveron 2016 for a recent discussion; *contra* Le Quellec 2013) based on the represented subject matter and archaeological data (e.g. Mori 1965, 1998; di Lernia 1999; Garcea 2001).

The Round Heads style was first identified and described by Breuil (1954) and Lhote (1958) in the Tassili n'Ajjer to designate all painted anthropomorphic figures with a round (or discoid) head. Since then, the recording of this kind of paintings has extended to the Tadrart Acacus, the Algerian Tadrart, the Djado plateau and the Wadi Aramat, while the presence of similar motifs is still highly debated in the Ennedi (Simonis et al. 2017) and the Jebel Uweinat/Gilf Kebir area (e.g. Muzzolini 1995). The paintings exhibit a variety of techniques, from simple contouring to full flat colouring and polychromy. Such variety led to conflicting stylistic classifications (e.g. Lhote 1958; Mori 1965; Sansoni 1994; Muzzolini 1995; for recent discussions, see Soukopova 2012; Gallinaro 2016; Tauveron 2016). The Round Heads anthropomorphs occur almost exclusively in rock shelters, in scenes interpreted as dances or ritual activities often involving animals and/or enigmatic motifs. This evocation of a sacred or symbolic world contributed to scholarly interest in this style (e.g. Sansoni 1994; Muzzolini 1995; Soleilhavoup 2007; Soukopova 2012).

Contextual analysis of the Round Heads paintings in the Tadrart Acacus (Gallinaro 2016) strongly supports the attribution of this style to the foraging groups locally named Late Acacus during the early Holocene (10.2–9.4 ka BP). The Late Acacus foragers had seasonal residential occupation and a broad-spectrum subsistence strategy that included hunting, fishing, gathering, food storage and the taming of Barbary sheep (*Ammotragus lervia*) (di Lernia 2001; Van Neer et al. 2020). Traces of ochre processing in the archaeological sites and the coherence between the represented motifs and the known subsistence strategies seem to confirm this attribution (di Lernia 1999; di Lernia et al. 2016). Quantitative analysis of the represented subjects also supports this association, showing a significant incidence of wild animals (mostly Barbary sheep and antelopes) and several elements that evoke gathering activities (Gallinaro 2016). The presence of a large quantity of bovids, including a small number of domestic animals (e.g. at Dobdobè, a site located on the northern part of the Acacus; Soleilhavoup 2007), suggests the persistence of this style during the first Pastoral period. This case contributed to the alternative proposal of a later dating of the Round Heads style during the full Pastoral period (e.g. Le Quellec 2013).

The distribution of Round Heads paintings in the Tadrart Acacus facilitates consideration of its cultural landscape. The paintings mostly occur in rock shelters or caves located on high ground along the courses of major wadis and tributaries, thus highlighting a deep knowledge and accurate use of the landscape, with clusters of sites showing intense economic and ritual activity in the region. The major concentrations—both in the number of sites and number and frequency of the represented motifs—are in the central and southern areas of the Acacus near the richest rock art sites of the Tassili. Although both areas show similar distribution patterns, the Tassili likely has a higher density of painted motifs (Fig. 22.2). The use of high-altitude sites has similarly been detected in the Tassili n'Ajjer although, in that case, it was interpreted as having sacred connotations (Soukopova 2012). In the later phases of Pastoral occupation, newer styles of rock art shift to abundant and rich lower altitude sites, with a strong link with wadi bottoms, in a changing environmental and economic trend (Gallinaro 2013; Gallinaro and di Lernia 2022).

Unfortunately, the understanding of links between rock art, occupation sites and landscapes is still hindered by the region's inaccessibility; only additional field research will ultimately improve chronological and contextual knowledge. However, recent development of online archives allows for a deeper and richer study of the Round Heads style (e.g. www.asasrtdata.eu; www.roundheadsahara.com) that can be further improved by the publication of recorded but thus far unpublished data.

22.3.2 The Pastoral Rock Art Landscape of the Messak

The Messak plateau comprises an impressive rock art landscape consisting almost entirely of petroglyphs found along the banks of networks of fossil wadis and on exposed bedrock of the plateau surface (Fig. 22.1b). The petroglyphs found in the Messak reach a quantity and quality unmatched in the central Sahara region. Thousands of motifs—from simple geometric signs to complex depictions with dozens of figures and intricate multi-layered

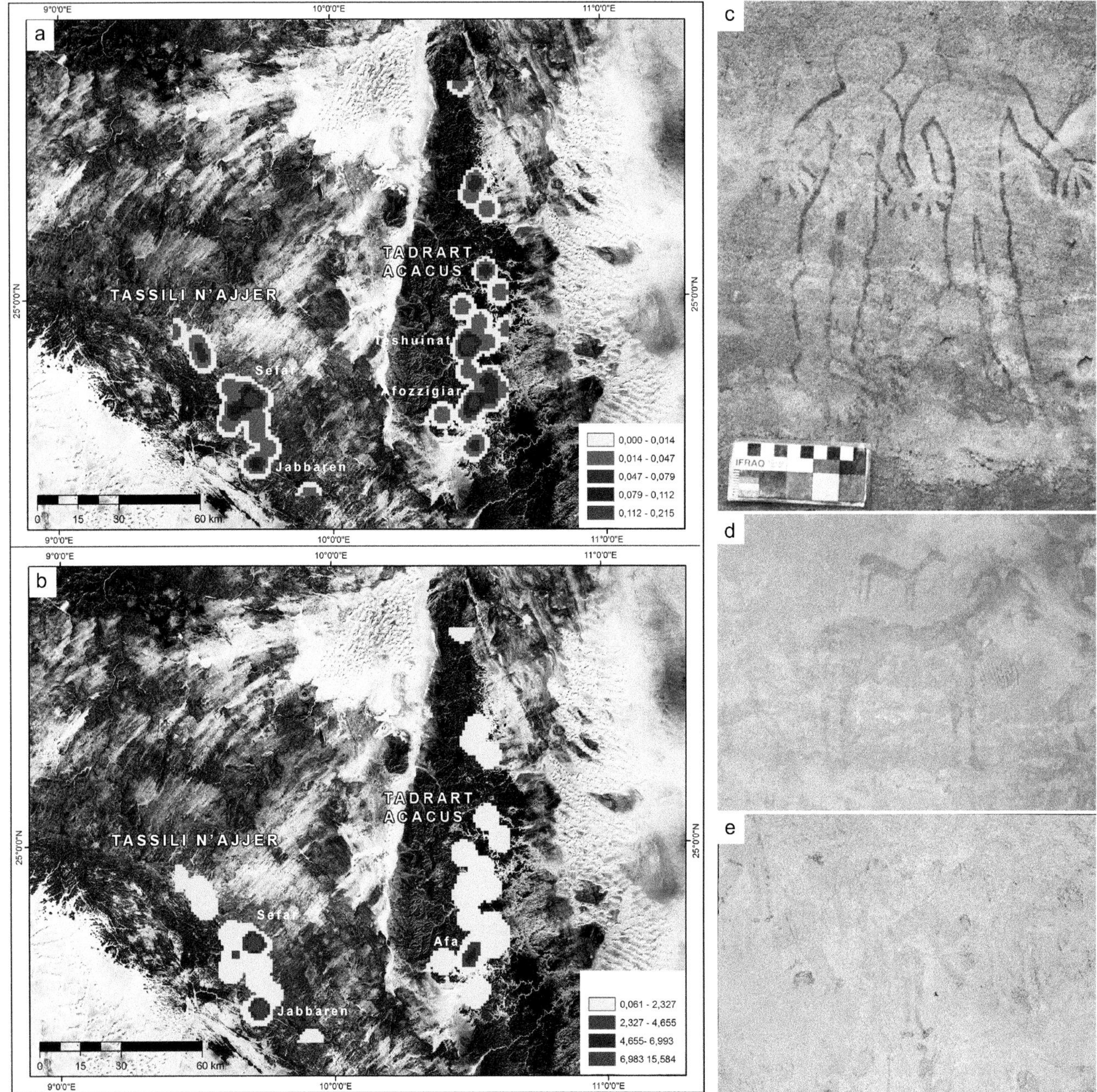

Fig. 22.2 **a** Density analysis of Round Heads sites in the Tadrart Acacus and Tassili n'Ajjer by location, **b** density analysis of sites by number of represented subjects (modified after Gallinaro 2016), **c** Round Heads anthropomorphs, Ghrub Cave, Tadrart Acacus, **d** Barbary sheep, Uan Afuda Cave, Tadrart Acacus, **e** plant gathering scene, Afozzigiar II rock shelter, Tadrart Acacus. (*Photographs* courtesy of The Archaeological Mission in the Sahara, Sapienza University of Rome)

palimpsests created in different styles and epochs—constitute a monumental archive and a challenging source of information (Mori 1965; Van Albada and Van Albada 2000). Numerous publications include systematic surveys of specific wadis and syntheses organised by themes and study areas (e.g. Frobenius 1937; Pesce 1967; Jacquet 1978; Jelinek 1984, 1985a, b; Lutz and Lutz 1995; Le Quellec 1998; Van Albada and Van Albada 2000; Graziosi 2005). Thanks to the generous support of Anne-Michelle and Axel Van Albada, an ambitious plan to manage the corpus of data from published and unpublished sources is currently ongoing in the ASArt-DATA project (Gallinaro 2019).

The petroglyphs of the Messak have been used for ecological reconstruction as they represent animals spanning from humid and savanna to arid environments. Animals such as crocodiles along the wadis, hippo, buffalo, elephants, rhino, giraffes and others are compatible with the humid phase of the Green Sahara, while gazelle, fennecs and camels correspond to the arid phases (e.g. Lutz and Lutz 1995; Le Quellec 1998; Mori 1998; Guagnin 2012) (see Knight, this volume). An outstanding feature of the Messak rock art is the theriomorphic motifs known as *Hommes-chiens* (e.g. Van Albada and Van Albada 2000). Despite debate over the chronology, scholars agree that the Pastoral Neolithic phase, and particularly the Middle Pastoral period (7.1–5.6 ka BP; di Lernia 2002), corresponds to the most intense phase of rock art production, given the frequency of cattle depictions and the presence of specific stylistic traits (di Lernia and Gallinaro 2010).

Contextual analysis of rock art and archaeological and environmental data permits insight into the specific cultural landscape of probably the first fully pastoral culture recorded in the central Sahara (di Lernia 2013). During the Middle Pastoral period, the Messak plateau was probably inhabited by groups of herders moving seasonally between the massif and the shores of temporary lakes in the dune fields of the Edeyen of Murzuq (di Lernia et al. 2013). The archaeological evidence of these occupations is scarce, owing to the limited presence of rock shelters and rare occurrence of preserved stratigraphies (Cremaschi 1994; Trevisan Grandi et al. 1998). The existence of ephemeral campsites located in endorheic depressions, suited to hunting and rearing cattle and small livestock, cannot be precisely dated (Perego et al. 2011; Biagetti et al. 2013; Gallinaro and di Lernia 2018). Nonetheless, there is a clear and well-dated record of ceremonial stone structures of the period (di Lernia 2006; di Lernia et al. 2013).

Pastoral rock art represents mostly domestic bovines and humans involved in a variety of activities such as milking, tending single cows or a herd and possibly transhumance, with probable ritual, convivial and hunting activities taking place. Some panels depict detailed representations of encampments with a wide range of animals and activities. Features can span over adjacent stone surfaces, permanently projecting the human experience onto the landscape (Fig. 22.3). A distribution analysis of about 100 typical scenes illustrating the pastoral world reveals a significant rock art landscape with definite trends across the massif (di Lernia et al. 2013). The scenes are disseminated all over the plateau, except for its northeastern side; their highest incidence occurs along the middle courses of the principal wadis, with clusters near the widest meanders. These locations are favourable areas for grazing and water supply in an otherwise harsh landscape. Significantly higher scene densities are recorded at some hotspots located in the central and southern areas of the massif facing the Edeyen of

Fig. 22.3 Site of Wadi Tiksatin, Messak. Example of a Middle Pastoral engraved scene of an encampment with a detail of a milking scene (1) and a detail of a herd of cow (2). Drawing modified after Van Albada and Van Albada (1996, p. 156) and Jelinek (2004, p. 157). (*Photographs* courtesy of A.M. Van Albada and A. Van Albada)

Murzuq. Recent scholarship highlights the connections between rock art clusters and ceremonial sites linked to the ritual deposition of slaughtered cattle in pits under megalithic stone structures (named *corbeilles*), dating precisely to the Middle Pastoral period (di Lernia et al. 2013).

Some of the *corbeilles* include standing or building stones with finely engraved cows. Three scenes depicting the slaughter of cattle are located inside one of the largest concentrations of *corbeilles,* and in one case, the slaughtering engraving has a topographic connection with a *corbeille* containing cattle remains (07/110 C1; di Lernia et al. 2013). A further hint at the complexity of associations linking human pastoral and symbolic landscapes is represented by the occurrence of incised scenes of probable rituals such as processions or dances of masked humans, quite close to one of the most interesting monumental *corbeille* sites.

22.4 The Figurative and Written Landscape of the Hyper-Arid Desert

Notwithstanding the desiccation of the central Sahara that began ~ 5.9 ka BP (Cremaschi and Zerboni 2011), the region continued to be inhabited, crossed and decorated from the last phases of the Pastoral Neolithic onwards. Here, we will focus on the rock art of the last 2500 years, generally classified as the Camel style. This consists of a representational world largely composed of depictions of stylised bitriangular humans, goat herds, isolated camels or caravans, palm trees, hunting and fighting. Scholars long ignored cameline rock art, favouring the earlier periods. However, scholarly interest in the area's protohistoric and historic cultural dynamics (e.g. Mattingly et al. 2003, 2017, 2020; Liverani 2005; Mori 2013) and ethnoarchaeological research (di Lernia et al. 2012, 2020; Biagetti 2014) also led to a more systematic recording of the later rock art (e.g. Zampetti 2008; di Lernia and Gallinaro 2011, 2022; Barnett 2019). In this period, the rock art landscape is deeply intertwined with the 'written landscape' comprised of painted and engraved ancient Libyan inscriptions in Tifinagh characters (e.g. Ait Kaci 2007; Biagetti et al. 2012, 2015). The chronological attribution and deciphering of these inscriptions are problematic. While the first North African use of this alphabet dates to the first millennium BCE, its presence in the central Sahara was dated to the first century BCE (Barnett and Mattingly 2003; Liverani 2005). The alphabet continued to be used thereafter and is still used today among the Tuareg. Most of the inscriptions relate short personal messages, epitaphs, riddles and names. Several toponyms and long genealogical lists have also been deciphered (Fig. 22.4).

Between the first millennium BCE and the first millennium CE, the first known Saharan state—the Garamantian kingdom—maintained an intensive, widespread trans-Saharan trade network centred in Garama (now Jarma), in the Wadi al-Ajal bordering the Messak. The Wadi Tanezzuft and the Tadrart Acacus formed the kingdom's southern frontier (see Mattingly et al. 2020 for a recent synthesis). Here, the principal settlements were in the oases of the Wadi Tanezzuft, where herding, hunting and intensive horticultural activities could be practised. A fortified system of castles that extended in the Tadrart Acacus also indicates a correspondence with the main mountain passes (Biagetti and di Lernia 2008; Mori 2013). However, sparse evidence suggests a more elaborate use of the landscape: the presence of a rich funerary landscape; thick dung deposits in rock shelters and caves in the central Wadi Teshuinat (e.g. Uan Muhuggiag rock shelter: Pazdur 1993; Uan Amil cave: Cremaschi and di Lernia 1998); and traces of rain-fed cultivation in the heart of the Tadrart Acacus, in use at least since the final phases of the Pastoral Neolithic period (di Lernia et al. 2012, 2020). The number of open-air sites with Tifinagh inscriptions and camel artworks is actually higher than any other rock art tradition in the area. They are abundant in the rock shelters located along the principal wadis

Fig. 22.4 Example of a boulder with Tifinagh inscriptions and camel style motifs, Tadrart Acacus. **a** General view, **b** detail of the panel. (*Photographs*: courtesy of The Archaeological Mission in the Sahara, Sapienza University of Rome)

of the massif, fitting with the occasionally patchy archaeological evidence and suggesting widespread occupation in the region even during the most arid phases (Gallinaro and di Lernia 2022). Additionally, the occurrence of sites near the principal mountain passes (known as *aqba* in Arabic) and ephemeral, rocky reservoirs of water (known as *guelta*) point to a landscape shaped by the human crossing of the massif (di Lernia et al. 2012, 2020). Thus, the figurative (cameline style) and written (Tifinagh inscriptions) landscapes of the last millennia reflect the region's changing pathways of human occupation and underscore rock art's strong potential to deepen our knowledge of chronological periods of the central Sahara that are still obscure.

22.5 Summary

The analysis of three cultural landscapes referring to the early, middle and late Holocene occupation of the Tassili n'Ajjer, the Tadrart Acacus and the Messak massifs, respectively (as the core of the central Saharan region), shows how the study of rock art integrated with environmental and archaeological data can add significant insights into interpretations of the past. Effort should be made to create shared comprehensive archives that can facilitate a deeper understanding of rock art styles, motifs and scenes situated in their environmental and social contexts. Only renewed field research can provide novel evidence for a sounder dating of rock art and improve our knowledge of human trans-Saharan connections.

Acknowledgements This research is part of the ASArt-DATA project funded by the European Union's Horizon 2020 research and innovation programme under the Marie Skłodowska-Curie grant agreement no. 795744. I wish to thank the Department of Archaeology in Tripoli and the Archaeological Mission in the Sahara from the Sapienza University of Rome, Italy. To Savino di Lernia go my warmest thanks for his comments on the manuscript. Special thanks go to Daniela Zampetti, Leader of the rock art teams from 2001 to 2009, and to Mauro Cremaschi and Andrea Zerboni for the geological and geomorphological study. I also wish to thank Donatella Calati, Adriana Ravenna and Anne-Michelle and Axel Van Albada for sharing their data with the ASArt-DATA project. The author is also grateful to Roberta Panzanelli for the improvement of English text.

References

Ait Kaci A (2007) Recherche sur l'ancêtre des alphabets libyco-berbères. Libyan Stud 38:13–37

Barnett T (2019) An engraved landscape. Rock carvings in the Wadi al-Ajal, Libya. Volume 1: Synthesis. Society for Libyan Studies, London, p 310

Barnett T, Mattingly DJ (2003) The engraved heritage: rock-art and inscriptions. In: Mattingly DJ (ed) The archaeology of Fazzan. Synthesis. The Society for Libyan Studies/The Libyan Department of Antiquities, London/Tripoli, vol 1, pp 279–326

Biagetti S (2014) Ethnoarchaeology of the Kel Tadrart Tuareg. Pastoralism and Resilience in Central Sahara. SpringerBriefs in archaeology. Springer, Cham, p 154

Biagetti S, di Lernia S (2008) Combining intensive field survey and digital technologies: new data on the Garamantian castles of Wadi Awiss, Acacus Mts., Libyan Sahara. J Afr Archaeol 6:57–85

Biagetti S, Ait Kaci A, Mori L, di Lernia S (2012) Writing the desert. The 'Tifinagh' rock inscriptions of the Tadrart Acacus (southwestern Libya). Azania 47:153–174

Biagetti S, Cancellieri E, Cremaschi M, Gauthier C, Gauthier Y, Zerboni A, Gallinaro M (2013) The 'Messak Project': Archaeological research for cultural heritage management in SW Libya. J Afr Archaeol 11:55–74

Biagetti S, Ait Kaci A, di Lernia S (2015) The 'written landscape' of the central Sahara: recording and digitising the Tifinagh inscriptions in the Tadrart Acacus Mountains. In: Kominko M (ed) From dust to digital. Ten years of the endangered archives programme. Open Book Publishers, Cambridge, pp 1–29

Breuil H (1954) Les roches peintes du Tassili n'Ajjer. Art et Metiers, Paris, p 161

Cremaschi M (1994) Le paléoenvironnement du Tertiare tardif à l'Holocéne. Doss l'Archéol 197:4–13

Cremaschi M, di Lernia S (1998) The geoarchaeological survey in the central Tadrart Acacus and surroundings (Libyan Sahara). Environment and cultures. In: Cremaschi M, di Lernia S (eds) Wadi Teshuinat. Palaeoenvironment and prehistory in south-western Fezzan (Libyan Sahara). CNR/All'Insegna del Giglio, Milano/Firenze, pp 243–296

Cremaschi M, Zerboni A (2011) Human communities in a drying landscape. Holocene climate change and cultural response in the central Sahara. In: Martini IP, Chesworth W (eds) Landscape and Societies. Springer, Heidelberg, pp 67–89

Davidson I, Nowell A (eds) (2021) Making scenes: global perspectives on scenes in rock art. Berghahn Books, p 352

di Lernia S (ed) (1999) The Uan Afuda Cave. Hunter gatherer societies of central Sahara. Arid Zone Archaeology Monographs, 1. All'Insegna del Giglio, Firenze, p 272

di Lernia S (2001) Dismantling dung: delayed use of food resources among early Holocene foragers of the Libyan Sahara. J Anthropol Archaeol 20:408–441

di Lernia S (2002) Dry climatic events and cultural trajectories: adjusting Middle Holocene Pastoral economy of the Libyan Sahara. In: Hassan F (ed) Droughts, food and culture. Kluver Academic/Plenum Publisher, New York, pp 225–250

di Lernia S (2006) Building monuments, creating identity: cattle cult as a social response to rapid environmental changes in the Holocene Sahara. Quat Int 151:50–62

di Lernia S (2013) The emergence and spread of herding in Northern Africa: a critical reappraisal. In: Mitchell P, Lane P (eds) Oxford handbook of African archaeology. Oxford University Press, Oxford, pp 523–536

di Lernia S (2018) The archaeology of rock art in Northern Africa. In: David B, McNiven IJ (eds) The Oxford handbook of the archaeology and anthropology of rock art. Oxford University Press, Oxford, pp 95–122

di Lernia S, Gallinaro M (2010) The date and context of Neolithic rock art in the Sahara: engravings and ceremonial monuments from Messak Settafet (south-west Libya). Antiquity 84:954–975

di Lernia S, Gallinaro M (2011) Working in a UNESCO WH site. Problems and practices on the rock art of the Tadrart Acacus (SW Libya, Central Sahara). J Afr Archaeol 9:159–175

di Lernia S, Massamba N'Siala I, Zerboni A (2012) 'Saharan waterscapes': traditional knowledge and historical depth of water management in the Akakus Mountains (SW Libya). In: Mol L, Sternberg T (eds) Changing deserts: integrating people and their environment. The White Horse Press, Cambridge, pp 101–128 di Lernia S, Tafuri MA, Gallinaro M, Alhaique F, Balasse M, Cavorsi L, Fullagar PD, Mercuri AM, Monaco A, Perego A, Zerboni A (2013) Inside the "African Cattle complex": animal burials in the Holocene Central Sahara. PLoS ONE 8:e56879. https://doi.org/10.1371/journal.pone.0056879

di Lernia S, Bruni S, Cislaghi I, Cremaschi M, Gallinaro M, Gugliemi V, Mercuri AM, Poggi G, Zerboni A (2016) Colour in context. Pigments and other coloured residues from the Early-Middle Holocene site of Takarkori (SW Libya). Archaeol Anthropol Sci 8:381–402

di Lernia S, Gallinaro M (eds) (2022) ATLAS of Tadrart Acacus rock art. A UNESCO World Heritage site in southwestern Libya. Arid Zone Archaeology Monographs, 9. All'Insegna del Giglio, Firenze, p 960

di Lernia S, Massamba N'Siala I, Mercuri A, Zerboni A (2020) Land-use and cultivation in the *etaghas* of the Tadrart Acacus (south-west Libya): the dawn of Saharan agriculture? Antiquity 94:580–600

Fowler PJ (2003) World Heritage Cultural Landscapes 1992–2002. UNESCO World Heritage Centre, Paris, p 133

Frobenius L (1937) Ekade Ektab. Die Felsbilder Fezzans, Ergebnisse der Diafe X (X deutsch-innerafrikanische Forschungsexpedition) nach Tripolitanien und Ost-Algier mit Ergnzungen der Diafe XII aus Zentral-Algier. Harrassowitz, Leipzig

Gallinaro M (2013) Saharan rock art: local dynamics and wider perspectives. Arts 2:350–382

Gallinaro M (2016) Rock art landscape of the central Saharan massifs: a contextual analysis of Round Heads style. In: Gutierrez M, Honoré E (eds) L'art rupestre d'Afrique. Actualité de la recherche. Editions L'Harmattan, Paris, pp 121–136

Gallinaro M, di Lernia S (2018) Trapping or tethering stones (TS). A multifunctional device in the Pastoral Neolithic of the Sahara. PLoS ONE 13:e0191765. https://doi.org/10.1371/journal.pone.0191765

Garcea EAA (2001) Uan Tabu in the settlement history of Libyan Sahara. Arid Zone Archaeology Monographs, 2. All'Insegna del Giglio, Firenze, p 256

Gallinaro M, di Lernia S (2022) Rock art styles in the Tadrart Acacus (SW Libya): a critical assessment. In: di Lernia S, Gallinaro M (eds) ATLAS of Tadrart Acacus rock art. A UNESCO World Heritage site in southwestern Libya. Arid Zone Archaeology Monographs, 9. All'Insegna del Giglio, Firenze, pp 77–97

Gallinaro M (2019) The ASArt-DATA Project: current perspectives on Central Saharan rock art. Quaderni di archeologia della Libya 22:157–166

Graziosi P (2005) Arte rupestre del Fezzan. Missioni Graziosi 1967 e 1968. Istituto Italiano di Preistoria e Protostoria, Firenze, p 199

Guagnin M (2012) From Savanna to desert: rock art and the environment in the Wadi al-Hayat (Libya). In: Huyge D, van Noten F, Swinne D (eds) The signs of which times? Chronological and palaeoenvironmental issues in the rock art of Northern Africa. Royal Academy for Overseas Sciences, Brussels, pp 145–157

Hachid M (2016) La chronostratigraphie des dépôts sédimentaires et les bandes pariétales de couleur sombre et claire du plateau du Tassili-n-Ajjer (Algérie): Leurs implications sur l'âge des peintures des périodes des Têtes Rondeset Pastorale ancienne. In: Gutierrez M, Honoré E (eds) L'art rupestre d'Afrique. Actualité de la recherche. Editions L'Harmattan, Paris, pp 65–106

ICOMOS (2007) Rock art of Sahara and North Africa: thematic study. ICOMOS, Paris, p 204

IFRAO (2000) Rock art glossary. Rock Art Res 17:147–156

Jacquet G (1978) Au cœur du Sahara libyen, d'étranges gravures rupestre. Archéologia 123:40–51

Jelínek J (1984) Mathrndush. In: Galghien. Two important Fezzanese Rock Art sites. Part I. Anthropologie, vol XXII, pp 237–268

Jelínek J (1985a) Tilizahren, the key site of the Fezzanese Rock Art. Part I—Tilizahren West Galleries. Anthropologie XXIII:125–165

Jelínek J (1985b) Tilizahren, the key site of the Fezzanese Rock Art. Part II—Tilizahren East, analyses, discussion, conclusions. Anthropologie XXIII:223–275

Jelínek J (2004) Sahara – Histoire de l'art rupestre libyen. Editions Jérôme Millon, Paris, p 560

Le Quellec J-L (1998) Art rupestre et préhistoire du Sahara: Le Messak libyen. Editions Payot et Rivages, Bibliothèque Scientifique Payot, Paris, p 616

Le Quellec J-L (2013) Périodisation et chronologie des images rupestres du Sahara central. Préhistoires Méditerranéennes 4. http://journals.openedition.org/pm/715

Liverani M (ed) (2005) Aghram Nadharif: The Barkat Oasis (Sha'abiya of Ghat, Libyan Sahara) in Garamantian times. Arid Zone Archaeology Monographs, 5. All'Insegna del Giglio, Firenze, p 520

Lhote H (1958) Peintures préhistoriques du Sahara; Mission H. Lhote au Tassili. Musee des Arts Decoratifs, Paris, p 64

Lutz R, Lutz G (1995) The secret of the desert: the rock art of the Messak Settafet and Messak Mellet. Golf Verlag, Innsbruck, Libya, p 177

Mattingly DJ (ed) (2003) The archaeology of Fazzan, vol 1, Synthesis. Society for Libyan Studies, London, p 454

Mattingly DJ, Leitch V, Duckworth CN, Cuénod A, Sterry M, Cole F (eds) (2017) Trade in the Ancient Sahara and Beyond. Cambridge University Press, Cambridge, p 466

Mattingly DJ, Merlo S, Mori L, Sterry M (2020) Garamantian Oasis settlements in Fazzan. In: Sterry M, Mattingly DJ (eds) Urbanisation and state formation in the Ancient Sahara and beyond. Cambridge University Press, Cambridge, pp 53–111

Mori F (1965) Tadrart Acacus. Arte rupestre e culture del Sahara preistorico. Torino, Einaudi, p 257

Mori F (1998) The great civilisations of the Ancient Sahara. L'Erma di Bretschneider, Rome, p 276

Mori L (2013) Life and death of a rural village in Garamantian times. Archaeological investigations in the oasis of Fewet (Libyan Sahara). Arid Zone Archaeology Monographs, 6. All'Insegna del Giglio, Firenze, p 405

Muzzolini A (1995) Les images Rupestres du Sahara. [Self published], Toulouse, p 447

Pazdur MF (1993) Evaluation of radiocarbon dates of organic samples from Uan Muhuggiag and Ti-n-Torha, southwestern Libya. In: Krzyzaniak L, Kobusiewicz M, Alexander J (eds) Environmental change and human culture in the Nile basin and Northern Africa until the Second Millenium BC. Poznan Archaeological Museum, Poznan, pp 43–47

Perego A, Zerboni A, Cremaschi M (2011) Geomorphological map of the Messak Settafet and Mellet (Central Sahara, SW Libya). J Maps 7:464–475

Pesce A (1967) Segnalazione di nuove stazioni d'arte rupestre negli Uidian Telìssaghen e Matrhandùsc (Messak Settafet, Fezzan). Riv Sci Preist 22:393–416

Sansoni U (1994) Le più antiche pitture del Sahara. L'arte delle Teste Rotonde. Jaca Book, Milano, p 328

Simonis R, Ravenna A, Rossi PP (2017) Ennedi, Tales on Stone. Rock Art in the Ennedi Massif. All'Insegna del Giglio, Firenze, p 288

Soleihavoup F (2007) L'art mystérieux des têtes rondes au Sahara. Faton, Paris, p 280

Soukopova J (2012) Round heads: the earliest rock paintings in the Sahara. Cambridge University Press, Cambridge, p 191

Taçon PSC, Chippindale C (1998) An archaeology of rock-art through informed methods and formal methods. In: Chippindale C, Taçon PSC (eds) The archaeology of rock-art. Cambridge University Press, Cambridge, pp 1–10

Tauveron M (2016) Art ancien du Sahara Central: les Têtes Rondes, état des recherches. In: Gutierrez M, Honoré E (eds) L'art rupestre d'Afrique. Actualité de la recherche. Editions L'Harmattan, Paris, pp 107–120

Trevisan Grandi G, Mariotti Lippi M, Mercuri AM (1998) Pollen in dung layers from rockshelters and caves of Wadi Teshuinat (Libyan Sahara). In: Cremaschi M, di Lernia S (eds) Wadi Teshuinat. Palaeoenvironment and prehistory in south-western Fezzan (Libyan Sahara). CNR/All'Insegna del Giglio, Milano/ Firenze, pp 95–106

Van Albada A, Van Albada AM (1996) Organisation de l'espace orné dan le Messak au Sahara Libyen. Anthropologie XXXIV:143–159

Van Albada A, Van Albada AM (2000) La Montagne des Hommes-Chiens. Art rupestre du Messak Libyen. [Self published], Paris, p 137

Van Neer W, Alhaique F, Wouters W, Dierickx K, Gala M, Goffette Q, Mariani GS, Zerboni A, di Lernia S (2020) Aquatic fauna from the Takarkori rock shelter reveals the Holocene central Saharan climate and palaeohydrography. PLoS ONE 15:e0228588. https://doi.org/10.1371/journal.pone.0228588

Zampetti D (2008) Le nuove ricerche sull'arte rupestre dell'Acacus: il senso del paesaggio, il senso della figura. In: di Lernia S, Zampetti D (eds) La Memoria dell'Arte. Le pitture rupestri dell'Acacus tra passato e futuro. All'Insegna del Giglio, Firenze, pp 339–349

23

Geoheritage and Cultural Heritage of the Central Sahara: Conservation Threats and Opportunities

Jasper Knight and Stefania Merlo

Abstract

The landforms and landscapes of the central Sahara are of considerable geoheritage and cultural heritage interest because of the unique geologic and climatic factors that have given rise to these landscapes and their physical and human properties, in particular over the Quaternary. Their value is recognised internationally by the inscription of a number of World Heritage Sites in the central Sahara region, supported by national scale and other international conservation designations across different central Saharan countries. However, geoheritage and cultural heritage conservation in the region is threatened by physical factors related to climate change and the survival of endangered species, in particular by human factors related to political instability, urbanisation and industrial development, oil and gas exploration, agricultural expansion, tourism and lack of effective management. The close connections between geoheritage and cultural heritage highlight not only their genetic relationships in the past, but also how, in combination, they can be valued and conserved in future.

Keywords

Heritage · Geoconservation · Management · World heritage sites · UNESCO · Rock art · Tourism · Climate change

J. Knight (✉)
School of Geography, Archaeology & Environmental Studies, University of the Witwatersrand, Johannesburg 2050, South Africa
e-mail: jasper.knight@wits.ac.za

S. Merlo
McDonald Institute for Archaeological Research, University of Cambridge, Downing Street, Cambridge CB2 3ER, UK
e-mail: sm399@cam.ac.uk

23.1 Introduction

Landscapes reflect the intersection of physical and human environments, properties and processes and, as such, exploitation of landscape resources, and therefore, landscape conservation and management requires consideration of both physical and human landscape components (Gray et al. 2013; Brilha et al. 2018). Although this is true for all landscapes on different spatial scales globally, it is especially the case in the central Sahara. Here, phases and styles of human activity during the Quaternary were affected by climate and geology that controlled the distribution of human occupation sites and the provision of water and food resources (Musch 2021), as well as how human activity, sociocultural practices and technology were used to exploit these environmental resources (Mattingly et al. 2007; Cremaschi and Zerboni 2009). Good evidence for this relationship comes from Garamantian-age (~2850–1450 BP) mortuary practices and necropolii (Biagetti and di Lernia 2008), land use practices (Mori 2018) and water management systems such as foggara (Dahmen and Kassab 2017; Wilson et al. 2020) that illustrate how prehistoric Saharan societies have exploited environmental resources. However, there are challenges in identifying and interpreting human–environment relationships in the Sahara, where evidence for past and present human activity, cultures and values may be hidden in the landscape such as by sand drift or just poorly managed due to their remoteness and transience in this vast and often inhospitable region (di Lernia 2006; Barker 2019). This is a limitation on both the understanding of past human societies in the central Sahara and in the interpretation of human–environment relationships such as adaptations to past climate variability. The latter can inform on likely impacts of future climate change on Saharan landscapes, resources and peoples.

Although there are strong reasons for valuing and then conserving geoheritage and cultural heritage in the central Sahara, many present challenges exist. There are still

J. Knight et al. (eds.), *Landscapes and Landforms of the Central Sahara*,
World Geomorphological Landscapes, https://doi.org/10.1007/978-3-031-47160-5_23

several elements of both the physical and human environments of the central Sahara that are poorly known. The central Sahara region includes many different countries with different legislative and management contexts, and this makes transboundary management of individual heritage sites problematic. Some parts of the Sahara are also difficult to access for field researchers and tourists (Kröpelin 2002).

23.1.1 Definition of Terms

The term *geoheritage* includes surface exposures of bedrock, the diversity of bedrock types present and their properties (lateral/vertical extent, structures, stratigraphy, fossils, minerals; termed their *geodiversity*), associated geomorphic features and evidence for certain physical processes or environments (Bronx and Semeniuk 2007; Erikstad 2013). Geoheritage also encompasses other heritage features that are genetically linked to the presence of bedrock types such as historic mining evidence, artefact knapping and human occupation within cave sites or exploitation of rock surfaces for rock art. This latter type of geologically dependent heritage therefore overlaps with *cultural heritage*, which refers to the evidence for past and present human cultures, practices and societal values. Cultural heritage can therefore include different types of archaeological evidence, built heritage of any age, as well as non-material cultural expressions such as language, music and folklore (Orton 2016; Esterhuysen et al. 2018). Many studies have examined both geoheritage and cultural heritage worldwide and their interconnections (Gordon 2012), although it is notable that geoheritage studies are not well developed in Africa, unlike cultural heritage studies that have strong palaeontological, archaeological and anthropological roots in the continent (e.g. Wurz and van der Merwe 2005; Keitumetse 2016; Brooks et al. 2020).

Despite the close relationships between the physical landscape and its properties (geoheritage) and evidence for the development of human activities and cultures within the landscape (cultural heritage), these different types of 'heritage' are commonly viewed as completely separate entities (Gray 2011) and are discussed and examined by different researchers who have very little interaction with each other. This is a significant limitation for areas such as the central Sahara where the co-relationships between different heritage types are well demonstrated and where palaeoclimate and water resources have exerted a strong control on prehistoric human occupation patterns. In an effort to overcome these limitations, this chapter discusses some overarching issues of geoheritage and cultural heritage in the central Sahara, highlighting their significance and management challenges. The chapter then identifies some future opportunities for geoheritage and cultural heritage site conservation, including the roles and limitations of sustainable tourism and ecotourism.

23.2 International Context of Heritage Conservation in the Central Sahara

In the central Sahara region as considered in this book, nine UNESCO World Heritage Sites (WHS) are present (Fig. 23.1). These include sites in Algeria, Chad, Libya, Mali and Niger that span the range of site types identified by UNESCO (as cultural, natural and mixed sites). The major properties of these nine sites and their general conservation status are given in Table 23.1. Although this is a

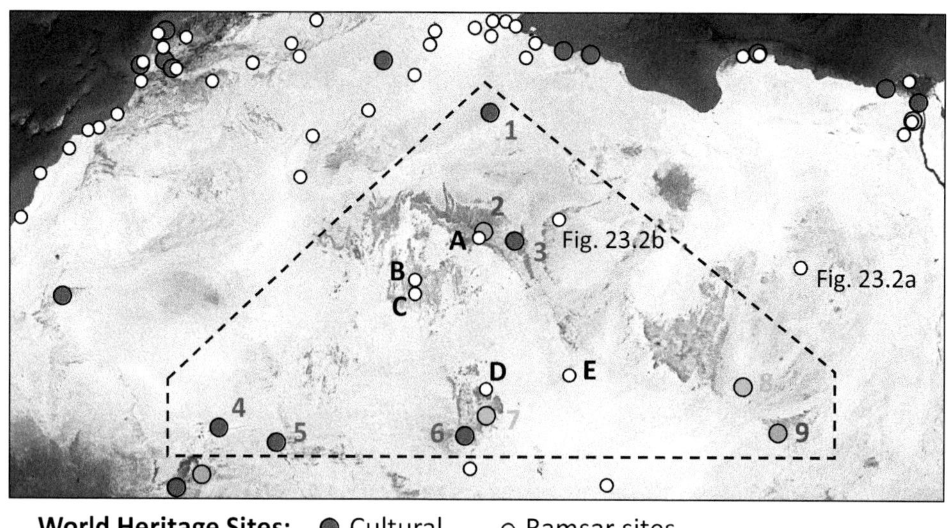

Fig. 23.1 Location of WHS and Ramsar sites in North Africa (correct as of 27 August 2023). Those located in the central Sahara region and discussed in this chapter are enclosed within the dashed polygon and listed in Tables 23.1 and 23.2. The locations of images in Fig. 23.3 are shown

Table 23.1 Issues affecting WHS integrity and site physical and cultural properties in the central Sahara region (sites are located on Fig. 23.1)

Site name and country (numbered site on Fig. 23.1) (year of inscription)	Site heritage type and the basis for inscription	Factors affecting the property (latest date of reporting)	Factors affecting the property identified in previous reports	Date of listing as 'in danger'
Old Town of Ghadamès (Libya) (1) (1986)	Cultural site. One of the oldest pre-Saharan cities and an outstanding example of a traditional settlement	War; water (rain/water table); fires; lack of maintenance of the property (2020)	Conflict situation prevailing in the country; torrential rain; fires	2016–present
Tassili n'Ajjer (Algeria) (2) (1982)	Mixed site. One of the most important groupings of prehistoric cave art in the world with more than 15,000 drawings and engravings	None listed	None listed	–
Rock Art Sites of Tadrart Acacus (Libya) (3) (1985)	Cultural site. Rock art scenes within caves including thousands of paintings in different styles and spanning the period 14–2 ka BP and depicting changes in climate and fauna/flora	Deliberate destruction of heritage; human resources; illegal activities; war (2020)	Vandalism; deliberate destruction of heritage; human resources; conflict situation prevailing in the country; illegal activities	2016–present
Timbuktu (Mali) (4) (1988)	Cultural site. Intellectual and trading centre from the fifteenth and fifteenth centuries; Islamic centre with three large mosques	Deliberate destruction of heritage; management systems/management plan; war; construction; pollution; desertification (2020)	Occupation of the property by armed groups; lack of management structure at the site	1990–2005, 2012–present
Tomb of Askia (Mali) (5) (2004)	Cultural site. Pyramidal structure built in 1495 and reflecting empire development and Saharan trade networks in the fifteenth and sixteenth centuries founded on salt and gold	Deliberate destruction of heritage; management systems/management plan; war; invasive species (2020)	Lack of site management; armed conflict; risk of collapse of the property	2012–present
Historic Centre of Agadez (Niger) (6) (2013)	Cultural site. Site developed in the fifteenth and sixteenth centuries; ancient street pattern still exist today; traditional mud brick structures and minaret	None listed (2015)	None listed	–
Aïr and Ténéré natural reserves (Niger) (7) (1991)	Natural site. Includes volcanic rocks of the Aïr massif and the Ténéré sand sea and related endemic plant and animal species	Civil unrest; erosion and siltation/deposition; forestry/wood production; human resources; identity, social cohesion, changes in local population and community; illegal activities; invasive/alien terrestrial species; livestock farming/grazing of domesticated animals; management activities; management systems/management plan; mining (2019)	Political instability and civil strife; poverty; management constraints (lack of human and logistical means); ostrich poaching and other species; soil erosion; demographic pressure; livestock pressure; pressure on forestry resources; gold panning; illegal activities (increase in poaching threats and timber harvesting); proliferation of the invasive exotic species (*Prosopis juliflora*); insecurity	1992–present
Lakes of Ounianga (Chad) (8) (2012)	Natural site. Interconnected saline, hypersaline and freshwater lakes in the Ennedi region	None listed (2015)	None listed	–
Ennedi massif: Natural and Cultural Landscape (Chad) (9) (2016)	Mixed site. Sandstone landforms reflecting wind and water processes, containing canyons, cliffs and natural arches; gueltas and rock art	None listed (2019)	None listed	–

Data source UNESCO WHS website (https://whc.unesco.org/en/list/), accessed 27 August 2023

Fig. 23.2 **a** Old Town of Ghadamès, **b** view of Timbuktu. All photographs reproduced under Creative Commons licence

very small sample size (9), five of these sites are classified by UNESCO as 'in danger', which at 56% is a far higher proportion than sites across Africa generally (15%) and globally (4.5%). At all sites, potential threats to site integrity and properties can be grouped under environmental, socioeconomic and management issues. Several localities in the central Sahara are also protected as Ramsar sites. Individual WHS are now examined in turn, informed by the site documentation available on the UNESCO WHS website (https://whc.unesco.org/en/list/). Information and photographs of some of these WHS are presented elsewhere. Tassili n'Ajjer rock art is discussed in Chap. 22. Rock Art Sites of Tadrart Acacus are discussed in Chap. 6 and presented in Figs. 6.3 and 22.2. Aïr and Ténéré natural reserves are discussed in Chap. 9. Lakes of Ounianga are shown in Fig. 18.4. The Ennedi massif is discussed in Chap. 5 and shown in Figs. 5.2 and 5.3. Photographs of several other sites listed in Table 23.1 that are not discussed elsewhere in this book are shown in Fig. 23.2.

Environmental issues relate to both climate change and its impacts on the physical environment and human activities such as industrial development that have potential to result in a deterioration in environmental quality. Climate change has a regional spatial context across the Sahara and is evidenced by both the persistence and likely amplification of aridity and high temperatures and an increasingly negative water balance (Adeniyi 2021). Several studies elsewhere from both humid and arid environments suggest that changes in climate (and microclimate within caves and rock shelters) can have negative effects on the preservation of rock art through rock spalling, salt weathering and deterioration of organic pigments (Giesen et al. 2014; Wright 2018; Taçon et al. 2021). Climate change in the Sahara is also likely to lead to potentially more erratic patterns of rainfall events (Vizy et al. 2013). These are significant because they can result in flash flooding and erosion of hillslopes and wadi systems and may in some instances trigger landslides and other mass movements. These events can negatively impact on built heritage features through destabilising walls and undercutting foundations. Environmental issues such as direct impacts of climate change, pollution and flooding/groundwater contamination are reported at several sites (Table 23.1).

Socioeconomic issues relate to the continued development of Saharan towns and cities and their associated footprints, in terms of resource use, water and air pollution, transport, infrastructure development and industrialisation in particular of oil and gas fields. Increased population numbers, density and resource requirements pose problems for the sustainability of Saharan cities. Tourism is commonly seen as a key socioeconomic opportunity, and the presence of WHS and similar conservation areas can help provide a basis for sustainable tourism (Kenrick 2019; Williams et al. 2020) but can also result is significant environmental impacts and the deterioration of destruction of cultural heritage (Keenan 2006).

Management issues relate to the detailed practices of WHS management, including national legislative and legal frameworks guiding site management structures, and engagement with stakeholders and local communities. Key gaps in site management identified in WHS reports include the convening, functionality and operation of management committees. Governmental support for sites is also often lacking, with the appropriate government ministers not taking responsibility for site management, or local management committees not empowered to take meaningful decisions. In many instances, there is a lack of stakeholder and community engagement—which is a requirement of WHS inscription—often because the structure or organisation for doing so does not exist. Funding and/or lack of training is often cited as limitations on effective management. External factors can also be significant, including civil unrest or insurgencies within individual countries (Keenan 2006).

Other international conservation designations are also present. UNESCO Global Geoparks are sites or regions that have international importance according to their geological and geomorphological heritage or scientific value. A key element of the designation of Global Geoparks is that they have a public outreach and access component, and it is this (and their specific geological focus) that distinguishes them from WHS. In total (as of the latest UNESCO Executive Board meeting of April 2021), 169 sites are designated as Global Geoparks, but only two of these are in Africa: M'Goun in Morocco (designated in 2014) and Ngorongoro-Lengai in Tanzania (in 2018). Neither of these are WHS. However, many natural WHS in Africa are inscribed under criterion VIII of the UNESCO World Heritage Convention as being an outstanding representative of Earth's history ('natural' sites on Fig. 23.1). A Geopark focusing on the Hoggar area of southern Algeria has also been mooted (Azil et al. 2020). Other locations are also protected internationally as Ramsar sites (Fig. 23.1) or Global Biosphere Reserves (e.g. Aïr and Ténéré, designated in 1997). This can help the conservation of IUCN Red List species that are found in the central Sahara region, such as the endangered Saharan cypress (*Cupressus dupreziana*), critically endangered Algerian fir (*Abies numidica*) and Nile crocodile (*Crocodylus niloticus*) that is found in desert gueltas. Ramsar sites in the central Sahara region, where the focus is on wetland ecosystem conservation, are listed in Table 23.2. Examples of guelta environments are discussed in Chap. 5.

23.3 Geoheritage in the Central Sahara

The central Sahara region covers a vast area, and as such, detailed field-based geological mapping has not been undertaken in all parts either in recent decades or with sufficient spatial detail (see Knight, this volume). Over large areas, surficial sediments also cover the bedrock surface. The best ground-truthed geological data, therefore, come from massifs such as the Tadrart Acacus, Tibesti and Ennedi where these have been well studied, mainly for applications to their rock art heritage (e.g. Gallinaro 2013). In a recent volume showcasing 'Africa's Top Geological Sites', out of 43 sites in total Viljoen (2016) included only three sites from the central Sahara region: Hoggar, Tassili and Tadrart Acacus (Algeria, Libya); Tibesti massif (Chad, Libya); and the Ennedi (Chad). These are all highland areas where bedrock tends to be well exposed (Azil et al. 2020). Specific studies of Saharan geoheritage are generally lacking, an exception being a description of oases in the Western Desert of Egypt (Plyusnina et al. 2016). There has been much more emphasis on the potential of different African sites for geotourism, with examples from Morocco (Bouzekraoui et al. 2018), Nigeria (Anifowose and Kolawole 2014), Algeria (Annad et al. 2017), South Africa (Knight et al. 2015;

Table 23.2 Ramsar sites in the central Sahara (marked in Fig. 23.1) and their properties

Ramsar site name (latter given on Fig. 23.1)	Site properties	Threats	Other designations
La Vallée d'Iherir, Algeria (A)	580 km^2 in area; 1100–1400 m asl; limestone plateau; intermittent streams, lakes, and marshes, freshwater springs, subterranean karst systems; habitats for lizards, snakes, four species of fish; threatened mammal species such as the Barbary sheep (*Ammotragus lervia sahariensis*), Dorcas gazelle (*Gazella dorcas*)	Tourism management; water pollution	Falls within part of a World Heritage Site; is a UNESCO Biosphere Reserve, National Park
Les Gueltates d'Issakarassene, Algeria (B)	351 km^2 in area; contains guelta ponds and wetlands; contains endemic fish species; threatened mammal species such the cheetah (Acinonyx jubatus), Dorcas gazelle and Barbary sheep' endemic plant species such as *Fagonia flamandii*, *Myrtus nivellei*, *Olea lapperini* and *Lavandula antinea*	Tourism management	National Park
Gueltates Afilal, Algeria (C)	209 km^2 in area; contains guelta ponds and wetlands; fish species include the Desert barbel; endemic plant species include Olea laperrinei, Rhus tripartita and Rumex simpliciflorus	Tourism management	–
Gueltas et Oasis de l'Aïr, Niger (D)	49,241 km^2 in area; permanent and temporary streams, oases and marshes; contains endangered species such as the cheetah, Barbary sheep and Dorcas gazelle and critically endangered Addax (*Addax nasomaculatus*); site for migratory birds, role of flood control	Desertification, development pressure	Falls within part of a World Heritage Site; is a National Park
Oasis du Kawar, Niger (E)	3392 km^2 in area; a complex of oases located between the Erg du Ténéré and the Erg de Bilma; date palm (*Phoenix dactylifera*) agriculture; refuge for mammals including the Cape hare (*Lepus capensis*), golden jackal (*Canis aureus*), Dorcas gazelle and Barbary sheep	Oil exploration, climate change	–

Data source Ramsar Sites Information Service website (https://rsis.ramsar.org/), accessed 27 August 2023

Penn-Clarke et al. 2020) and Angola (Tavares et al. 2015). Several studies have also been concerned with geoheritage loss (e.g. AbdelMaksoud et al. 2019).

Good examples of Saharan geoheritage are meteorite impact structures that are present across the central Sahara (mapped and reviewed by Reimold and Koeberl 2014). The reasons for this are that impacts can exhume mantle rocks from deeper levels, making them more accessible for scientific observation; rocks are tilted and differentially weathered, exposing strata in outcrop; high-energy and high-temperature impact often results in unusual minerals being formed; and impact structures often have a clear topographic expression in regional landscapes, with broken concentric rings of ridges separated by valleys. The topographic expressions of meteorite impacts on the land surface in humid areas also in turn influence patterns of rivers, soils, agriculture, settlements and rocks and thus later human activity (Gibson and Reimold 2015). This demonstrates the close relationship between geology and geodiversity, the physical environment and its resources and development of patterns of human activity and thus cultural heritage.

23.4 Cultural Heritage in the Central Sahara

The best documented aspect of cultural heritage of the central Sahara is rock art, and this is central to much research activity over recent decades; WHS inscription of the sites of Tassili n'Ajjer, Tadrart Acacus and the Ennedi; and the focus of much tourist activity (see Gallinaro, this volume) (Fig. 23.2). Rock art, by definition, is developed on rock faces and include both painted scenes and petroglyphs, but the relationship of these art forms to specific rock types, mineralogy, structure or facing aspect is not well understood. Much rock art is therefore interpreted in a cultural context (e.g. Ben Nasr and Walsh 2020) or with respect to activities such as pastoralism or migration taking place across the landscape by prehistoric groups (Barnett and Guagnin 2014; Honoré 2019). Other cultural heritage elements are also present in the central Sahara region, such as funerary monuments ranging from prehistoric megalithic tombs to Garamantian necropolii (Biagetti and di Lernia 2008; Biagetti et al. 2017); Roman, Arab and colonial military forts; prehistoric and historic settlements; mining and smelting sites; churches and mosques; and water/agricultural systems near and around oases (Biagetti and di Lernia 2008; Mori 2018). Mattingly and Sterry (2013) describe the progressive development of urban settlements in central Libya, spanning the period 3000–1000 BP, for which archaeological evidence is still retained in the landscape. Likewise, the urban development of Timbuktu (a WHS) from 1000 BP onwards took place in concert with changes in wadi and Niger River dynamics (Park 2010), highlighting the connection between the development of cultural heritage and its landscape and climatic context. Traditionally, vehicle-based archaeological surveys have been used to locate, visit, record and monitor such highly visible archaeological monuments and sites. More recently, the availability of high spatial resolution and free satellite imagery via platforms such as Google Earth has facilitated the discovery and monitoring of thousands of such traces in the central Saharan landscape (Biagetti et al. 2017; Nsanziyera et al. 2018; Rayne et al. 2020). Monumental and visible archaeology has been the main focus of monitoring and protection from national authorities and international organisations in the Sahara. It is increasingly recognised by archaeologists, however, that this is just one part of the rich archaeological record of the Sahara (al-Rimayh 2006). In fact virtually every square kilometre of the Sahara contains small-scale ephemeral archaeology created over thousands of years by the activities of ancient and more modern pastoralist and hunter-gatherer societies: small spreads of spears and stone piles, traces of tent footings and animal pens, fireplaces and field walls (Barker 2006). These discontinuous spreads of artefacts, that can be distributed over hundreds of metres, known in archaeology as 'off-site' or 'non-site' archaeology, are poorly suited for easy location and study, both via vehicle-based and remote sensing-based survey. This, coupled with the need of detailed geomorphological surveys (rare in the Sahara), alongside detailed on-foot archaeological surveys, has made it particularly difficult to map, interpret and protect this type of cultural heritage (Barker 2006). This archaeology is particularly vulnerable to damage or disturbance by travellers and vehicle traffic, mostly from lack of awareness of its potential importance (Keenan 2006).

Anthropological studies also describe cultural practices, traditions and behaviours of peoples such as Berbers and Tuareg, and this can also be included as examples of cultural heritage (e.g. Hincker 2004; Chalcraft 2015, 2016; Biagetti 2017).

23.5 Discussion

Although the central Sahara region is of significant interest for its geoheritage and cultural heritage and despite a number of international conservation designations existing, there are significant challenges to effective heritage conservation and management at protected area sites (Table 23.3). These issues cut across all sites and correspond to the combination of environmental, socioeconomic and management issues discussed above.

Both the physical landscape and cultural heritage features can be impacted by modern human activities in the

Table 23.3 Challenges for protected area management in the central Sahara

Management challenge	Nature of these issues
Security	Internal conflict and insurgency; internal political instability
Climate change	Climate hazards (floods)
Tourism	Lack of tourist infrastructure; pollution, environmental degradation, erosion associated with high tourist numbers; under-developed sustainable and ecotourism
Industry development and resource exploitation	Oil and gas exploration and development is often prioritised ahead of heritage conservation, water/air pollution
Urbanisation	Housing development, transport infrastructure
Pollution (air, water)	Waste management, groundwater abstraction, transport infrastructure
Lack of institutional management capacity	Lack of compliance to WHS management guidelines
Loss of biodiversity	Invasive species, competition, agriculture, climate change, pollution
Agricultural intensification	Intensive pivot irrigation agriculture; groundwater depletion and pollution; soil salinisation

Sahara. Prospection for oil mineral deposits and consequent opening of oil extraction operations has significantly affected the preservation of both monumental and ephemeral cultural heritage (Anag et al. 2002; Biagetti et al. 2013). Countries such as Libya have adopted new heritage management legislation ('Antiquities Law', 1994), aimed at preventing further damage to its archaeological and natural landscapes caused by oil prospecting. Archaeological assessments have also been conducted since the late 1990s in advance of any oil operation (Biagetti et al. 2013). As a result, thousands of archaeological sites have been documented; however, scarring left by roads and seismic survey lines in the broader landscape of the Sahara will remain forever (Fig. 23.3a). Likewise, pivot-irrigated agricultural systems have potential to significantly change soil, groundwater and microclimate systems (Hadeid et al. 2018) (Fig. 23.3b).

Protected area management is also set in a wider context of ongoing change to both the physical and human environments. For example, the Sahara has already expanded in area by 11–18% over the twentieth century as a result of climate change (Thomas and Nigam 2018) and is predicted to increase still further by one-third by the end of the twenty-first century (Zeng and Yoon 2009). This is the outcome of both enhanced heating of desert surfaces through albedo feedbacks (Vizy et al. 2013) and by changes in precipitation regime over the region (Adeniyi 2021). Previous studies have also highlighted the close historical connection between climate and environmental change and sociocultural processes in the region (e.g. Gatto and Zerboni 2015), and it is likely that these interconnections will continue in future. Political, demographic and socioeconomic changes in the Sahara (e.g. Bredeloup 2012; Freire 2014; Brachet 2016; Collier 2017) also pose challenges for site management structures, safety of development of tourist facilities

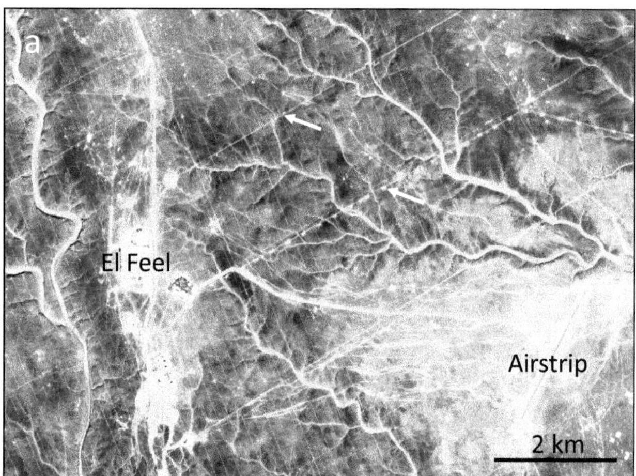

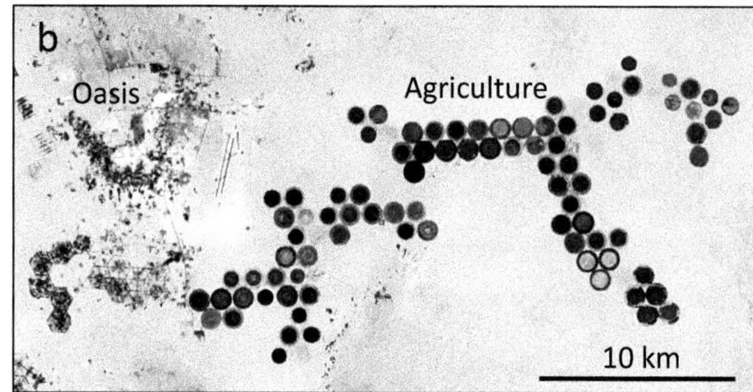

Fig. 23.3 Contemporary human activities modifying the Sahara landscape at a regional scale. **a** Straight and parallel oil prospection lines (white arrows) and El Feel (elephant) oil extraction operations infrastructure in the Messak Settafet, Libya; **b** the Al Khufrah Oasis in southeastern Libya (near the Egyptian border) is one of Libya's largest agricultural projects. Locations of these sites are shown in Fig. 23.1

and the survival of traditional cultures and trading networks (Kohl 2009; Biagetti 2014). Monitoring and protecting archaeological sites from negative impacts of climate change, vandalism, neglect or destruction have been noted from across the region (Remini et al. 2011; Menozzi et al. 2017; Mugnai et al. 2017).

The potential for cultural heritage tourism and geotourism as strategies for sustainable socioeconomic development in the Sahara region has been identified in several studies (e.g. Kröpelin 2002; UNESCO 2003; al-Rimayh 2006; Keenan 2006), but the infrastructure to support such activity (airports, tourist visas, transport, guides, hotels, etc.) is generally lacking (Chalcraft 2016; Ngwira 2019). Keenan (2006) argues that tourists are only one of many actors that can potentially cause damage to in particular rock art but also other cultural heritage artefacts that are subject to damage, looting and vandalism. Keenan (2006, p. 242) estimates that 2 million artefacts such as lithics and potsherds have been looted from the Tassili region (Algeria) alone. However, cultural heritage tourism and geotourism should not be seen as a panacea for either socioeconomic development at a national scale or for guaranteeing the maintenance or survival of cultural heritage or geoheritage sites (Keenan 2006; Williams et al. 2020). Chalcraft (2016) argues that urbanisation and high tourist use of sites has marginalised local people economically and also threatened the cultural identities and traditions that gave rise to site inscription in the first place. Education, training and active involvement of local people in tourism and conservation strategies are needed, but this should also be set in an international, transboundary context (UNESCO 2003; Keenan 2006).

A framework guiding the conservation of geological heritage in Africa is the Declaration of Antananarivo (2019). This Declaration was signed by representatives of the PanAfGeo project 'Geoscientific knowledge and skills in African Geological Surveys'; the Organisation of African Geological Surveys (OAGS); the European Association for the Conservation of Geological Heritage (ProGeo); the African Association of Women in Geoscience (AAWG); and the African Geoparks Network (AGN). This Declaration states among other things that geodiversity and geological heritage in Africa should be considered in the management of protected areas; that African countries should have a legislative framework to achieve this; and that local development strategies should consider geoconservation as an element of management, planning and impact studies. However, this Declaration is aspirational only, and it is unclear to what extent national governments and other stakeholders have an appetite for geological issues when faced with socioeconomic development challenges. In South Africa, for example, national legislation for geoconservation exists, but its scope and practical implementation are contested and often unclear (Cairncross 2011; Knight et al. 2015). Other African countries may be even less prepared.

23.6 Summary

The central Sahara region offers a number of sites of geodiversity and cultural heritage importance, but, despite the conservation status of these sites, they are commonly under pressure as a result of climate change, political instability and lack of effective management. Conservation as WHS, Ramsar sites, National Parks and others does not guarantee protection, and this highlights the sometimes precarious nature of the survival of geoheritage and cultural heritage sites across the central Sahara and in North Africa more generally. More rigorous management frameworks are needed at the local level, and these need to be accountable to both national governments and international institutions such as UNESCO. Tourism is often seen by managers and governments as the only activity that can lead to local socioeconomic development and site conservation, but this is not applicable to all sites, and there are many limitations of this narrow approach that are not considered by managers or politicians. This is a limitation of this most common approach towards heritage site conservation in Africa.

Acknowledgements We thank Stefano Biagetti and Amanda Esterhuysen for their comments and advice on this chapter.

References

AbdelMaksoud KM, Al-Metwaly WM, Ruban DA, Yashalova NN (2019) Sand dune migration as a factor of geoheritage loss: evidence from the Siwa Oasis (Egypt) and implications for geoheritage management. Proc Geol Assoc 130:599–608

Adeniyi MP (2021) Possible influence of climate change on water balance over West Africa under the global warming levels of 2 and 3 °C. J Water Clim Ch 12:1684–1673

al-Rimayh TF (2006) The natural and human essentials for desert tourism in South-west Libya. In: Mattingly D, McLaren S, Savage E, al-Sasatwi Y, Gadgood K (eds) The Libyan desert: natural resources and cultural heritage. Society for Libyan Studies Monograph 6, pp 253–259

Anag G, Cremaschi M, di Lernia S, Liverani M (2002) Environment, archaeology, and oil: the Messak Settafet rescue operation (Libyan Sahara). Afr Archaeol Rev 19:67–73

Anifowose AYB, Kolawole F (2014) Appraisal of the geotourism potentials of the Idanre Hills, Nigeria. Geoheritage 6:193–203

Annad O, Bendaoud A, Goria S (2017) Web information monitoring and crowdsourcing for promoting and enhancing the Algerian geoheritage. Arab J Geosci 10:276. https://doi.org/10.1007/s12517-017-3061-6

Azil C, Rezzaz MA, Bendaoud A (2020) Aspiring Hoggar and Tidikelt geoparks in Algeria. Arab J Geosci 13:1078. https://doi.org/10.1007/s12517-020-06017-y

Barker G (2006) The archaeology and heritage of the Sahara. In: Mattingly D, McLaren S, Savage E, al-Sasatwi Y, Gadgood K (eds) The Libyan desert: natural resources and cultural heritage. Society for Libyan Studies Monograph 6, pp 9–26

Barker G (2019) Libyan landscapes in history and prehistory. Libyan Stud 50:9–20

Barnett T, Guagnin M (2014) Landscapes in the Wadi al-Ajal, South-West Libya. J Afr Archaeol 12:165–182

Biagetti S (2014) Ethnoarchaeology of the Kel Tadrart Tuareg. Pastoralism and Resilience in Central Sahara. Springer Briefs, Cham, p 154

Biagetti S (2017) Resilience in a mountain range: the case of the Tadrart Acacus (Southwest Libya). Nomadic Peoples 21:268–285

Biagetti S, di Lernia S (2008) Combining intensive field survey and digital technologies: new data on the Garamantian Castles of Wadi Awiss, Acacus Mts., Libyan Sahara. J Afr Archaeol 6:57–85

Biagetti S, Cancellieri E, Cremaschi M, Gauthier C, Gauthier Y, Zerboni A, Gallinaro M (2013) The 'Messak Project': archaeological research for cultural heritage management in SW Libya. J Afr Archaeol 11:55–74

Biagetti S, Merlo S, Adam E, Lobo A, Conesa FC, Knight J, Bekrani H, Crema ER, Alcaina-Mateos J, Madella M (2017) High and medium resolution satellite imagery to evaluate Late Holocene human-environment interactions in arid lands: a case study from the Central Sahara. Remote Sens 9:351. https://doi.org/10.3390/rs9040351

Ben Nasr J, Walsh K (2020) Environment and rock art in the Jebel Ousselat, Atlas Mountains, Tunisia. J Mediterran Archaeol 33:3–28

Bouzekraoui H, Barakat A, Touhami F, Mouaddine A, El Youssi M (2018) Inventory and assessment of geomorphosites for geotourism development: a case study of Aït Bou Oulli valley (Central High-Atlas, Morocco). Area 50:331–343

Brachet J (2016) Policing the desert: the IOM in Libya beyond war and peace. Antipode 48:272–292

Bredeloup S (2012) Sahara transit: times, spaces, people. Pop Space Place 18:457–467

Brilha J, Gray M, Pereira DI, Pereira P (2018) Geodiversity: an integrative review as a contribution to the sustainable management of the whole of nature. Environ Sci Pol 86:19–28

Brocx M, Semeniuk V (2007) Geoheritage and geoconservation—history, definition, scope and scale. J Roy Soc Western Austral 90:53–67

Brooks N, Clarke J, Ngaruiya GW, Wangui EE (2020) African heritage in a changing climate. Azania 55:297–328

Cairncross B (2011) The National Heritage Resource Act (1999): can legislation protect South Africa's rare geoheritage resources? Res Pol 36:204–213

Chalcraft J (2015) Metaculture and its malcontents: world heritage in southwestern Libya. In: Mármol C, Morell M, Chalraft J (eds) The Making of heritage: seduction and disenchantment. Routledge, London, pp 22–43

Chalcraft J (2016) Decolonizing the site: the problems and pragmatics of world heritage in Italy, Libya and Tanzania. In: Brumann C, Berliner D (eds) World heritage on the ground: ethnographic perspectives. Berghahn, Oxford, pp 219–247

Collier P (2017) Africa's prospective urban transition. J Demograph Econ 83:3–11

Cremaschi M, Zerboni A (2009) Early to Middle Holocene landscape exploitation in a drying environment: two case studies compared from the central Sahara (SW Fezzan, Libya). C R Geosci 341:689–702

Dahmen A, Kassab T (2017) Studying the origin of the foggara in the Western Algerian Sahara: an overview for the advanced search. Water Sci Tech: Water Supp 17:1268–1277

di Lernia S (2006) Cultural landscape and local knowledge: a new vision of Saharan archaeology. Libyan Stud 37:5–20

Erikstad L (2013) Geoheritage and geodiversity management—the questions for tomorrow. Proc Geol Assoc 124:713–719

Esterhuysen A, Knight J, Keartland T (2018) Mine waste: the unseen dead in a mining landscape. Progr Phys Geogr 42:650–666

Freire F (2014) Saharan migrant camel herders: Znaga social status and the global age. J Modern Afr Stud 52:425–446

Gallinaro M (2013) Saharan rock art: local dynamics and wider perspectives. Arts 2:350–382

Gatto MC, Zerboni A (2015) Holocene supra-regional environmental changes as trigger for major socio-cultural processes in Northeastern Africa and the Sahara. Afr Archaeol Rev 32:301–333

Gibson RL, Reimold WU (2015) Landscape and landforms of the Vredefort Dome: exposing an old wound. In: Grab SW, Knight J (eds) Landscapes and landforms of South Africa. Springer, Cham, pp 31–38

Giesen MJ, Ung A, Warke PA, Christgen B, Mazel AD, Graham DW (2014) Condition assessment and preservation of open-air rock art panels during environmental change. J Cult Heritage 15:49–56

Gordon JE (2012) Rediscovering a sense of wonder: Geoheritage, geotourism and cultural landscape experiences. Geoheritage 4:65–77

Gray M (2011) Other nature: geodiversity and geosystem services. Environ Conserv 38:271–274

Gray M, Gordon JE, Brown EJ (2013) Geodiversity and the ecosystem approach: the contribution of geoscience in delivering integrated environmental management. Proc Geol Assoc 124:659–673

Hadeid M, Bellal SA, Ghodbani T, Dari O (2018) L'agriculture au Sahara du sud-ouest algérien: entre développement agricole moderne et permanences de l'agriculture oasienne traditionnelle. Cahiers Agric 27:15005. https://doi.org/10.1051/cagri/2017060

Hincker C (2004) Inaden's identity and craft: the social value of techniques among the western Tuareg. Homme 169:127–151

Honoré E (2019) Prehistoric landmarks in contrasted territories: rock art of the Libyan Desert massifs. Egypt. Quat Int 503:264–272

Keenan J (2006) Tourism, development and conservation: a Saharan perspective. In: Mattingly D, McLaren S, Savage E, al-Sasatwi Y, Gadgood K (eds) The Libyan desert: natural resources and cultural heritage. Society for Libyan Studies Monograph 6, pp 241–252

Keitumetse SO (2016) African cultural heritage conservation and management. Springer, Cham, p 227

Kenrick P (2019) Supporting cultural tourism in Libya—a brief history. Libyan Stud 50:51–57

Knight J, Grab SW, Esterhuysen AB (2015) Geoheritage and geotourism in South Africa. In: Grab SW, Knight J (eds) Landforms and landscapes of South Africa. Springer, Switzerland, pp 161–169

Kohl I (2009) Beautiful modern nomads. Bordercrossing Tuareg between Niger, Algeria, and Libya. Dietrich Reimer Verlag, Berlin, p 142

Kröpelin S (2002) Damage to natural and cultural heritage by petroleum exploration and desert tourism in the Messak Settafet (Central Sahara, Southwest Libya). In: Jennerstrasse 8 (eds) Tides of the desert—Gezeiten der Wüste. Contributions to the archaeology and environmental history of Africa in honour of Rudolph Kuper. Heinrich-Barth-Institut für Archäologie und Geschichte Afrikas, Universität zu Köln, pp 405–423

Mattingly D, Lahr M, Armitage S, Barton H, Dore J, Drake N, Foley R, Merlo S, Salem M, Stock J, White K (2007) Desert Migrations: people, environment and culture in the Libyan Sahara. Libyan Stud 38:115–156

Mattingly DJ, Sterry M (2013) The first towns in the Central Sahara. Antiquity 87:503–518

Menozzi O, Di Valerio E, Tamburrino C, Shariff AS, d'Ercole V, di Antonio MG (2017) A race against time: monitoring the necropolis and the territory of Cyrene and Giarabub through protocols of remote sensing and collaboration with Libyan colleagues. Libyan Stud 48:69–103

Mori L (2018) Urbanisation in the central Sahara in Garamantian times: a look from the south. Origini 42:193–209

Mugnai N, Nikolaus J, Mattingly D, Walker S (2017) Libyan antiquities at risk: protecting portable cultural heritage. Libyan Stud 48:11–21

Musch T (2021) Exploring environments through water: an ethnohydrography of the Tibesti Mountains (Central Sahara). Ethnobiol Lett 12:1–11

Ngwira PM (2019) A review of geotourism and Geoparks: is Africa missing out on this new mechanism for the development of sustainable tourism? Geoconserv Res 2:26–39

Nsanziyera AF, Rhinane H, Oujaa A, Mubea K (2018) GIS and remote-sensing application in archaeological site mapping in the Awsard area (Morocco). Geosciences 8:207. https://doi.org/10.3390/geosciences8060207

Orton J (2016) Prehistoric cultural landscapes in South Africa: a typology and discussion. S Afr Archaeol Bull 71:119–129

Park DP (2010) Prehistoric Timbuktu and its hinterland. Antiquity 84:1076–1088

Penn-Clarke CR, Deacon J, Wiltshire N, Browning C, du Plessis R (2020) Short report: geoheritage in the Matjiesrivier Nature Reserve, a world heritage site in the Cederberg, South Africa. J Afr Earth Sci 166:103818. https://doi.org/10.1016/j.jafrearsci.2020.103818

Plyusnina EE, Sallam ES, Ruban DA (2016) Geological heritage of the Bahariya and Farafra oases, the central Western Desert, Egypt. J Afr Earth Sci 116:151–159

Rayne L, Gatto MC, Abdulaati L, Al-Haddad M, Sterry M, Sheldrick N, Mattingly D (2020) Detecting change at archaeological sites in North Africa using open-source satellite imagery. Remote Sens 12:3694. https://doi.org/10.3390/rs12223694

Reimold WU, Koeberl C (2014) Impact structures in Africa: a review. J Afr Earth Sci 93:57–175

Remini B, Achour B, Albergel J (2011) Timimoun's foggara (Algeria): an heritage in danger. Arab J Geosci 4:495–506

Taçon PSC, May SK, Wesley D, Jalandoni A, Tsang R, Mangiru K (2021) History disappearing: the rapid loss of Australian Contact Period rock art. J Field Archaeol 46:119–131

Tavares AO, Henriques MH, Domingos A, Bala A (2015) Community involvement in geoconservation: a conceptual approach based on the geoheritage of South Angola. Sustainability 7:4893–4918

Thomas N, Nigam S (2018) Twentieth-century climate change over Africa: seasonal hydroclimate trends and Sahara Desert expansion. J Clim 31:3349–3370

UNESCO (2003) The Sahara of cultures and people. Towards a strategy for the sustainable development of tourism in the Sahara, in the context of combating poverty. UNESCO, Paris, p 77

Viljoen R (ed) (2016) Africa's top geological sites. Struik Nature, Cape Town, p 312

Williams MA, McHenry MT, Boothroyd A (2020) Geoconservation and geotourism: challenges and unifying themes. Geoheritage 12:63. https://doi.org/10.1007/s12371-020-00492-1

Vizy EK, Cook KH, Crétat J, Neupane N (2013) Projections of a Wetter Sahel in the twenty-first century from global and regional models. J Clim 26:4664–4687

Wilson A, Mattingly DJ, Sterry M (2020) The diffusion of irrigation technologies in the Sahara in antiquity: settlement, trade and migration. In: Duckworth CN, Cuénod A, Mattingly DJ (eds) Mobile technologies in the Ancient Sahara and beyond. Cambridge University Press, Cambridge, pp 68–114

Wright AM (2018) Assessing the stability and sustainability of rock art sites: insight from Southwestern Arizona. J Archaeol Method Theory 25:911–952

Wurz S, van der Merwe JH (2005) Gauging site sensitivity for sustainable archaeotourism in the Western Cape Province of South Africa. S Afr Archaeol Bull 60:10–19

Zeng N, Yoon J (2009) Expansion of the world's deserts due to vegetation-albedo feedback under global warming. Geophys Res Lett 36:L17401. https://doi.org/10.1029/2009GL039699

Index

J. Knight et al. (eds.), *Landscapes and Landforms of the Central Sahara*,
World Geomorphological Landscapes, https://doi.org/10.1007/978-3-031-47160-5